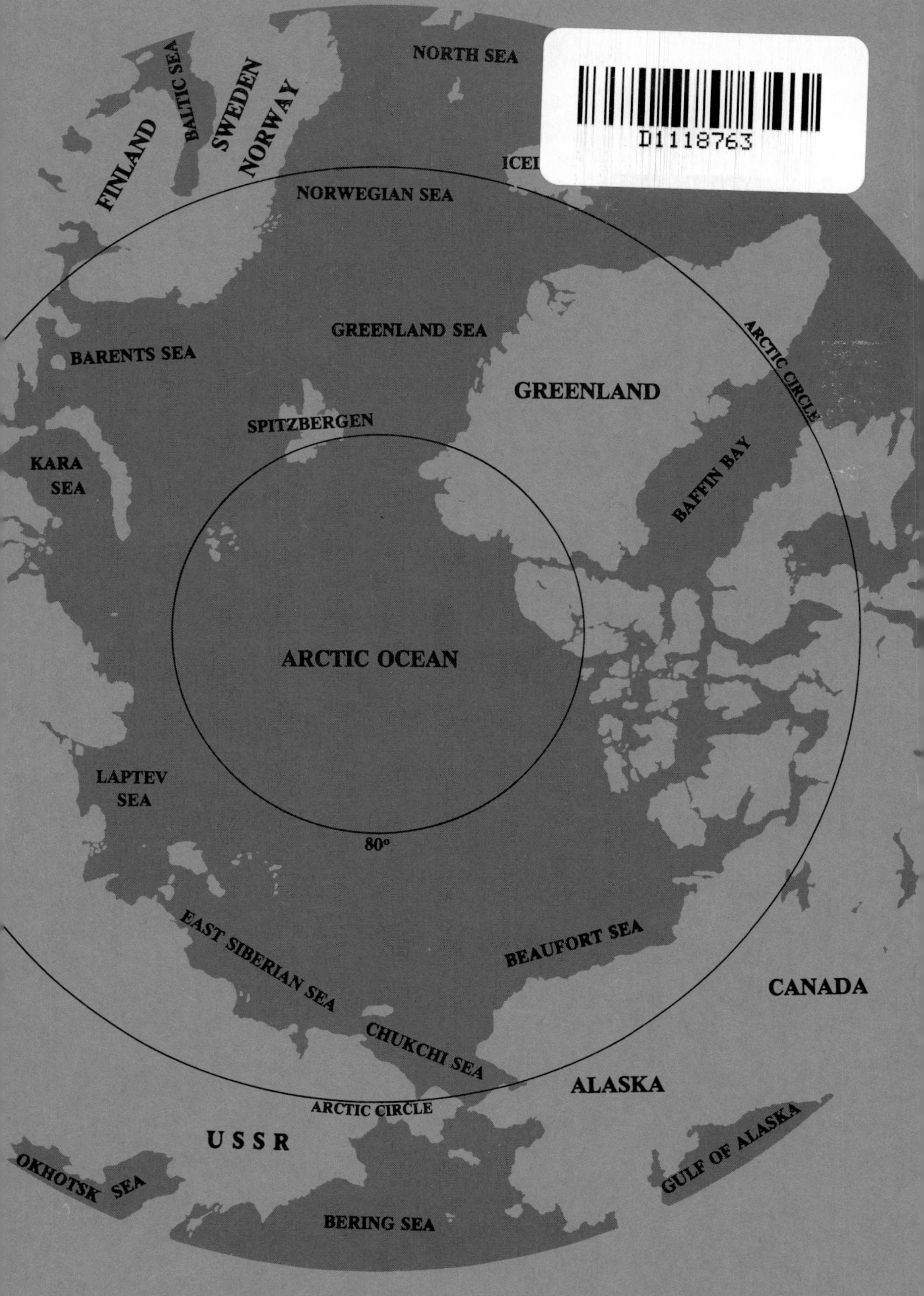
NORTH SEA
BALTIC SEA
FINLAND
SWEDEN
NORWAY
NORWEGIAN SEA
BARENTS SEA
GREENLAND SEA
GREENLAND
ARCTIC CIRCLE
SPITZBERGEN
KARA
SEA
BAFFIN BAY
ARCTIC OCEAN
LAPTEV
SEA
80°
EAST SIBERIAN SEA
BEAUFORT SEA
CANADA
CHUKCHI SEA
ALASKA
ARCTIC CIRCLE
USSR
OKHOTSK SEA
GULF OF ALASKA
BERING SEA

To my good friend Tim
Parsons from
Don Hood
Feb. 14 (Valentines Day) 1977

ASSESSMENT OF THE ARCTIC MARINE ENVIRONMENT
selected topics

ART CREDITS:

Inside front cover: Polar delineation adapted from «PHYSICAL MAP OF THE ARCTIC» American Geographical Society of New York, 1930

Following page: « WINTER SUN» from stoneware tile design by Wenzel Studio, 109 West Spruce, Louisville, Colorado

ASSESSMENT OF THE ARCTIC MARINE ENVIRONMENT
selected topics

Edited by D. W. HOOD AND D. C. BURRELL

Occasional Publication No. 4
Institute of Marine Science
University of Alaska • Fairbanks

ELEANOR KELLEY
Technical Editor

Based on a symposium held in conjunction with
Third International Conference on
Port and Ocean Engineering under Arctic Conditions

POAC—75

under financial sponsorship of the U.S. National Science Foundation; the Alaska Sea Grant Program; and the Institute of Marine Science, University of Alaska

with the support of the American Association for the Advancement of Science, Alaska Division; Engineering Committee on Ocean Resources; The Arctic Institute of North America; Canadian Society for Civil Engineering; Marine Technology Society; Norwegian Institute of Technology, Division of Port and Ocean Engineering; U. S. Office of Naval Research, Naval Arctic Research Laboratory; and the Geophysical Institute, University of Alaska

ISBN 0-914500-07-4
Library of Congress catalog card number: 75-43209

Printed and bound by The Maple Press Company
in the United States of America

foreword

Assessment of the environmental components in the high latitudes of the World Ocean is a challenge and commitment as relevant as man's future. Resource extraction in the north country is of critical priority for development in all nations which hold dominion over these lands. The long-established sources for these vital commodities to meet the worldwide needs of modern machines and industry are rapidly diminishing. Extraction operations in the North are frustrated by major problems of logistics and a general lack of historical data on this unique environment. There is often a call for new technology — or new application of old technology — to cope with the very different engineering problems posed by an obstinate climatic condition.

Covered with ice and snow most of the year, the northern regions would appear to be relatively free of the many concerns associated with heavier populated, highly developed areas. Further insight reveals, however, a sensitive living system whose natural elements function near the threshold of their tolerance. In the marine ecosystem, plankton are the driving force of productivity, and here they take on a fundamental role. As major marine providers, these organisms must also adapt to long periods of low sunlight, frequent ice cover and highly mixed seas — all of which factors are generally adverse to optimum growth. Despite these apparent handicaps, the plants of the plankton community demonstrate a capability of producing food materials under near-freezing conditions at rates comparable to those found anywhere else in the world.

Man's impact on this growth is not yet fully understood, but it is clear that the metabolism of the phytoplankton is influenced by small amounts of foreign hydrocarbon components. If other organisms are more sensitive, the evidence is not clearly indicative. Certainly one of the prime technological problems to be solved in developing the North is to determine the response of key organisms at the critical period of their life history to the effects of components that may be derived from man's activities.

The symposium "Environmental Assessment under Arctic Conditions," to which this book is addressed, is a first attempt to broadly appraise the environmental problem of developing the northland. Continued efforts of similar sensitivity will go a long way in gathering environmental knowledge essential to the optimal welfare of both man and the arctic marine resources upon which his future depends.

D. W. H.

preface

Assessment of the Arctic Marine Environment is based on topics selected from a symposium held 11-15 August 1975 in Fairbanks, Alaska, in connection with the Third International Conference on Port and Ocean Engineering under Arctic Conditions (POAC-75).

The official proceedings of the conference, consisting of 108 papers exclusive of the contents of this book, have been compiled separately in two volumes available from the same publisher (Institute of Marine Science, University of Alaska, Fairbanks). The reader is referred to the POAC-75 proceedings document for discussion of such engineering and mechanical-physical aspects as ice dynamics and related problems of arctic offshore structures and technology, transportation, the seabed and sub-bottom investigations.

The heavy emphasis directed at this conference to the arctic environment prompts a summary of the background behind so great an effort to understand and solve problems associated with the northern regions. The POAC conferences were initiated in response to a growing need for establishing closer cooperation between "port" and "ocean" interests in an environment where any industrial development is faced with great difficulties and risks unique to severe climatic conditions. These very challenges suggest a natural bond between the mutual concerns of all who work in the arctic environment — engineer and scientist alike. It appears now that we have made a significant advance in successfully bridging the usual but unserving dichotomy of port and ocean objectives in this icy domain.

Norway was selected host for the first POAC conference, a choice due partly to the country's geographical location on the rough and cold waters of the North Sea and the North Atlantic, extending up into the ice-ridden seas of the Arctic. Then, too, there was the timely North Sea circumstance of still-accelerating gas and oil exploration and exploitation, a development requiring extraordinary efforts in the related fields of port and ocean engineering. Understandably, much emphasis at the 1971 conference was placed on structures and how they are affected by the forces of wind, wave, current and ice activity. Various coastal engineering aspects were also included. The conference was attended by 300 participants and resulted in 110 papers, the proceedings of which were published in 1972 by the Norwegian Institute of Technology. A similar pattern, including an increased interest in ice, was followed at POAC-73, held in Iceland in 1973, with 150 conferees presenting 70 papers. Proceedings were published by the University of Iceland in 1974.

With the rapid development of oil and gas exploration in Alaska and its surrounding waters, it was only appropriate that the third POAC conference (POAC-75) should be hosted by the University of Alaska in Fairbanks, gathering 300 participants from 15 northern countries. In the sense that the North Sea is one of the most hostile seas in the world with respect to winds, currents and waves, the Alaskan environment presents one of the most adverse combinations of low temperatures, ice in the ocean or on land, and the associated problems with development and transportation. In addition, the environment has proved, by hard experience, to be very sensitive to man's interference. Development in Alaska is faced with unusual problems, and efforts to solve them are no less dramatic. The Manhattan cruise, the trans-Alaska pipeline, and the Cook Inlet platforms are just examples. The engineering tasks being undertaken in the arctic marine environment rank undoubtedly among some of the most difficult and glorious in the world.

This book is by no means intended as a complete treatise on the assessment of the arctic environment, but it does approach this important topic with thoughtful consideration, a discussion of present activities, and some detail on methodology and findings in selected areas of study. In a scope uncommon to a single publication, the contents demonstrate an interdisciplinary approach to a complex problem. If perchance the pages that follow help guide the way to a sensible approach to assessment and rational utilization of the arctic marine environment, our efforts are shown to be in the right direction.

PROF. PER BRUUN
Secretary General, POAC

Chairman, Division of Port and Ocean Engineering
The Norwegian Institute of Technology
Trondheim, Norway

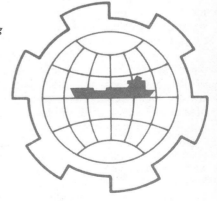

contents

PART I / ASSESSMENT CONCEPT AND PROJECT APPROACH

Section 1. THE ARCTIC PREMISE

Section 2. REGIONAL PROGRAM PERSPECTIVES

PART II / CONTEMPORARY TOPICS

Part I

ASSESSMENT CONCEPT
AND
PROJECT APPROACH

section 1

the arctic premise

CHAPTER **1**

Introduction:
A statement of the problem

D. W. HOOD [1]

> *You who . . . the High North is luring . . .*
> *Honor [her] . . . ever and ever,*
> *Whether she crown you, or whether she slay;*
> *Suffer her fury, cherish and love her —*
> *He who would rule must learn to obey.*
>
> **Robert W. Service (1874-1958)**

ENVIRONMENTAL FACTORS
UNIQUE TO ARCTIC ASSESSMENT

The diverse high-latitude regions of the World Ocean are characterized in common by extreme light and temperature conditions, a widely splayed phenological spectrum dominated by seasonal ice and associated low temperatures.

The assessment of life processes at all trophic levels is greatly complicated by the cold-stressed environment prevailing in the northern domain. Not only does the presence of ice impede conventional access to the oceanic regions of the world, but it represents ecologically a little-understood phase in the ocean medium. It is apparent that ice conditions exert important influences on the birds and mammals of the Arctic, acting at a gross level as a movable partition between distribution of those species that normally reside within the boundaries of the ice pack and those which are excluded from the same area when ice is present. In the spring of the year, the primary productivity which occurs in the sea ice is vital to grazing organisms and possibly serves as a source of detritus for benthic biota underlying the ice flows. An understanding of the dynamics which drive this finely tuned

---

[1] *Institute of Marine Science, University of Alaska, Fairbanks, Alaska 99701.*

system requires detailed information on not only the ice regime — but knowledge of the underlying water masses throughout the entire year. The difficult logistic implications of such studies have so far been approached only by occasional use of helicopter, light aircraft and icebreaker observations, as such surveillance opportunities have been made available.

The behavior of spilled oil in ice and the effects of oil on the associated ecosystem are largely unknown. Although relatively little experience has been gained in this medium, certainly the problems attendant with clean-up of oil accidents are complicated by the presence of ice and severe cold conditions.

An additional environmental factor of the north is the existence of indigenous coastal populations who continue to depend upon the resources of the sea for a significant portion of their subsistence.

Historically, the northern regions have proved rich in biological resources and have been heavily harvested for many centuries (Hunt 1975). Transportation corridors have been sought through the arctic regions to link east and west commerce since well before the fifteenth century; yet there remains to be developed a satisfactory passage through the Arctic Ocean to connect the Atlantic and Pacific shipping lanes. In response to increasing urgency for removal of oil, gas and minerals from resource-rich areas of the Arctic to balance world energy deficits, probably such a passage will at last arise during the next score of years.

Assessing the Arctic in its relatively undeveloped state may provide a clearer focus than in the lower latitudes, where high population density has imposed ocean-use demands for waste disposal, recreation, industrial centers, and competition for beach properties. Fundamentally, it appears that assessment of the arctic marine environment means reconciling mineral resource extraction with maintenance of the natural biological resources. Although development must inevitably advance, resource management decisions must be made with an awareness that a finite stress limit exists in any environment. The ultimate alternative to this line of consciousness spells hardly less than extinction of the marine ecosystem and removal of that part of the earth's life support system. The human proposition in such a sequence of events puts man himself under perpetual stress to look out for his very existence. There was an apparent wisdom in the lyrics of a popular Canadian balladeer who saw, through the dazzle of an earlier resource boom, nature's gentle admonition: "He who would rule must learn to obey" (Service 1958).

Renewable resources

The ice-covered, windswept high-latitude seas, so often considered to be barren of living things and of little significance for renewable resource extraction, in fact represent some of the world's richest areas of biological productivity.

Although ice cover and long periods of darkness limit photosynthesis during the winter months, the summer — with long days of light accompanied by ideal growing conditions — supports a primary production regime comparable to any in the world. In support of primary productivity, abundance of nutrients and stability of the water column at the time of plant growth are perhaps often more important factors than is the availability of light.

From a depth of a few hundred meters to the sea bottom, the water column of the ocean is rich in the essential plant nutrients phosphorus, nitrogen and silicon. The surface waters, however, are depleted of these nutrients for much of the year, especially in the lower latitudes, where stratification of the water column is so intense that mixing with the nutrient-rich deep water occurs only on a time scale of many years. In high latitudes, on the other hand, the nutrient-rich waters are very near the surface (30-100 m) even during high periods of photosynthesis, and mixing with the deeper waters occurs at least annually through the action of thermal cooling and heavy storms. Complete mixing to depths below the critical level for plants is detrimental to the photosynthetic process. Consequently, in high latitudes, the major bloom of phytoplankton occurs in the spring, after only minor stratification of the surface water has taken place as a result of ice melt, surface runoff or solar insulation. Once stratification occurs, primary production is quickly initiated and remains high until the nutrients are depleted (see Chapter 10, Fig. 10.10, this volume). After this burst, the level of production then drops to that supported by recycling of nutrients or input of nutrients from convective sources.

Even more impressive than the primary productivity level of the high-latitude seas is the associated secondary production which prevails in the great standing stocks of birds, mammals, fish and invertebrates. Migrating birds and mammals draw the most public attention because they are highly visible and also may visit southern climates during part of their life cycle. Birds, in particular, rear their young in the long sunlit days of the far northern summer, where it is critical that the offspring mature to competent flying age before the ensuing migration. In the delicate balance of the Arctic, a few days' slippage due to bad nesting conditions or early freeze-up can be catastrophic to a migrating species. However, few places on earth can match such high production of food sources that are readily available to these organisms.

Some of the world's great fisheries are found in this cold-dominated area. Many examples of these biologically rich northern seas can be given, but perhaps the most productive is the Bering Sea, separating the Asian and American continents.

The Bering Sea is both temperate and arctic in its characteristics. Although much of its waters lies below 60° N latitude, nearly all of the extensive Bering shelf becomes covered with ice each winter. The Aleutian Islands reach south to a latitude of 50° N, and the temperate Alaskan Stream flows along the south side of the island chain in an east-west direction. The island passes serve as transport channels for exchange of water between the North Pacific and the Bering Sea. The penetration of the Bering Sea waters by Pacific water through the shallow eastern passes gives rise to the upwelling phenomenon, which is very important in the support of the very large Bering Sea biomass. In the west, the passes both broaden and deepen where the major oceanic exchange occurs.

Over the past several years, the Bering Sea ecosystem has been under extensive consideration by oceanographers from the Soviet Union, Japan, and the United States. A summary of these studies is given in the book, *Oceanography of the Bering Sea*, edited by Hood and Kelley (1974). More recently, scientists of the United States initiated an ecosystem study under the acronym PROBES (Processes and Resources of the Bering Sea Shelf). This study, briefly discussed by McRoy, Muench and Elsner in Chapter 8 of this volume, is an effort to produce an ecosystem model describing the energy transfer from the primary to the top trophic levels on the southwest Bering Sea shelf (known as the Golden Triangle area). The Alaskan pollock, in all its life stages, is to be used as a biological tracer in this study. Present evidence indicates that the level of primary productivity estimates for the Bering Sea shelf is inadequate to support the remarkable biomass endemic there unless a greater efficiency is realized in trophic dynamics than has been found in warmer seas. Part of this efficiency may come from a higher net production in colder water because the photosynthetic processes are less affected by cold than are functions of respiration. It also appears that fortuitious timing of biological events and shorter food-chains may be major factors.

Since renewable resources represent the primary use of this region, it is with this emphasis that Bering Sea assessment must lie. Man's exploitation of particular animals of the northern seas has drastically changed the abundance of some species. Most notably affected was the sea cow (*Rhytina stelleri*), probably extinguished in the early 18th century, and the sea otter (*Enhydra lutris*), which is now recovering from near-extinction after enormous rehabilitation efforts in its behalf. Other mammal populations have been reduced and threatened, as have the fish stocks of herring (*Clupea harengus pallasi*), yellowfin sole (*Limanda aspera*), Pacific halibut (*Hippoglossus stenolepis*), ocean-run salmon (*Oncorhynchus* spp.), Pacific cod (*Gadus macrocephalus*), Pacific ocean perch (*Sebastes alutus*), and other commercially important species over the past few decades. Meaningful assessment of the effects of man's activity on the ecosystem through the extraction of commercial fish is not presently possible, since little information on the early life history is available for any of the species except halibut and salmon. The only analysis possible is on the exploited portion of the population for which good catch statistics have been collected.

Over 2,000,000 metric tons of fish flesh are removed each year from the southeast Bering Sea alone (Pruter 1973). This must without doubt have an effect on the marine ecosystem. Assessment of this effect, however, is complicated by rapidity of the development and resulting over-exploitation of the commercial fishery. The ubiquitous pollock (*Theregra chalcogramma*) fishery, representing over 80 percent of the total Bering Sea catch, was started by the Japanese only as recently as the early 1960s. Indications are that the pollock are now overfished, and collapse of the fishery is predicted within the next few years if the harvesting effort remains unchecked.

Fisheries data for the Bering Sea show that most species of fish being harvested are over-exploited at the present time. This indicates that nearly all of the biological energy fixed in this ocean that man has found economically feasible to extract and utilize is now being harvested. Unfortunately, this is being done at a rate in excess of its sustainable yield. This type of exploitation places the ecosystem to which these commercially desirable species belong under stress. Not only are there fewer adult fish, but the numbers of juveniles in the reproductive cycle are also accordingly reduced, leaving food resources available. Other species which are present may fill the niche for food vacated by the depleted species. The effect of fishing efforts on the biological energy resources of the Bering Sea is that of directly extracting about 250,000 metric tons of fixed carbon from the system each year. This idea can be developed further. If pollock are considered to feed on the second trophic level, and if efficiencies of 10 percent are assumed for each level, the fish taken annually would have consumed about 25,000,000 metric tons of carbon fixed by photosynthesis. Total primary productivity of the Bering Sea is estimated to be about 274,000,000 metric tons carbon per year (McRoy and Goering, in press). The amount of total fixed energy directly removed by fishing — about 0.1 percent — is insignificant. However, the fact that nearly 10 percent of the total productivity must be utilized to provide this level of harvest is indicative of the inefficiency pervading this kind of extractive industry. The opportunity for improving efficiency lies in such practices as harvesting lower trophic levels, by fish culturing to give less diversity and therefore more efficiency, and by application of better management methods. These alternatives appear very attractive not only in the Bering Sea, but in all other productive oceans as well.

Non-renewable resources

Very promising oil and heavy-mineral provinces lie yet untapped on the continental shelf portions of many arctic seas. These are sometimes associated with mineral deposits which also exist on land regions often accessible only by sea and in submerged lands which will be developed through underwater placer mining techniques or by oil drilling. The extraction of these resources is imminent and will impinge upon habitats and ecosystems of the renewable resources.

The philosophy of need versus desire for ever-increasing requirements for fossil fuel by modern society has long been debated. There seems to be little question but that the individual nations of the world, especially the United States, have set an economy and a way of life which is most inefficient in the use of energy and materials. Maintenance of the highest standards of living on earth is largely at the expense of energy reserves and mineral resources. The fact that the United States is also the bread basket of the world compensates somewhat for the extravagance of its technology. One is set back, however, when he realizes that, on an energy basis, our farmers use more energy in fossil fuels than they produce in food, whereas the Chinese coolie produces about 30 times his energy consumption in food. It is an interesting conjecture as to where this pattern of operation will lead.

The industrial nations are dependent on the heavy use of fossil fuels until other energy substitutes can be obtained. Water power, atomic energy, better use of coal, the ocean, wind, and solar radiation are all promising sources of exploitable energy. All have their side effects and environmental impacts. Hopefully, sufficient effort will be brought to bear to find relief for oil and gas as primary energy sources before their reserves are depleted. In the meantime, there seems to be no realistic way to slow down oil production significantly.

The impact of non-renewable resource extraction in northern seas, as well as elsewhere, will probably be of less importance to the visible higher trophic levels than to the lower trophic levels; yet, perhaps unwisely, the concern of the public is concentrated on the higher life forms. The term "sensitive species" has been coined with reference to organisms for which the public has special concern. Into this group fall the birds and marine mammals, salmon and other commercial fish species. An oil spill is the most likely environmental insult which would affect the adult forms of the "sensitive species." Depending upon the circumstances, losses due to an oil spill would probably be most serious in those organisms which become directly exposed to it, such as some species of birds, marine mammals and beach organisms. Adult fish have not been found to be greatly affected by floating oil.

A FUNDAMENTAL DEBATE

Renewable versus non-renewable resource extraction

Although harvesting of renewable resources inevitably takes a certain toll on the entire marine ecosystem, it should rarely affect the productivity of the lower trophic levels if technology is well managed. Conversely, non-renewable resource extraction has a potential, due largely to accidents, for directly reducing the numbers of organisms in the higher trophic levels, but its effects on the lower trophic levels and early life stages of harvestable species are more suspect and largely unknown.

The dilemma, then, is primarily a question of economic morality in treating the needs of an energy and material glutted world consumer. Oil or other *non-renewable* resources can be extracted radically with relatively quick financial gain to the producer. Or, there is the alternative approach of a long-term but less dramatic return from the *renewable* living system. In simplest terms, it is a matter of heat versus food — both elements essential to man's basic survival. Unfortunately, it is the very polarization of opinion on this problem that prolongs a solution to it. On the one side, there are those who deny any benefits to be gained from extraction of non-renewable resources and who claim that such action will lead to the demise of the biological system. This point of view is not rationally compatible with evidence that regions where heavy oil development have occurred in the world do not show a detectable decline in biological populations. Then there is the other faction who, because of political and financial pressures, would extract and process energy at any environmental cost.

Although petroleum-derived hydrocarbons have been identified throughout the ecosystem where petroleum-oriented industries are located, the long-term effects of low concentrations of these substances will not be known until several generations of biota are allowed to develop. Far greater impact has been observed as a result of growth of the associated industries and municipalities that have accompanied major oil-field developments. Differentiation between these direct and associated effects is usually difficult, since both activities tend to occur together in most areas of oil extraction. In the high-latitude regions, however, there is a tendency to move crude petroleum components to more densely populated areas in the south for the refining and manufacturing processes. Although this practice may not favor full employment in the area of resource extraction, it should be of great value ecologically in preserving the renewable resource base for future generations to use and enjoy.

Those concerned with assessment of the arctic environment are faced with serious difficulties brought on by an often-shortsighted interest in reducing the oil dependency of highly industrialized nations. This has resulted in rapid development of some coastal and offshore regions of the Arctic where oil has only recently been found. Despite the occurrence of active harvest since the early 1700s, there is little information available on the biological processes and resources of the north. This absence of an adequate environmental data base severely limits existing scientific capability in gaining useful information on a time-scale responding to the public urgency for energy from oil. Assessment of the standing stocks of a given area on a short time-scale is feasible; however, many years would be required to obtain data which would allow a determination of a perturbation in the system due to non-renewable resource extraction, because of the natural fluctuations which occur in most marine communities. Meaningful early life history and process studies also demand time and highly qualified scientific manpower. Satisfactory techniques necessary for assessment of many life processes, particularly in the second trophic level, have not yet been sufficiently developed.

The argument could be pursued that development of high-latitude areas for non-renewable resources should proceed without environmental assessment because of the urgency of the energy requirement and the lack of existing baseline data on important elements of the ecosystem. Such an argument is fallacious and dangerous. Historically, man has often proved himself out of phase with needed information, but this observation has only emphasized the urgency of his conditions and incited him to all-out efforts to recover lost ground. Society now seems to be faced with a uniquely pressing problem of finding both energy and food resources. Certainly both are essential, and both must be acquired simultaneously. As rapidly as resources will permit, measures must be taken to assimilate existing pertinent information and to formulate a well-planned program for a meaningful understanding of the arctic marine ecosystem.

Two kinds of oil spills are of particular concern as potentially hazardous to the "sensitive species." An extensive spill may cover a large area that is difficult to clean up and which involves large numbers of organisms. Or, the timing and location of an oil spill may be such as to directly impinge on a major portion of a population at a critical time in the life cycle. Potential victims of the latter type spill might include nesting populations of arctic terns or other wildfowl, a seal rookery, or a beach population of salmon eggs or larvae.

The effects of non-renewable resource extraction on "sensitive species" of the ecosystem will not be seen so much at the adult stages as in the juvenile portion of their life cycles. In the case of fish and vertebrates, the juveniles populate a lower trophic level, beginning with the immobile egg stage, pass through the zooplankton phase, and finally emerge as fingerling fish in the top trophic level. The early life stages of most commercial fish in high-latitude regions are poorly understood, with the possible exception of salmon and halibut. It is known, however, that many juveniles spend their early lives in the plankton-rich nearshore water and estuaries. Since it is probable that the impact of oil development will be more critical to the nearshore component of the marine ecosystem, this region should therefore receive corresponding priority in the planning of assessment studies.

Acknowledgment

Contribution no. 294, Institute of Marine Science, University of Alaska.

REFERENCES

HOOD, D. W., and E. J. KELLEY (Eds.)

    1974 *Oceanography of the Bering Sea*. Occas. Publ. No. 2, Inst. Mar. Sci., Univ. Alaska, Fairbanks, 623 pp.

HUNT, W. R.

    1975 *Arctic passage*. Charles Scribner Sons, New York, 395 pp.

McROY, C. P., and J. J. GOERING

  (In press) Annual budget of production in the Bering Sea. *Marine Science Communications*, Vol. 2.

PRUTER, A. T.

    1973 Development and present status of bottomfish resources in the Bering Sea. FAO Technical Conference on Fishery Management and Development. Vancouver, B. C., Canada.

SERVICE, R. W.

    1958 Men of the High North. In *Collected poems of Robert Service*. Dodd, Mead & Company, New York, p. 79.

Assessment of the Arctic Marine Environment: Selected Topics
Copyright © 1976 by Institute of Marine Science, University of Alaska, Fairbanks

CHAPTER **2**

Man in the polar marine ecosystem

M. J. DUNBAR [1]

THE HUMAN SCOPE

An ultimate perspective

The primary and the ultimate problem of mankind is to reconcile its commercial and industrial activities with the maintenance of the natural system; for if man manages to upset the natural system beyond the point of recovery, he extinguishes himself. It has been suggested, I think by Sir Fred Hoyle, that the probability of the presence, now and in the past, of other civilizations besides our own in the universe is high, but that contact between them is limited or rendered improbable by time. Communication between points separated by many thousands of light-years is rendered difficult by the probability that civilizations have limited life-spans, that they probably extinguish themselves some 10,000 years after they begin. This is a chilling thought; one wonders where to put the start of our civilization in the past, and from that date to speculate on how long we have to run.

The global problem

Let us come back to earth, therefore. One of the paths toward extinction is that of the global atomic war, which many science fiction writers have treated very effectively. Another is the failure of the original source of energy in the system, that is, of the sun — which does not concern us here and which we can do nothing about. A third is the destruction of the natural global ecosystem, a problem which we can break down into geographic units for greater ease of handling, and in which the crucial thing is to establish with confidence how tough or how vulnerable the ecosystem is. Within this global problem, our immediate interest is the arctic marine environment. Are we, or are we not, in danger of destroying it?

[1] *Marine Sciences Centre, McGill University, Box 6070, Station A, Montreal, Quebec.*

11

MAN'S IMPACT ON THE MARINE ARCTIC ECOSYSTEM

Environmental stamina

The word "perturbation" is easier to understand and to define than is the word "stress." My understanding of what is meant by perturbation in the ecological context is a numerical or quantitative change in some part of a system, usually a rather sudden change in any of a number of time scales, and one from which most systems recover. This is a change which in turn causes stress of one kind or another, either on individuals in the communities, or on species, or on the system as a whole. Thus, very high lemming populations in "lemming years" are perturbations of the system, and they cause stress on individual lemmings, and low lemming populations cause stress on the system, resulting in starvation or emigration of predators, and so on. Margalef (1968) spoke of "stresses resulting from the lack of equilibrium between environment and population."

Both perturbation and stress, within limits, are normal to the system, but it is not proper, using these definitions, to speak of environmental "rigor" as stress. The low temperatures of polar regions and the high temperatures of deserts are both normal properties of the environments to which the living elements are adapted, and to consider life in the deep water of the Arctic Ocean as stressful is decidedly anthropocentric. I take issue with the view, therefore, that tropical environments are biologically stressful and that polar environments are physically stressful. Both types of environment are normal to the regions in which they are found, and therefore form parts of adjusted ecosystems. On the other hand, conditions beyond the normal range of environmental extremes may legitimately be considered as perturbations to which the system is not adapted to deal with in the normal way.

Marine ecosystems are buffered, by the very nature of water, against climatic extremes. In fact, abnormal perturbations, and therefore stressful conditions, in the sea must be rare, whether physically or biologically determined. Extraordinary perturbations in the natural process must likewise be only infrequent with the exception of local and occasional events such as submarine eruptions and population outbreaks such as that of the crown of thorns. I can see no reason to believe that polar marine environments are any more or less vulnerable or "fragile" than temperate or tropical environments. The activities of man, however, from the beginnings of agriculture and fishing, have consistently altered the natural ecosystem of the world, and since the industrial revolution have done so with continued acceleration.

The whaling industry and ecological balance

The destruction or disappearance of a specific population (a whole species) is undoubtedly a perturbation, but it does not necessarily lead to stress, except possibly to man himself. The destruction of a "key industry" species in a simple ecosystem (such as lemmings) can cause great stress to the system, as already mentioned, and could in theory destroy it (Dunbar 1973). But most ecosystems can absorb the loss of a specific member without much trouble. Take as examples any of the recent known extinctions, from passenger pigeons to Labrador ducks; extinction has been a very common event indeed in the history of life on earth. There is, however, an aspect of serious reductions of specific populations by human activity that has not been given much attention, and that is the change in ecological balance which may make recovery of the population in question more difficult than might be considered on ordinary principles of conservation. The blue whale of the Antarctic has been reduced to a comparatively small proportion of its original numbers, and as a result this reduction has given rise to various estimates of surplus euphausiids (krill) in the Antarctic Ocean. This event in turn has started two of the largest whaling nations to turn their practical attention to the direct harvesting of the krill, the assumption being that if the pressure is eased on the blue whale by reduction or stoppage of the whaling industry, the whale population will recover and can once again take over the consumption of the krill. This does not necessarily follow. It might be expected that other predatory groups such as fish, seals, and seabirds would move in on the krill thus made available to them. At the Antarctic Ecological Conference held at Cambridge University in 1968, this possibility was briefly discussed but discarded for lack of evidence. At the Polar Oceans Conference at McGill University in 1974, however, there were several suggestions that this was in fact what was happening, particularly with respect to the fish populations. Indeed, it may well be that the historical dominance and great success of the sea mammals in polar regions has much to do with the comparative lack of success of the fishes.

If alternative predators on the krill increase in population numbers in this way, the system as a whole will suffer no damage; life will go on without the whales. But it must be expected that the recovery of the whales would be a very long process, perhaps impossible, for the whales must now compete, at a disadvantage, with the new surging fish and bird population. Perhaps the direct harvesting of the Antarctic krill, furthermore, may be just the wrong thing to do if we hope to encourage the recovery of the blue whale.

I suggest that this may be what has happened to the bowhead whale in the Canadian eastern Arctic. The bowhead was considered almost extinct in the Baffin and Hudson bay regions in the early years of this century, and it was afforded complete protection from commercial exploitation. The bowhead population is still very low (Mitchell 1973), however, and it is very doubtful whether this can be attributed to Eskimo hunting in that area. The story of the right whale in the northeastern Atlantic region is much the same. One is tempted to wonder why the recovery has been so slow, and the answer may lie in the mechanism described above. Unfortunately, except for documentation of well-known changes which can be ascribed to climatic change, the records do not allow us to come to conclusions about any changes that may have occurred in the numbers of competitors for the same food supply, such as caplin, other fish, seals, or birds.

All this for the moment is speculation, but it may serve to underline the need for research along these particular lines of population dynamics and ecological competition. Certainly the replacement of one competitor by another is a normal ecological phenomenon under many sets of circumstances, and the case of the Antarctic krill can therefore serve as an illustration of the necessity of obtaining real understanding of the ecological rules before undertakings go forward in the commercial fisheries or in industry development which might jeopardize any particular species or group of species, such as sea mammals and birds, in the face of oil development and transportation.

Significance of impact studies

This means that really effective impact studies must go into very much greater detail than they do at present, and this in turn means that they should have been begun 50 years ago. One of my university colleagues, in the Department of Economics, has said that economics is an extremely complex discipline, not a simple matter like ecology — "just a bunch of rabbits." I think part of the trouble we are in, environmentally, derives from this "bunch of rabbits" view of ecological science on the part of engineers and the bureaucrats. The fact is, of course, that we are a very long way from the level of understanding ecological interrelations required for adequate answers to the questions put to the ecologists by the industrialists. And when this is made clear, the reaction is impatience, leading to the signal to go ahead and never mind the consequences.

Beaufort Sea studies. The Beaufort Sea Project is one of the better examples in which a real effort is being made to understand an extremely complex system before the signal to drill for oil is given. A great deal of information has been, and is being, accumulated on a wide range of subject matter, and there is no doubt that the project will add significantly to basic knowledge of the southern Beaufort Sea and will help to solve certain engineering problems. But the time available for the study (about 18 months) is far too short for the development of decisive judgements concerning the feasibility or hazards of drilling for oil in that region. This is most apparent in the ecological aspects; it has become disconcertingly evident, for instance, that our knowledge of fish populations and their life cycles is in a primitive stage of development, little more advanced than the preliminary natural historical survey. This makes the whole project look like a five million dollar sop to that group known in the trade by the almost derogatory term of "environmentalists," and leads to the cynical view that no matter what the final report of the project looks like, the oil development will be put into action anyway.

Other marine impact studies in arctic Canada are far less comprehensive than the Beaufort Sea study, and there seems to be a tendency to forget that most of the year in the north is winter, and that what happens in the sea in winter is quite different from the events in the short summer during which most impact studies are made.

Myth of the fragile Arctic. On the other side of the coin, there is the constant repetition of the "fragile Arctic" story. I have discussed this briefly elsewhere (Dunbar 1973), and have come to the conclusion that it is based on two things — the vulnerability of permafrost tundra terrain to heavy vehicles in summer, and the fact that ecosystems in high latitudes are simpler (less diverse) than in the tropics. The first point is straightforward enough, but once the fact is known, steps can be taken to avoid doing the damage. The second point seems to have been seriously misunderstood. The simpler, oscillating system, given certain safeguards, is likely to be much tougher and more resilient than the highly diverse "stable" tropical systems, particularly the rain forest (Gomez-Pompa et al. 1972). So far, work on this problem has been at the theoretical model level, and it is impressive. Tropical experiments have been carried out, but so far there has been no attempt to design experiments in the arctic marine environments, except to study directly the effects of oil on life on a small scale, which really tells us nothing about the main issue — namely, the large-scale effects of serious specific depletion.

STATUS OF MAN'S UNDERSTANDING OF ARCTIC MARINE ECOLOGY

Incomplete and comparatively primitive though our arctic marine ecological knowledge may be, we can at least point to the most serious gaps that exist in it, with particular reference to the impact of industry, always remembering that it is important to discover the weakest or most vulnerable links in the biological cycle, the breaking of which could seriously disturb the whole system beyond the point of recovery within a reasonable time. After all, the most durable of our arctic resources are the renewable resources, by definition, and it would be foolish to destroy them in the short-term process of extracting the non-renewable resources.

Microbiology. A comparison of the arctic marine ecological literature with that of the world marine scene reveals some remarkable gaps in the arctic work. We seem to know almost nothing at all, for instance, about cold water marine bacteria, and in fact very little about the whole of the microbiological part of the biological cycle. Some work is now beginning on the metabolism of the bacteria involved in the degradation of oil in cold water, something that has now become an urgent practical problem of short-term importance; but we know nothing of the rate of bacterial breakdown of organic detritus in cold water, the process of nitrification that produces nitrates and nitrites, or of the parallel activities of the producers of phosphates and silicates. Upon these inorganic fertilizers depends the whole of the annual biological production cycle. And there has been no work at all done on the denitrifiers in arctic regions, the bacteria which convert nitrite and nitrate to elemental nitrogen, or on the nitrogen fixers, bacteria and blue-green algae, which use elemental nitrogen in the formation of their own proteins. In fact, it is only comparatively recently that bacteriologists as a group have come to realize that bacterial activity can even take place at all under arctic temperatures. The lack of this research means that we suffer from the severe handicap, in estimating the environmental impact of industrial activity, of not knowing the rates at which these microbiological processes take place, for it is certainly not legitimate or reasonable to assume that in such estimations we can simply apply the Q-10 rule and extrapolate from temperate or laboratory rates.

Exploitable marine species

The population dynamics of large numbers of species in temperate and tropical seas have been worked out to the point where good predictions can be made of the consequences of different levels of exploitation pressure. This work has been done in response to the need to regulate and conserve the species concerned. Very little such work has been done in the arctic marine environment, in contrast to the considerable attention paid to terrestrial arctic animals such as the furbearers and their food supply. Again, the difference reflects the levels of commercial exploitation. In the marine systems, only a few species have been studied from this point of view, such as the ringed seal, the walrus and the Arctic char. In all three instances, this work cannot be said to have passed the preliminary stages nor with adequate geographical coverage. This makes the presence of industrial attack upon the system the more alarming; it should be noted also that in this case the commercial exploitation aims not at an individual species with renewable populations, but at non-renewable resources, the extraction of which is likely to be hazardous to the life of the region: In other words, the commercial interest in the conservation of the renewable resources does not exist, with consequent lack of motive to take the biological investigation seriously.

A knowledge of the population dynamics and life tables of marine species at higher trophic levels is necessary not only in the interests of rational exploitation, but for the reasons discussed above, namely the prediction of the consequences to be expected following a serious depletion of one or two key species from whatever cause, as illustrated by the cases of the blue whale in the Antarctic and the bowhead whale in the north. Directly related to this problem is that of the limits of ecosystem tolerance, or the ability to withstand perturbations beyond the natural amplitudes. There are therefore two problems: that of the effects of perturbations on individual species, and that of the ability of the system to adjust to the loss of a key species. Here again, we need the application to the arctic scene of techniques developed for the study of temperate ecosystems, because the first attack on these highly complex problems must be in the form of intelligent modeling, which can show where field work and direct measurements are needed, and which provide a conceptual framework in which the research can progress. A good example of such work is that of Steele (1974). Again the specifically arctic trouble here is that the oil and gas industries have developed too fast in the north for the scientific world to prepare its defenses or to catch up with the rapid Northward Course of Empire.

Ice biota

Another field of investigation that started late is that of the ice biota, chiefly diatoms, living within the sea ice itself. This turns out to have certain practical applications of possible immediate importance. This study began only 15 years ago, apart from a few references to the occurrence of this within-ice life without any attempt to measure it or to record its annual cycle. Recent reviews of this work have been given by Horner (in press) and by Grainger (in press), and there is no space to go into detail here. I would refer, however, to two papers of special interest, those of Buinitsky (in press) and of Hoshiai (in press). Buinitsky reports that the strength of sea ice is approximately cut in half by the intense layers of plant growth, and Hoshiai finds in the Antarctic that the horizontal distribution of the plant growth in the ice is highly variable, or patchy. These two points taken together clearly indicate the need for some sort of mapping technique for this phenomenon, and the working out, if possible, of a means of estimating the plant growth rapidly and over large areas, perhaps by remote sensing.

Suspended inorganic matter

Ice drilling operations, like large rivers, pour much suspended inorganic matter into the sea, and the behavior of this suspended matter is interesting, important, and somewhat neglected. The Beaufort Sea Project, in one of its studies, has produced the surprisingly high estimate for the sediment load carried by the Mackenzie River into the Beaufort Sea of 15×10^6 tons/yr. Certainly the plume of high turbidity visible from the air extends very far to the east of the river mouth. The process of flocculation in seawater has not been given the attention it deserves, except in the Netherlands. In Canada there is only one worker contributing to this field of research, namely Dr. Kranck of the Bedford Institute, whose papers are attracting the attention of biologists and chemists as well as marine geologists (see Kranck 1975). The flocculation process, a phenomenon very largely of estuarine and river-mouth regions, forms larger flocs from smaller particles, and deposits them on the sea-floor. This has several very practical results: It builds up sediment in coastal and estuarine regions, thus calling for dredging in many areas; it forms a boundary region between two ecosystems, the freshwater and the saltwater systems; it clears the oceanic water seaward of a great deal of suspended matter; and it must also remove from the coastal water much particulate pollutant material, including such things as oil droplets and the solid effluent from pulp and paper plants. Particularly in the Alaskan coastal waters and the southern Beaufort Sea, therefore, but in fact in any region where coastal industrial development occurs, this process should be given very close attention. The process is by no means yet fully understood. It is not clear, for instance, whether it is dependent primarily on the mixture of fresh and salt water, with attendant changes in the behavior of electric charges, or upon sheer concentration of suspended matter, or upon some other effect. Nor is it known what the effects of high turbidity and of flocculation are upon life in the sea and on biological production. One might reasonably suppose, for instance, that the presence of a concentration of flocs of from 10 to 75 microns diameter would be unfavorable to the life of the microplankton. For many of these points I am indebted to K. Kranck (personal communication).

Climatic change

While paying close attention to what mankind does to natural ecosystems, one should not forget to keep a watchful eye on what nature does to them. It is remarkable how unwilling we are to accept the truth of that old Hippocratic conclusion that "change itself is the only reality." I am not as familiar with the Bering Sea and Alaskan arctic regions as I am with the eastern Arctic and Greenland waters, but it is quite clear in the east that the arctic-subarctic-boreal boundaries today are not the same as they were 30, 50, or 70 years ago. There has been a climatic upswing in west Greenland since about 1915 which brought the Atlantic cod up the coast in commercial quantities, and it appears that the relaxation of that upswing has brought the Atlantic salmon. The lesson to be learned, of course, is that one must not expect the living resources of the north to stay in the same place all the time, and this makes the assessment of man's activities all the more complex. In a given region, what is the state of the system, what are the key species, how will their distribution shift, and in what direction is the ecosystem going to change? All these questions need much more attention than they are given, and all of them involve marine climatic change. We need a CISCC, to coin an acronym, a Continuing International Study of Climatic Change, to give us the best possible information on the probable future events in the marine climate. Something of this sort is part of the program of the POLEX operation, but only for a limited period and only in a limited region.

EFFECTS OF MAN ON HIS OWN HUMANITY

Arctic native peoples

There is an obvious and justified misgiving on the part of the native northern peoples concerning the present activities of industrial humanity. The native peoples of Canada and Alaska clearly place much greater importance on the survival of the renewable resources, of which they are a part, than upon the extraction of the non-renewable. Obviously they must do so, for their way of life is dependent upon the living resources, and they wish it to remain so. In the case of the Eskimos, or Inuit, the living resources of the sea are particularly important, and it is the marine resources with which we are concerned here. We are thus faced with a basic question of values, the choice between immediate financial gain offered by oil and gas exploitation (and gain for whom? – one might ask), and the long-term survival of a living system. It may be, possibly, that this choice will not have to be made, and that is what impact studies are all about; but the past history of industrial man is not encouraging in this respect.

Bureaucratic pollution

The last point I would raise is a more general one, and perhaps a fairly new one. We have a new kind of pollution among us, in the form of bureaucratic growth in the organization of science. Parkinson's Law has been with us for some time, but it was invented, or discovered, in terms of government departments other than those devoted to science, or so we were given to hope. It always threatened the scientific departments, and now the threat is translated into reality. The growth of a heavy matrix, or connective tissue mass, of administrative offices occupied by men and women whose main function is to make work for themselves and to build little sub-empires within bigger empires, is stifling the scientific productivity and initiative of this and other countries. And this cancerous growth is not only encouraged by the politicians and senior civil servants; it is planned and supervised by them. We were warned, of course. Several years ago Hans Zinsser warned us that the administrative camel would crowd the intellectual rider out of the tent (Moe 1951). Now it is happening, with special reference to the scientific rider, and the process is both increasing the cost of scientific research and decreasing the scientific returns.

In government scientific departments with which I have had contact, I believe it is the rule that in years immediately following the last war there were perhaps two administrative levels separating the summer student from the Deputy Minister. Now there are often six or seven. The levels inserted during the years contribute nothing to the scientific productivity of the department, and they are not intended to do so. They are "administrators," define that as you will, and their status and their pay are proportional to the number of staff members they have working "for" them. It is to the financial advantage of the younger members of these departments to leave productive scientific work as soon as they can and start climbing the administrative ladder. This process is foisted upon the public (41 percent of the Canadian National Income is spent on the civil service) in the name of efficiency. Obviously, it is a sham; it is quite clearly the reverse of efficiency, since it absorbs energy and time unproductively and on an enormous scale.

The growth of this administrative monster continues daily. It is not the place here to go into detail, although detail is available in abundance. There is time only to point out that scientific research suffers increasingly from this growth, that it is upon this situation that the framers of "science policy" should be focusing their attention, and that it is this burden that has handicapped, perhaps more than anything else, the progress in ecological science in the Arctic, of which we stand so desperately in need today. This may seem to have only passing relevance to the subject of this chapter, but I do not think so. We are heavily dependent on government science in northern research, for obvious reasons, and it is in government science that this problem has become overwhelming. There is a growing sense of despair among the younger men and women of science who wish to continue in productive scientific research rather than to give in to the siren of the administrative career.

CONCLUSION

At the beginning of this paper I posed the question, "Are we or are we not in danger of destroying the arctic marine environment?" The answer is, of course, that we don't know, and I think we have to admit that we are not really doing the best we can to find out.

Conference Discussion

McAuliffe — How do you explain why the blue whale is slow to come back after being depleted as the food source (krill) increased? Seals and fish have come in and now appear to compete for krill. Initially, the blue whale was most efficient in obtaining krill compared with predation by seal and fish. As the krill supply is now high, why can't the blue whale again compete?

Dunbar — It appears that, in the evolutionary view, the homotherms had the advantage in the use of the polar regions during the pliocene cooling; so that the greatest competition for the whales might be expected to come from the other homotherms rather than from the fish — but the Antarctic fish do seem to be increasing. But I appreciate your main point, and I agree that ultimately the original balance should be restored, whales and all; but this competition may make it slower than expected as has happened in the case of the bowhead whale in the North.

Korringa — In July 1975 the International Commission for Exploration of the Seas (ICES) held a symposium in Aarhus on long-term changes in the sea. The main conclusion was that the changes observed in the composition of fish stocks and in plankton should not be ascribed to climatological or hydrological fluctuations, which are minor, but rather to man's fishing activities.

Norton — You have materially contributed to ecological theory by your distinction between non-oscillatory stability of systems (e.g., tropical rain forests) and stability in the sense of resiliency in the face of regular perturbations (e.g., North American tundra systems). Does this geographic distinction hold for strictly marine systems? And, are there practical ways in which industrial activity in northern regions might reduce its ecological effects by mimicking the natural disruptive forces that give the biota its character?

Dunbar — Two very interesting questions, but I am not sure that I can answer either of them satisfactorily. I doubt whether anything strictly similar to the tropical rain forest condition exists in the sea. In the rain forest, the ratio of production to biomass has been reduced to a minimum. In the most non-oscillating marine environment, which is probably the deep water of the Arctic Ocean, the biomass is very low indeed, and nothing much is known about the production rate — it must be low also. But no marine system builds up the great mass of standing tissue that is found in the rain forest. In temperate and high latitudes, in surface waters, the seasonal environmental oscillation is well developed, and the system is well adjusted to it. But it seems that the tropical marine systems also oscillate, judging from recent results from our station in the Caribbean, so there may be no true parallel to the rain forest anywhere in the sea. On the second point: the idea is ingenious, but I think it is a matter too delicate for reply. Certainly the native population would object strongly!

REFERENCES

BUINITSKY, V. K.

(In press) Oceanic life in sea ice. Proc. SCOR/SCAR Polar Oceans Conference, May 1974, Montreal.

DUNBAR, M. J.

1973 Stability and fragility in Arctic ecosystems. *Arctic* 26(3): 179-185.

GOMEZ—POMPA, A., C. VAZQUEZ-YANES, and S. GUEVARA

1972 The tropical rain forest: a nonrenewable resource. *Science* 177: 762-765.

GRAINGER, E. H.

(In press) Primary production in Frobisher Bay, Arctic Canada. Synthesis volume on marine productivity, International Biological Programme.

HORNER, R. A.

(In press) History and recent advances in the study of ice biota. Proc. SCOR/SCAR Polar Oceans Conference, May 1974, Montreal.

HOSHIAI, T.

(In press) Seasonal change of ice communities in the sea ice near Syowa Station, Antarctica. Proc. SCOR/SCAR Polar Oceans Conference, May 1974, Montreal.

KRANCK, K.

1975 Sediment deposition from flocculated suspensions. *Sedimentology* 22: 111-123.

MARGALEF, R.

    1968   Perspectives in ecological theory. University of Chicago Press, Chicago, 111 pp.

MITCHELL, E. D.

    1973   The status of the world's whales. *Nature Canada* 2(4): 9-25.

MOE, H. A.

    1951   The power of freedom. *Pacific Spectator* 5(4): 435-448.

STEELE, J. H.

    1974   The structure of marine ecosystems. Blackwell Scientific Publications, Oxford and London, 128 pp.

CHAPTER **3**

Ecological assessment as criterion in the rational exploitation of natural resources

P. KORRINGA[1]

INTRODUCTION

Man's concept of nature

It is not easy to define the concept "nature." Perhaps one can come as close as this: "Nature is man's perceptible terrestrial reality in the sense that it came into being without his interference." Whatever the case, we can rest assured that man has for ages considered nature as a source of food and has devised religious concepts which tell him that nature has been put especially at his disposal to be taken advantage of. And that is what he did — first by collecting edible items in the wild from a great diversity of plants and animals; later, when he realized that not so many mouths could be filled that way, his interference reached a gross scale and reduced the number of species drastically by means of what is now called agriculture and animal husbandry. The areas unsuitable for these activities were long considered hostile to man and were called "waste;" accordingly, plants and animals of no conceivable use to man were by definition labeled "weeds" and "vermin."

Nature's redefinition of man's place

In due course, man saw that indiscriminate interference with nature could lead to disaster, to a complete loss of productivity. As woods were felled, erosion would wash away the valuable topsoil that before had served as a sponge during heavy rains. Or, once destroyed by man or his cattle, the protective mat of vegetation was simply blown away by the wind. Several once-fertile areas became derelict this way, and desert type conditions may

---

[1] *Netherlands Institute for Fishery Investigations, Box 68, IJmuiden-1620, Netherlands.*

now prevail where long ago one found centers of civilization. To compensate for such losses, man made an effort to put more and more "waste" to "good use" by reclamation, irrigation and manuring — driving back what was left of "nature," which had taken refuge in the poorest grounds.

It took a very long time for man to realize that nature should not be looked upon as something hostile to him. He had to see that he could not do without nature, that he was not a sort of supernatural creature transcending nature. And finally he came to realize that he arose from nature and belongs to it, being just one of many mammals competing for food and space with other animals.

Recognizing the cost of his abusive activity in terms of irreparable erosion and extermination of several species of larger animals, man felt now inclined to be more careful in his attack on what was left of nature. Public opinion has now become aroused in such a way over environmental impact that it is a sheer political necessity to take the environment into account in planning. Ecological assessment is therefore essential in our time if one wishes to exploit areas which thus far have belonged to the domain of nature.

A question of duty

There is considerable confusion in the basic philosophy of the protection of nature. Where the Christian background prevails, one is often inclined to ascribe to mankind the duty to manage this world, to protect all its denizens big and small against indiscriminate attacks which might lead to their extermination. One feels that man simply does not have the right to eliminate any species of plant or animal, even if they seem to be hostile to him. But in a more critical perspective, such an arrogant attitude is not tenable. Margalef (1968) once said that one can have different views on man's duty on earth, but he seldom encounters someone who claims that man should be the director of the zoo. In reality, it is only the utilitarian arguments which count; one distinguishes primarily what nature does for man's prosperity and only secondarily considers how it affects his more fundamental well-being. Man's prosperity is at stake in his exploitation of nature, not only through agriculture and animal husbandry, but also in his extraction of minerals and development of water power. In all these cases, man should be aware of the danger in over-exploiting limited resources. Since agriculture and animal husbandry are cases of applied biology, the techniques of which have been developed both by practical experience and by scientific investigations, research carried out in the natural environment to further improve these practices must be considered of great importance. One should therefore reserve fair sections of "nature" for that purpose. Nature's wilderness, not exploited by man, is also of considerable practical importance for its regulatory capacities: it serves as a "gene-pool" from which man can select in his efforts to improve the hereditary qualities of his cultivated plants and cattle; it acts as a buffering agent, supplying predators and parasites when something gets out of control in the arable fields; it contributes to the global supply of oxygen in the air, biodegrades certain waste products, and also improves the climate.

A matter of necessity

More difficult to define is the need of nature for man's well-being. Man expects more from nature than the purely material contributions just listed. He needs nature for his "pastoral recreation" and wishes to have the possibility to enjoy nature by visiting nature reserves in which he can enjoy both the landscape and its denizens. The scarcer real nature becomes in a given country, the greater the need for this type of recreation.

Some ecologists claim that man has no right to destroy nature for his materialistic needs. According to them, man should refrain from any further attacks on nature, since one cannot foresee the various and often complex effects of such impacts. This attitude makes discussion impossible and reveals at the same time its weakness: one fears evidently deleterious effects for man himself, not for nature as such.

Ecologists are now consulted before one undertakes any further attacks on what remains of nature, and they should try to use the same kind of "hard" arguments as the planners from other disciplines. This means that they should try to express their conclusions in figures, if possible in money; otherwise they will lose their case too easily when interests clash.

One must be realistic and should comprehend that there is not a very strong argument to be had in nostalgia for conditions that once prevailed when the human population was much sparser than it is today. Further, the "biological equilibrium" so often cited is not as labile as is assumed by common opinion. The greater the diversity of species, the more stable the entire population composed of plants and animals. A complex population with long food-chains is interesting as such, but it has a very modest net production in comparison to its primary production of organic matter provided for by the green plants. Farmers have understood this for a long time — hence their practice of growing crops in monoculture, however vulnerable these isolated crops may be to attack by predators, parasites and disease. The equilibrium observed in nature is not static, but truly dynamic — all participants trying to sustain themselves as broadly as possible. It does happen in nature also that species disappear from a population. But this does not mean that the biological equilibrium has failed; ecological balance cannot be characterized as a house of cards which collapses when one card is taken away somewhere. It is rather like a kaleidoscope that can show various types of equilibria, even when the number of participating units is reduced, which means in nature that fewer species participate in maintenance of the biological equilibrium.

DETERMINATION OF ECOLOGICAL VALUES

Ecological assessment should determine the value of a given piece of nature; it should show what man has to lose if he decides to use this area, or parts thereof, for other purposes.

Abiotic versus biotic features

In assessing the value of a given area, one should first of all distinguish the abiotic factors (such as the nature of the bottom deposits, climate, hydrology, and relief, which together determine the habitat potential for communities of plants and animals) from the biotic factors (flora and fauna). Since the nature of the vegetation determines largely which animals can thrive in a given area under consideration, more attention should be paid to the flora than to the actual fauna in the biotic assessment. In this assessment, one has to make a clear distinction between the flora (the list of plant species encountered) and the vegetation (the spacial distribution of the individual plants in relation to their habitat). In terrestrial ecological assessment, one usually pays far more attention to flowering plants than to cryptogams, to birds and some of the larger diurnal mammals than to species of all the other groups of the animal kingdom.

Diversity factor

In making an ecological evaluation, one attaches great value to the factors "rarity" and "diversity," with respect to both abiotic and biotic factors. A type of soil or climate of rare occurrence is just as interesting as rare species of plants and animals, in which one has to take into consideration whether the item is rare on a regional, national, or global basis. The factor "rarity" is easily overemphasized. For small plants and animals, easily disseminated, a verdict often holds that "rare species do not exist; the point is where to look for them." For species with larger individuals that encounter barriers in spreading, one may attach great value to their occurrence within national borders, whereas they may abound in neighboring countries where ecological conditions happen to be more conducive to them. This diminishes the true value of the observed rarity. Rarity and threatening extinction often indicate great vulnerability of a species, usually caused by its narrow ecological range, which may easily lead to its disappearance by minor climatological or other environmental changes and ensuing alteration in vegetation. This could also happen without man's interference, and it did happen on a large scale before man's appearance on this planet. Still, biologists cannot help being fascinated by the rarity of a species and often attach undue value to it.

Diversity of landscape, of flora and fauna, is usually considered as a factor which should weigh more heavily in the opinion of the ecologists than rarity alone. An area with a varied landscape is considered to be more attractive than a rather monotonous area, and great numbers of species of plants and birds indicate the biological importance and stability of the area under consideration. In terrestrial ecological assessment, one therefore attaches great value to the list of species recorded, if possible made up in the various seasons, and indices are used by ecologists to classify the terrians studied. But here, too, one easily fails: some very important bird sanctuaries (such as the Dutch Wadden Sea or sites where wild geese hibernate) are very rich in food but show a limited number of species. Just as on arable land, the greatest production of food is to be expected where the number of species is small and the food-chains short.

Other considerations

Yet other factors are taken into consideration in the ecological assessment of an area, especially when one considers its allocation as a nature reserve or national park as part of humanity's legacy. These values include ecological maturity (climax phase of natural succession in the vegetation), how much time nature would require to restore an area after a calamity of some sort, the size of the area, and its accessibility. Nature reserves completely out of reach for recreational purposes are of limited value to a nation.

DATA CLASSIFICATION

Pricing the parameter

Serious efforts have been made to classify the data for each separate parameter and to use multiplication factors for the various parameters, depending on the importance attached to them for the entire scheme. Such efforts should finally produce a price tag for ecological assessment in an effort to negotiate with planners from varied disciplines on a common basis. An interesting example of this can be found in Helliwell (1969). It is clear that such a procedure is open to serious criticism, both as regards the arbitrary way of classifying data for each separate parameter, and considering the subjective nature of the multiplying factors used to indicate the relative importance of each parameter. That it is virtually impossible to evaluate nature in hard figures became evident in the attempt of a commission set up by the Dutch government to reconsider a proposed enclosure of the Oosterschelde estuary within the framework of the Delta Project (Anonymous 1974). It was possible to calculate the value of the oyster and mussel resources, based on a series of production figures, worked out to the level of national economy. However, no realistic price tag could be attached to the value of the Oosterschelde as a hibernation site for millions of wading birds. From a purely scientific point of view, an assessment in which one makes notes on the abiotic factors and further on the flowering plants and singing birds encountered, forgetting about all the rest, could not possibly lead to a clear insight of the biological value of a given terrain. Several generations of biologists of the Leyden University have carried out investigations on flora and fauna in a section of dunes north of the city of The Hague, called Meijendel, taking into account the environmental conditions. Fifty years of investigation in this area have been summarized in a well-illustrated book that offers a wealth of information on many groups of plants and animals (Bakker et al. 1974). However, if the question is posed of whether a good quantitative understanding has now been reached of the biological economy of this limited area, or of the various interspecific relationships, the reply must be that the still rather scanty and fragmentary data cannot possibly give a complete picture of life in Meijendel.

'Pastoral recreation' - the terrestrial condition

There is still a surprising abundance of nature left, despite the many densely

populated countries and the erosional ravages of climate, relief and hydrology. It is just as surprising that the rather superficial ecological assessments
made thus far have often led to a wise choice of areas to be reserved for
man's enjoyment, either to be protected completely against human interference or established as a national park where only minor disturbances are
allowable. The reason why assessment results are usually relatively good
(despite the limited value of the figures one produces in ecological assessment in comparison with figures for agricultural or industrial exploitation) is
that the observations made cover exactly what is of special interest to those
who wish to use nature for recreation. For special investigations of a purely
scientific character, it is often more practical to travel to areas where man
has hardly treaded than to try to maintain a limited terrain in a densely
populated country completely free from human interference.

What does the average nature lover want? A natural landscape with ample
acreage and natural contours — adorned, if appropriate, with the trees which
belong there. In that landscape he wishes to find a diversity of flowering
plants, further — a variety of birds, perhaps some butterflies, and possibly
some larger mammals. In reality, he does not care about the cryptogams and
other plants without distinct flowers; about mollusks, crustaceans, and spi-
ders; about bats and mice. In short, his concern is not with all sorts of
animals which lead a nocturnal life or are otherwise inconspicuous. Therefore, there is no real need to grasp the entire flora and fauna in all their
elements if one wishes to advise on the value of a given terrain for the
purpose of what is called pastoral recreation.

What worth the sea

What about the need for environmental assessment of the sea? In fact, the
same sort of philosophy applied for the land holds good for the marine
environment. Man wishes to use the sea in various ways. For ages it has
served his transportation and fishery needs; in our time, it is also heavily
used for recreation, extraction of minerals, and waste disposal. The oceans
may be vast and seem to be invulnerable, but they are not so homogeneous
as they may seem. Only where the seas are quite shallow (the continental
shelf) or where upwelling of nutrient-rich water takes place is there a
concentrated fishery. About half of all the catches made in the whole world
come from under 0.1 percent of the surface of the seas. The edges of the
sea are often of considerable importance as nursery grounds for fishes and
shrimps and also for shellfish farming (Korringa 1967, 1973a). Therefore,
every intended impact by pollution or engineering works, land reclamation
included, should be carefully planned with the fisheries authorities if one
wishes to avoid deleterious effects in an often far greater area than where
one plans to operate. Fishery biologists in coastal countries study fluctuations in fish stocks as affected by environmental factors and by man.
Though natural factors do affect the degree of success in reproduction of
many species of fish and shellfish, it often turns out that man exerts a far
greater influence on the events under water by his fishery operations than
he is inclined to believe. The total fertility of the section of the sea under

consideration, expressed in nutrient salts or primary production, is usually a parameter beyond man's control. But just as on the land, a great diversity of species and long food-chains will lead to a noteworthy stability of the community but at the same time results in modest net production at the end of the food-chains. Where large stocks of old fishes occur, feeding voraciously but hardly growing at all, even the net production will be meager. Reducing the average age of the fishes increases the net production of the sea by elimination of specimens with poor conversion rates. A well-loaded table will not remain untouched for long in nature. When many of the older fishes are removed by the fishery, young specimens of the same or of other species will make use of the available food. On the other hand, too great an intensity of fishing leads to overfishing — the consequences of which include: reduced average size of fish caught, a smaller number of year-classes in the catches (and hence greater fluctuations in the size of the annual catches), and a declining catch tonnage per unit of effort. It is the fishery biologist's duty to study these matters and to advise his government on ways and means to come as close as possible to what is called the optimum sustainable yield. Further it is the fishery biologist who should study the life cycle of the commercially important species, locating their spawning areas and nursery grounds. This knowledge makes it possible for him to indicate the inshore areas which are nursery grounds or spawning sites of noteworthy importance for the stocks under consideration (Korringa 1973a). Recommendations on these matters should also be made available for those who are considering use of these inshore sites for other purposes such as for gravel dredging. On this basis, it is possible to reach a sound decision at a higher level on the best use to be made of an area under consideration for the benefit of the participating country and as much as possible without endangering other national interests. In a sense, this research that is carried out by the fisheries biologist on behalf of the fishing industry could be called ecological assessment.

The sea as pastoral recreation

Is there also a need for ecological assessment, in the sense discussed for terrestrial conditions, to advise on reserving parts of the sea for recreational purposes? In a way there is, especially when it concerns animals, which are a joy for the outdoor recreationist. As such, one can mention a variety of seabirds and seals, though they all live primarily on the land and only use the sea as source of food; dolphins and other creatures of the whale family; the tropical coral reefs with their colorful beauty and great diversity of species; and the fishes and variety of invertebrates. For such cases, ecological assessment is essential, for all those creatures can suffer losses and occasionally be driven almost to extinction by indiscriminate hunting or destruction of rookeries; pollution can destroy the beauty of coral reefs, vulnerable also to explosives and crowbars used in search of rare specimens to be sold as collectors items. Protection of nature is surely a good and necessary thing for these groups of animals. But what about the truly marine species in temperate and arctic water — the fishes, the mollusks, the crustaceans, the worms, the sponges, the ascidians, and the many small zooplankton and phytoplankton creatures. What kind of protection do they need against man's impacts, and what ecological assessment is required to pave the way to wise decisions in case of a clash of interests?

Fish and shellfish

For species of fish and shellfish of commercial importance, it is the fisheries biologist who carries out research and gives advice to government authorities. Just as in the case of terrestrial production, a limited number of species of fish leads to far greater net profits on the basis of a given fertility rate of the sea than a great diversity of species. For non-commercial species of no great value in the food-chains which lead to man, protection seems hardly necessary: thus far no species of fish or marine invertebrate has been exterminated by man nor will ever be exterminated, given the great range of their natural occurrence and the limited area of man's activities. Only few people, even very few naturalists, can call to mind more than 100 species of fish. Still, the number of species of fish has to be counted in the tens of thousands. The same holds good for the invertebrates of the benthos living on or in the seabed and for the planktonic species: only very few specialists know them, and their diversity of species is often great. Nobody can grasp them all. There seems to be little chance that man will ever eradicate some of the tens of thousands of species in the sea, and on the other hand there is virtually nobody who can claim rightfully that he derives part of his happiness from observing and admiring them in their natural habitat. The fisheries biologist will say that the number of species is far too great already in the sea and that larger catches could be made by the fishery if the fauna of the sea approached to some extent the monoculture pattern so characteristic on arable land. Cultivation of mussels and oysters already shows such greatly increased yields.

Of course, one should be careful with pollution by dumping or discharging waste containing ingredients which are both toxic and non-degradable. Thus far, though, pollution of the sea has been mostly both localized and reversible. Since coastal waters in particular are liable to pollution, it is the shellfish beds and nursery grounds for fish and shrimps which are the most vulnerable of all. In waters such as the North Sea, however, surrounded by densely populated and highly industrialized countries, there is as yet no sign of reduction in the fish stocks through domestic or industrial pollution (Korringa 1973b). Rumors about dying oceans, often based on sweeping statements of laymen, are greedily absorbed by the masochistic public and should be regarded with considerable care and criticism.

CONCLUSION

The conclusion must be that ecological assessment in the marine environment should on the one hand be left to the fisheries biologist, who is responsible for the rational exploitation of the living resources of the sea and thus for man's prosperity. Complementing this function is the assessment of the ecologists, advising on the well-being of birds, seals, whales, and coral reefs — all of considerable importance in the recreational sector, thus responsible for man's well-being. They should focus their attention on rare sets of environmental conditions with an interesting fauna appealing to the general nature lover and should try to protect such areas when a clash of interests might occur. Here, too, the ecologist will experience the disadvantage of finding it virtually impossible to affix a price tag to his object, the more so since he deals with collective properties and not with private goods. As in the terrestrial consideration, we have to admit frankly that ecological assessment is not carried out to protect foremost the interests of the multitude of species of animals and plants competing for a place on this earth. Principally we think in terms of the interests of man himself, who wishes to exploit natural resources as rationally as possible and to enjoy the pleasures of observing a well-defined and limited number of denizens of the sea during his pastoral recreation.

Conference Discussion

Dunbar — Are there not two important types of "stability:" the steady-state, highly diverse pattern of the tropical rain forest which is stable but very vulnerable to perturbation and the oscillatory "stability" of mid- and high latitudes which can respond elastically to perturbations and return to status quo. So the most vulnerable ecosystem is the tropical rain forest, not the arctic tundra.

Remember also "Margelef's Paradox" — that nature evolves to produce a greater diversity, thereby ultimately increasing its own vulnerability.

Korringa — A tropical rain forest is composed of hundreds, if not thousands, of species of trees, and is inhabited by many species of animals, large and small. Predation, parasites, or disease endanger some species of trees— but never the forest as a whole. The same holds for the invertebrates and for the animals of a higher level in the food-chain which live there. Climatic extremes likewise endanger certain species, but do not easily threaten the forest as a whole. And the same can be applied to a sea containing a high diversity of marine species: the biomass remains very stable despite fluctuations in species abundance.

Gerritsen — The philosophy on which you base your viewpoints seems basically pragmatic or utilitarian. Would it be possible to use a different philosophy which is more basic?

Korringa — Of course, one can advance other philosophies than the pragmatic one I just presented, but my duties as a fisheries biologist force me to be practical rather than purely theoretical. I am strongly convinced that nature's protection should have a utilitarian basis in which man takes the central position — not nature itself.

REFERENCES

ANONYMOUS

1974 Rapport Commissie Oosterschelde. Staatsuitgeverij's—Gravenhage, 199 pp.

BAKKER, K., ET AL.

1974 Meijendel, duin-water-leven. Meded. Meijendel Comite', N. S. 28. W. van Hoeve, B. V. Den Haag/Baarn, 272 pp.

HELLIWELL, D. R.

1969 Valuation of wildlife resources. Regional Studies, Vol. 3, Pergamon Press, pp. 41-47.

HILL, M.

1968 A goals-achievement matrix for evaluating alternative plans. *J. Amer. Inst. of Planners* 34-1.

KORRINGA, P.

1967 Estuarine fisheries in Europe as affected by man's multiple activities. *Amer. Assoc. Adv. Sci.* 83: 658-663.

1973a The edge of the North Sea as nursery ground and shellfish area. In *North Sea science*, edited by E. D. Goldberg. MIT Press, Cambridge, Mass., pp. 371-382.

1973b The ocean as final recipient of the end products of the continent's metabolism. Pollution of the oceans: Situation, consequences and outlooks to the future. In *Ökologie und Lebensschutz in internationaler Sicht*, edited by H. Sioli. Rombach Freiburg, pp. 91-140.

MARGALEF, R.

1968 *Perspectives in ecological theory*. Univ. of Chicago Press.

CHAPTER **4**

Ecological principles for guiding engineering design in arctic regions

BEATRICE E. WILLARD [1]

When is man going to realize that land is a community to which he belongs rather than treating it as a commodity he can exploit.   (Aldo Leopold *1945*)

INTRODUCTION

The pressures to obtain more fuels have brought about much well deserved attention and investigations in the arctic and antarctic regions of the world. Most view this fact with a certain degree of ambivalence, since we all know well the special risks, sensitivities, and problems of operating in an arctic environment in contrast to temperate or tropical environments. Therefore the challenge is great for the scientific community to exert its best efforts in bringing expertise to bear in solving, minimizing, and mitigating problems, before they arise as much as possible.

Terminology

Following are definitions of a few terms that are well known, but which have come into broad and imprecise usage recently.

Environment. Sum total of the physical and chemical factors in which living things live; for man, environment also includes economic, social, political, and cultural factors.

Ecology. The science that investigates the nature and operation of ecosystems as entities. *Eco* is derived from the Greek root *Oikos*, meaning home

[1] *Executive Office of the President, Council on Environmental Quality, 722 Jackson Place, N.W., Washington, D.C. 20006.*

or habitat, which is the same root as in economics; this is not mere coincidence, in Ancient Greek, economics and ecology were synonymous terms, used interchangeably. In modern application, we find that the two disciplines widely complement each other in function and results, when their congruency is recognized and sought.

Ecosystem. Any of the recognizable units of the landscape such as white spruce forests, Sitka spruce forests, or tundra stands. Ecosystems are composed of environment factors, organisms, and the dynamic interactions that operate among organisms and environment factors. The components and processes vary from one ecosystem to the next, so each ecosystem type has to be looked at individually.

Environmentalism (conservation). A philosophy of management. Teddy Roosevelt coined the term "conservation" and defined it as "wise use." It leads to developing an ethic about use and attitudes toward land, air, water, living things, people, ecosystems. People of all types of background are environmentalists.

Growth. This is defined by the *New Webster Dictionary* as a "gradual development toward maturity," i.e., perfection. It is interesting that this definition says nothing about increase in size or numbers.

Assumptions

In considering the balance of energy, environment, and growth, it is important to recognize basic assumptions. I am assuming that for the purposes of this discussion, the varied benefits of technology, which require adequate energy, are appreciated by those of us concerned with the total environment. We favor growth, as stated above — "a gradual development toward maturity" — not mere increase in number and size. We enjoy and wish to retain the attractive features and productivity of the arctic environment. We want to bring our activities into harmony with our environment, with those ecosystems of which we cannot help being a part.

These assumptions imply an interest in achieving a balance among energy, environment, and growth where none of the three dominate the other two.

NATURAL GUIDELINES FOR BALANCING DEMANDS ON ENVIRONMENT

Scientists and engineers are trained to look for and apply knowledge of natural principles, such as gravity and Charles and Boyles "laws" — i.e., "principles." We know that there are innate earth forces with which we cannot argue, but to which we must adapt, with which we must harmonize human activities. This is a recognition distinguished from the outlook typical of lawyers and politicians, whose attention is riveted on the process of human law-making and law enforcement, all of which can be changed by further acts of humans.

For decades, natural scientists have dismayedly watched their peers in other disciplines take actions diametric to natural principles, actions that would set in motion whole sets of processes for which later generations must pay the costs.

A poignant example of using man-made versus natural "laws" is drawn from the experience of an eminent geologist at the University of California in Los Angeles. This man had studied the Mono Basin of eastern California for years. He had aged its moraines, its shorelines, its volcanic features. He had studied in detail the processes by which each of these was made. In the early 1930s, he became aware that the engineers at the Los Angeles Metropolitan Water and Power District planned to route the aqueduct from the basin through the Mono Craters, which had erupted within the previous 2000 years. So he went to the engineers and explained his scientific findings, only to be told summarily that they were not interested in his findings and intended to go ahead as planned, in spite of the fact that an easy route around the craters existed. And they did. Within a few months, they literally were in hot water inside the Mono Craters, and this part of the project had to undergo redesign. The oversight cost the project three years' delay, untold thousands to millions of dollars, and man-years.

In the 1960s a similar thing happened to construction on Interstate 70 west of Denver, where the engineers failed to heed the geologists who pointed out that Straight Creek Tunnel was routed directly through a major fault in the Continental Divide. The scientists counseled the engineers to move north or south a few hundred feet, so as to avoid this zone of mountain movement. But the engineers did not do this. Several years and millions of dollars were expended because of this refusal to accept the facts about operation of natural principles and forces in the human environment. An innovative tunneling machine was caught in the tunnel and crushed by the forces of the mountain movement.

These examples are cited not to breed fear, caution, and dissention but to catalyze healthy, active respect for natural principles and their operation. When we move in rhythm with them, they work for us. Today, the great challenge of engineering science is no longer "how much of the earth's forces can we overcome," but how can we harmonize necessary human activities with the dynamics and processes of the earth's ecosystems, with the least alteration of these systems, so as to foster the well-being of all mankind and the environment for generations to come. This is not to say means cannot be devised for extending human capabilities where needed. The point is that we need to exercise the attitudes and techniques outlined in our National Environmental Policy Act, a document which incorporates natural principles into human law:

Sec. 101. (a). The Congress, recognizing the profound impact of man's activity on the interrelations of all components of the natural environment, particularly the profound influences of population growth, high-density urbanization, industrial expansion, resource exploitation, and new and expanding technological advances, and recognizing further the critical importance of restoring and maintaining environmental quality to the overall welfare and development of man, declares that it is the continuing policy of the Federal Government, in cooperation with State and local governments, and other concerned public and private organizations, to use all practicable means and measures, including financial and technical assistance, in a manner calculated to foster and promote the general welfare, to create and maintain conditions under which man and nature can exist in productive harmony, and fulfill the social, economic, and other requirements of present and future generations...(of humans).

Thus, Congress wrote natural ecological principles directly into this law of our land, making natural principles the human law. Following is an examination of several of these principles and how they operate in ecosystems within the framework of choices policymakers have before them.

Basic ecological principles

Interrelationships. The principle of primary importance is that everything on earth is connected to and affects everything else, directly or indirectly. Nothing operates in isolation. A prime example of this principle in operation is the fact that DDT and radionuclides have been found in the tissues of all organisms tested to date, even in flightless antarctic penguins. Since penguins must obtain these substances from outside the Antarctic Continent, for none of either has been introduced directly to that region by man, it is obvious that these substances have traveled to penguins through a series of food-chain links.

Ecosystems and niches. Earth is covered with a vast array of ecosystems, large and small, which interact with each other. The composite makes up the ecosphere. Within these systems, each type of organism has a role to play — a "niche." To man, these roles of other organisms are sometimes obscure, negligible, or detrimental. Ecological research demonstrates that such organisms are nevertheless significant in the total functioning of ecosystems. A prime example is a cypress swamp bordering the Savannah River between the nuclear plant and the river, which was considered negligible in usefulness to man until recently. Research of the material flow pattern in that swamp showed that, under the specific conditions prevailing (a long-undisturbed system with water pH 7), kaolinite of the soil was absorbing all the ^{123}Cs being released by the plant; therefore, radioactive cesium did not reach the river. But if the pH level of the water were changed by adding small quantities of nearby well water, cesium was released from the kaolinite into swamp water and eventually into the river.

Material cycling and energy flow. Chemical substances in ecosystems cycle through and among systems at varying rates. For this reason they are available for re-use in geologic or shorter time. Conversely, energy follows a one-way downhill path, sometimes circuitous, always dissipating eventually as heat.

Limiting factors. Within all ecosystems, specific features of the environment interact with the genetically controlled nature of the organisms to restrict or limit the functioning of these organisms. These interactions define the operating boundaries of that system and the organisms within it. Frequently, numerous physical and chemical factors interact with a group of species to describe the limiting factors of the system.

Precise knowledge of limiting factors is especially important in the case of arctic ecosystems, which generally are operating closer to the outer limits for life than in other parts of the earth. Very small changes in even one factor can tip the balance between life and death; not that there is no resilience in tundra ecosystems, but the margins are narrower. The relative simplicity of the arctic ecosystem, due to the smaller number of species present, causes the total system to respond more strongly to perturbation within a single species or population.

Carrying capacity. The composite processes of environmental factors and organisms interact to produce a dynamic optimum operating budget commonly referred to as the ecosystem's "carrying capacity." This is the same principle that operates in engineering design — enabling transmission systems to carry a given electrical load, computers to handle a given information load, and ice-breaker hulls to withstand certain forces. Engineers are familiar with what happens to these systems when their loads are exceeded and they cease to function properly or productively. So it is with ecosystems; but because they are living, they have more built-in dynamic resilience. On the other hand, living systems may not indicate that the carrying capacity has been exceeded until the actual point is well past in time, numbers, and balance.

Ecosystem development. Over a period of geologic time, ecosystems develop from especially simple systems on biologically naked surfaces of rock, sand, or water to progressively more complex systems. The nature of each of these systems changes through this evolution until a stable, permanent, and highly complex system is reached. Natural processes exist for healing various naturally occurring ecosystem disruptions such as fire, landslides, and insect infestations. These processes are generally slow, but sure. They can heal perturbations introduced by man, especially when the perturbations are introduced in a manner that complements or is integrated with existing processes. But if processes are introduced that are diametric to existing ecosystem processes, they can be as toxic to ecosystem operation as cyanide to the human system.

For example, surfaces in arctic tundra systems can take centuries or more to heal. But tundra is very easily transplanted with little or no visible reduction in vitality of the plants. Furthermore, styrofoam sheeting acts as excellent insulation for maintaining permafrost and is the most economic, efficient, and effective medium for transferring tundra turf from a site where it˙ will be disturbed to where it is needed to cover a barren surface. It is important, however, that wet snowbed and dry fellfield tundra are transplanted to the same type sites, respectively.

Specialization, diversity, stability. Ecosystem development processes result in increased numbers of species inhabiting a system and absorbing the existing ecological roles (niches). Competition for air, water, food, and space brings increased specialization, creating added niches and augmenting diversity. As in economic and social systems, diversity creates stability by buffering the influence of any single perturbation in the system. Therefore, the simpler the system, the more responsive to and far-reaching is the perturbation; the more diverse the system, the less responsive it is to perturbation. An example of this occurred during 1972 and 1973 winter ice storms in the Southeast. Natural forests sustained breakage proportional to their inclusion of trees that had brittle wood. Thus, plantations of long-leaf pine sustained the greatest damage, both because of the brittle wood, and long needles that held considerable ice.

APPLICATION OF ECOLOGICAL PRINCIPLES TO ARCTIC ENGINEERING

One fact is highly important in all applications of ecological principles: the most complex system created by man is simple in comparison to the simplest life system — the single cell. And when numerous cell systems are compounded into the variety of types of organisms that comprise arctic ecosystems (terrestrial and marine), that complexity is at least an order greater than in man-made systems.

The most important single thing anyone doing research and construction in any environment can do is to develop individual and collective skills in conceptualizing the implications of this single fact. For people operating in the arctic environment, these skills are at least twice as important, because life systems are operating closer to their existence thresholds, are more finely tuned to the environment, and they have less resilience and leeway for error. This is not to say there is not considerable variability in arctic ecosystems; it is to say that changes reverberate through them more directly and with greater force, because of their simplicity and because of the stresses set up and maintained by the extremes of environment factors they endure.

Value of reconnaissance studies

These facts demand that we have more complete and accurate information about the operation of arctic ecosystems than we may need for activities in other latitudes. Meeting this demand is made easier by the simplicity of the arctic ecosystems. These facts also necessitate that we utilize information and understanding about arctic ecosystems with a high degree of skill and accuracy. They require that we undertake careful, complete planning and design before entering the field. They necessitate meticulous organization to take into account new facts that come to light as construction progresses.

If we do these things, the results are rewardingly positive. For example, in Colorado I had several occasions to use my knowledge of the dynamics of alpine tundra ecosystems to assist engineers in project design. The process was quite like an experienced mathematician applying the principles of calculus to solve a complicated equation. I would go into the field with the engineers, describe the processes that operated in each alpine ecosystem, set out the parameters of the ecosystem operation, and then predict what would happen from the particular project being proposed. The engineers quickly saw that certain types of ecosystems could tolerate certain kinds of activities much more readily than other ecosystems; therefore those were the places to construct power lines and highways or to locate pipelines and build recreational facilities.

For example, a gas line was built across the Continental Divide, west of Denver. It was routed in old rights-of-way and under existing bridges, so as to minimize disturbance of ecosystems. On the tundra, it was buried by a carefully thought-out process. Plastic was laid on either side of the site. The turf was lifted by backhoe and laid upright on the plastic. The soil was removed by the hoe and put on the other side. The pipe was laid in, covered with subsoil, then topped with the soil. The turf was replaced by hand, settled in, and watered. Within a year, the presence of the line was barely discernible. And this was far cheaper in time, money, and manpower than re-seeding the ditch would have been, especially in view of the fact that it is difficult to establish seed in arctic regions due to a wide variety of physical and biological hazards such as needle ice uprooting, desiccation, animal foraging, drought, blizzards, and short growing seasons.

In another instance, I participated in the helicopter reconnaissance of the general area where a transmission line might be routed. I ranked the types of ecosystems as to their susceptibility to damage from construction, and the engineer located the tower sites accordingly. The same procedure was used on Trail Ridge for the siting of visitor facilities.

This approach, together with an engineering knowledge of the problems associated with maintaining a tailing line at high elevation in arctic winters, was used in a recent tunnel operation through 10 miles of Precambrian granites and metamorphics of the Continental Divide to bring molybdenum ore to a site acceptable for milling and tailing disposal.

The cost of making reconnaissance and prediction studies described above was not expensive in comparison to what it would mean to lose the health of those ecosystems for humans in future generations. In fact, it was usually only a fraction of a percent of the engineering costs.

Prime principle in ecosystem assessment

A device contrived for environmentally sound engineering is called the "3E's tripod." One leg stands for engineering, one for economics, and one for ecology. With a tripod, all legs must be present, functioning, and used to make the "sighting operation" operable and valid.

A functional plan or project cannot emerge in any walk of life if one integral component of it remains undeveloped or inoperative throughout the period of its applicability. And yet, some engineers continue to insist that they cannot apply ecological science until their planning, design, and budgeting are complete. The errors of such an attitude can be analyzed on the basis of the ecological principles represented by the 3E's tripod and some common sense.

First, to gather data on ecosystems so as to objectively evaluate alternative project approaches requires one to three years, so that the range of variations can be better estimated. Our present data base is so geographically spotty, as well as fragmented in comprehensive knowledge of ecosystem processes, distribution, and variability, that time is required to do the needed field and library research. Leaving ecological research to the last minute leaves insufficient time to accomplish this process.

Second, telescoping ecological investigations provides no opportunity for developing a sense of the "carrying capacity" of ecosystems and regions for facilities.

Third, the limiting factors operating within a system are not always obvious; sometimes the major ones operate for only a few days per decade. This is true, for example, of the Gulf of Mexico upslope snowstorms that limit deciduous trees from the natural ecosystems of the high plains. At infrequent intervals, these storms deposit heavy snow and ice along the east base of the Rocky Mountains during early fall, when deciduous trees usually still have their leaves. The snow is trapped by the leaves and branches until an intolerable load is released by massive breaking of the branches. One could live in the high plains for years without observing this phenomenon, thus missing a prime feature controlling the nature and operation of the system. The longer the observation period, the greater is the chance for accuracy of prediction.

Fourth, application of the prime principle "that everything affects everything else" prompts early ecological reconnaissance as an environmental problem analysis. Only intensive studies over several years of time will reveal the vast variability of arctic conditions.

For example, for 17 summers I have intensively studied 200 square feet of alpine tundra on Trail Ridge in Colorado to see how these ecosystems recover from impacts of visitors' feet. In one year, I saw the initial response of plants to impact release. In four years I saw differences of seedling establishment, as well as differences in plant responses to wet and dry seasons or to short and long seasons. Observations over 17 seasons revealed several waves of succession of plants and their destruction, the invasion of mosses and lichens, and the demise of some mature plants.

A corollary to applying the prime principle is that the human computer needs time — several field checks and re-runs — to assure that a high degree of validity has been attained from available data and that all appropriate connections have been made. Projects must be examined from all possible angles in the context of ecosystems and their variability. This is impossible with a once-through, rapid preparation. Each feedback loop must be re-traced, each input questioned. Then, all possible imaginable changes in functioning of the system must be postulated and analyzed for probabilities. Different disciplines must raise different questions and analyze the same data from different standpoints.

After that, the trends suggested by the analyses must be viewed from all conceivable standpoints. By so doing, we frequently discover new and unexpected ways of integrating man's activities with those of the whole ecosystem, so as to harmonize man with nature.

Fifth, small ecosystem perturbations, such as a plant uprooted or an animal killed, will heal in a short time relative to the time involved to produce the original ecosystem. Large perturbations can take systems back to primary development (bare surfaces) and then require centuries or even millenia for them to reheal.

Construction workers who completely clear a site and bury all usable tundra turf leave behind an often impossible chore of reestablishing vegetation. Owners must telescope ecologic time to try to restore a productive ecosystem. Yet all this could have been avoided by a few simple measures.

Sixth, knowledge and understanding of the whole system enables us to recognize those conditions necessary to maintain species diversity. This diversity is a major indicator of ecosystem vitality; as it declines, so does ecosystem health — often in subtle ways that are unseen at first. There is no quick way of predicting all possible combinations of interactions that can culminate in ecosystem deterioration.

Seventh, straightforward materials and resource commitment and potential recycling can be relatively easily calculated. More complex are the cumulative commitments of resources and the progressive effects of approaching each project as though it were the first and last to enter the systems under consideration. The off-site impacts of increased demands for energy, commodities, and utilities attendant with population growth are rarely analyzed with more than an inventory approach. Often overlooked is adequate consideration of changes in the social and cultural structure of native peoples, community maintenance costs, and changes in land commitment.

Eighth, assessment of short-range versus long-term resource commitment demands an understanding of ecosystem dynamics. For example, the upper part of a coastal redwood watershed in California was committed to clear-cutting of trees about 10 years ago. In the short-range period, this commitment netted dollars of profit for the company, salaries for employees, and lumber for building — all good things. But the long-range commitment of those resources converted a viable, diverse forest, that held soil in place and tempered effects of winter storms on the landscape, into a barren hillside that rapidly and progressively eroded with each rain. Redeposited downstream, this displaced soil is slowly smothering the roots of giant redwoods growing on the floodplain and gradually killing the forest in Humboldt State Park. An alternative approach that would have distributed resource commitment differently, but would have yielded many of the same human benefits, would have been to selectively remove mature trees throughout the forest. By this means, the watershed-control function of the rest of the forest could have been retained.

A second example is the commitment of prime agricultural crop land to housing developments, shopping centers, highways, and industrial facilities. One million acres of prime soil are covered each year in our nation. Rocky areas, where soil is poor, are seldom used for building; yet millions of people in the world are undernourished most of their lives.

I trust a third short-term commitment will not be made — i.e., turning the arctic tundra regions of the world to industrial development before the long-term commitments are thoroughly evaluated.

SUMMARY

Seek to think, feel, smell ecology just as you think, feel, smell engineering. Identify with engineering. Start ecological-environmental research even before projects are conceived. Approach environmental matters from the holistic viewpoint — for the region, the program, or the flowsheet of a process. Use the 3E's as a unit — engineering, ecology and economics, supporting each other. Establish baselines of ecosystem information, including the human components of community features. Seek new and better ways of five-dimensional analysis of effects of projects on ecosystem processes (the other two are time and evolution). Develop a repertoire of potential alternatives; seek to augment this with time. Increase your data base about these alternatives with time. Involve the public in your thinking, planning, and deliberations. They are a vital, helpful, constructive ecosystem component, if so viewed and amalgamated into your work. Document your data sources so others can know where the information came from. Take nothing for granted. Things which seem so self-evidently harmless to you must be patiently proved to the uninitiated.

Arrange to monitor projects when in operation so as to check your accuracy of prediction and to augment your data base, increasing your ability to predict accurately in the future.

. . .We in this generation must come to terms with nature. . .
We are challenged as mankind has never been challenged before,
to prove our maturity and mastery, not of nature but of ourselves.

(Rachel Carson *1962*)

section 2

regional program perspectives

CHAPTER **5**

The North Sea problem

H. C. BUGGE [1]

INTRODUCTION

In the last few years the North Sea has become an important area of offshore petroleum exploitation. A substantial gas find off the coast of the Netherlands at the end of the 1950s indicated the possibility of oil and gas in other parts of the North Sea bordered by the United Kingdom (UK), the Netherlands, the German Federal Republic, Denmark, Sweden, and Norway. These states agreed, mostly bilaterally, on delimitations of the shelf south of the 62nd parallel as based on the median line principle (Fig. 5.1).

Norwegian and British explorations

In 1966 exploratory drilling started in the Norwegian sector. Since then, some 500 wells have been drilled in the whole area, most of them in the British sector. So far there have been 40 oil and gas finds of interest in the British sector, 21 in the Norwegian sector and 10 in the other sectors (Fig. 5.1).

Production has started from one site, the Ekofisk field, situated in the southwestern part of the Norwegian sector. Oil and wet gas from Ekofisk will be transported to Teeside in the UK by pipeline; the dry gas will be transported by pipeline to Emden in Germany.

An important gas field, Frigg, is situated on the 60th parallel, straddling the Norwegian and British sides of the median line. A final decision has not yet been made on how to transport the gas. One possibility is a pipeline to the Norwegian coast, most probably to the Karmøy area.

In 1974 the most important oil field so far, Statfjord, was found even further north, close to the median line. The Norwegian state oil company *STATOIL* has a majority interest in that field with its total production capacity estimated at 2 billion barrels.

The estimated total exploitable reserves on the Norwegian continental shelf south of the 62nd parallel amounts to 1 to 2 billion tons (about 10 billion barrels) of oil and the same oil equivalent of gas reserves.

[1] *Department for Pollution Control, The Royal Ministry of Environment, N-Oslo-1, Norway.*

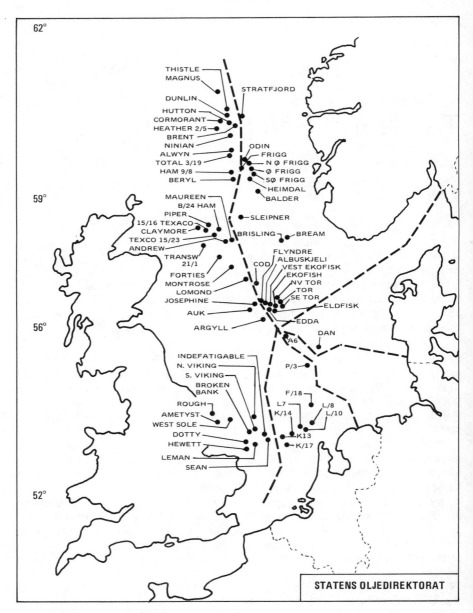

Fig. 5.1 Locations of North Sea oil and gas fields.

In the British sector, production is on the point of starting from a number of fields. Several pipelines to the British coast are under planning construction.

Economic significance

Together, the reserves found so far in the North Sea are more than twice as big as the total known reserves in the Gulf of Mexico. This resource will make Norway, and later the UK, a net exporter of oil and gas. In fact, the Norwegian yearly production may be 10 times the national consumption in the beginning of the 1980s.

For Norway, the North Sea is only a beginning. We have not yet started drilling north of the 62nd parallel, where the continental shelf covers large areas (Fig. 5.2). The question of delimitation towards the deep sea is still open, pending the outcome of the Law of the Sea Conference. Negotiations are going on between Norway and the Soviet Union on the delimitation of the shelf in the Barents Sea. The delimitation of the shelf around the Spitsbergen Islands presents particular problems related to the international status of those islands. The Norwegian view is that the Norwegian shelf continues up to and beyond Spitsbergen. The geology indicates that there may be enormous oil and gas reserves in these areas.

Clearly, offshore petroleum activity will be a major factor in the future development of Norway for a very long time.

IMPACT ASSESSMENT

National policy

Work presently being carried out in Norway for the purpose of assessing the impact in different fields of the petroleum industry is not limited to environmental effects in the stricter sense, but focuses also upon economic and other social consequences. In Norway, these aspects are seen as very much interrelated. Present Norwegian oil policy is based on a comprehensive assessment of social effects, of which the environmental aspect is only one element.

In the present legal and administrative system in Norway there is no formal procedure to assess environmental effects. There does not exist for example, the environmental impact statement recognized in the United States system.

Of course, from the very beginning, the oil industry and public agencies have carried out those studies of the environmental conditions which have been necessary from a technical and economic point of view.

But it was not until after 10 years of industrial activity that the first assessment of social effects and hazards to the environment was made and presented to the Parliament (Norwegian Government 1974). Such an assessment had then become a political necessity. The social and environmental issues had become crucial in the public debate over further development on the continental shelf, in particular on the question of where and when to start drilling north of the 62nd parallel.

Social issues

This Parliamentary report provides a comprehensive assessment of effects and draws up basic objectives and principles for Norwegian oil policy.

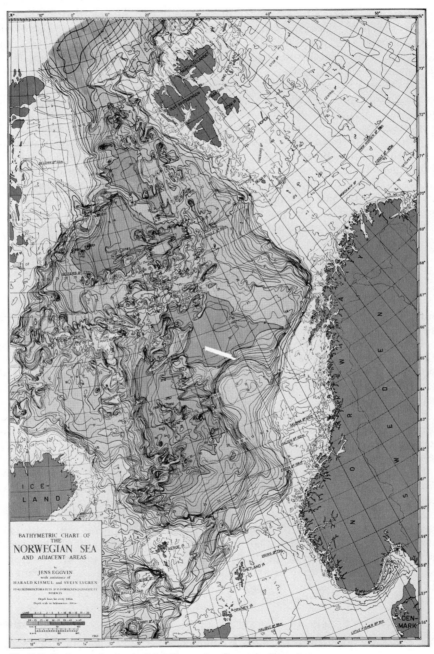

Fig. 5.2 Bathymetric chart of the Norwegian Sea and adjacent areas.

In this document, several problem areas of socio-economic consequence are identified: e.g., the impact of oil industry on the national economy; the impact on existing production, employment, and settlement patterns in Norway; undesirable social effects; and possible hazards to the marine environment. The report states what is obvious: that the oil will make Norway a wealthier nation. The question is how to be richer, even very rich, without acquiring at the same time many new problems, and without losing important social values and traditions appreciated today. This is a more difficult dilemma than it might seem in a country where social stability has been a predominant feature.

Since there is already full employment in Norway, the oil activity and use of the associated revenues will necessarily encroach upon — and threaten — other sectors. There will be a domestic cost squeeze. People will leave traditional activities and go to jobs on offshore installations or platform yards on shore. Too rapid change of this kind is not considered desirable in Norway, where the maintenance of the existing pattern of settlement is a primary political goal.

The principal conclusion in the report is that the oil may be very beneficial to Norwegian society. It will make it possible to create a better society, to improve the quality of life for the Norwegian people and to strengthen their efforts towards the developing countries. However, in reaping these fruits of the oil activity, certain negative effects must be avoided. To cope effectively with these deterrents, a firm policy and new legal administrative mechanisms are needed. This means research in many areas, and it means time.

The report then draws up basic principles in Norwegian oil policy: moderate speed in oil exploration and production; strict public control of activities; prime allocation of income for public needs; substantial state participation, mainly through the state oil company *STATOIL*; decentralization of on-shore activities; and minimization of pollution.

The government has based its long-term economic policy on these principles. In a 1975 report to Parliament on Norway's natural resources and economic development, the government proposes a slower economic growth during the coming years than would have been possible with more rapid development in the oil industry. The government will use a very substantial part of the new income for public needs — social and environmental policies being among the most important concerns.

Environmental problems

Physical features. The North Sea ranges in average depth from under 100 to 200 m. The bottom is flat and mostly sandy, especially on the British side. Conditions for pipe-laying are good, and it is possible to bury the pipe for long distances. Further north the bottom is deeper and very rough and rocky. The possibility of pipe-laying in these areas is still uncertain.

The weather in the area is generally very bad. The North Sea is known as a particularly stormy area. On the other hand, there is no ice problem. The North Sea, the Norwegian Sea, and most of the Barents Sea up nearly

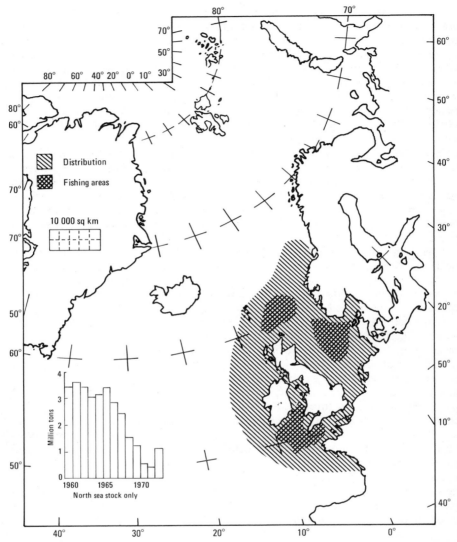

Fig. 5.3 Locations of mackerel distribution and fishing areas on the Norwegian continental shelf.

as far as Spitsbergen are free of ice year-round and have no icebergs. Ice becomes a problem if oil is found on or off the Spitsbergen.

The main current in the North Sea is circular and anti-clockwise. Along the Norwegian coast, there is a strong and stable northward current which has its origin in the Baltic Sea.

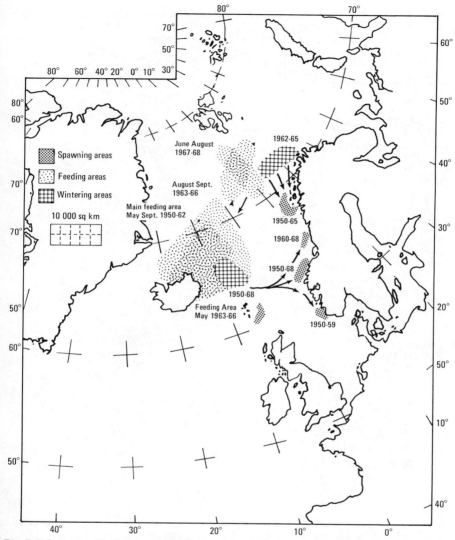

Fig. 5.4 Spawning, feeding, and wintering areas for Atlanto-scandian herring in the North Sea.

Commercial fisheries. The North Sea and the other sea areas off the Norwegian coast and the Barents Sea are some of the richest fishing grounds in the world. Here are found the principal spawning grounds and, to a certain extent, the maturation area for the most important types of fish in the North Atlantic. The annual fish catch on the Norwegian continental shelf is approximately 6 million tons, taken by an international fishing fleet.

Locations of main areas for some species are shown in Figures 5.3, 5.4 (Atlanto-scandian herring), and 5.5 (cod). It is important to note the partic-

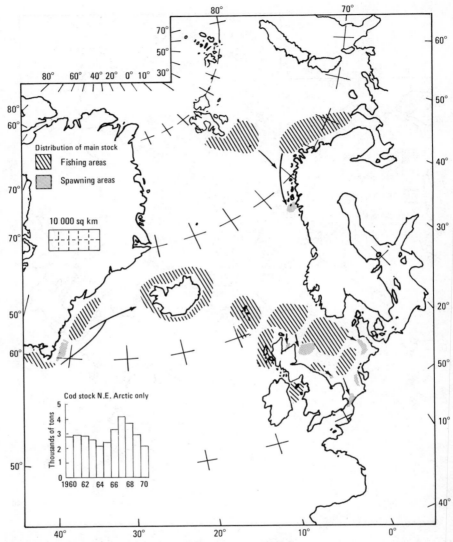

Fig. 5.5 Fishing and spawning areas for Northeast Arctic cod.

ular and very limited spawning areas. The cod, which is the most important resource, has its spawning grounds in Lofoten and its maturation area in the Barents Sea. The young cod travels from the spawning grounds to the maturation area with the current, and they pass the coastal banks off Troms and Finnmark at their most sensitive stages.

The problem of possible threat to these resources from oil activity can be summed up in the following way: pollution from ordinary oil operations on the continental shelf will probably be insignificant if reasonable measures are taken. The risk of a major accident with a substantial oil spill is small, but such a possibility must be taken into account, especially because of the deep waters and the very rough weather conditions in the area.

As to the probable effect of an oil spill, there is not yet any certain answer. The effects of oil on marine organisms differ very much from one type of organism to another, with different types of oil, and under various environmental conditions. No certain conclusions can therefore be drawn from studies in other sea areas. To get an exact answer to the area in question, regional studies will have to be carried out.

In spite of the fact that some crude-oil components are toxic, have a high persistence and may be taken up in the food-chain, Norwegian scientists think it is quite unlikely that oil pollution can be a really significant factor in the long-term development of the fishery resources on the Norwegian continental shelf.

Oil pollution may have important local effects, however. Crude oil, even in very low concentrations in seawater, has a deadly effect on primary organisms such as fish eggs and larvae. Eggs and larvae of many important species drift in the surface layers. A major oil spill in an important spawning area would therefore be expected to have a particularly serious effect. Under extreme conditions, a major oil spill might reduce substantially a whole year-class of type of fish in question. The oil also has a tainting effect on fish. This was proved in 1974 when a small tanker spilled fuel oil in a fishing area in northern Norway. Local fishermen had to give up fishing for about two months because of the bad taste and smell of the fish.

Pollution control

These and other facts are serious to Norwegian fishermen and to the government, especially because the most promising areas for drilling north of the 62nd parallel — the banks — are also the important fishing and spawning areas. The general conclusion reached in the Parliamentary report on this question is that, provided moderate speed and reasonable precaution in exploration and exploitation, oil pollution will not create greater problems than can be dealt with in a satisfactory way. However, special restrictions and limitations may be necessary in particularly vulnerable areas.

The Parliament has stated that it is desirable to start drilling north of the 62nd parallel only if such activity can be carried out within an acceptable risk level. The report draws up an action program to keep pollution at a minimum, including such measures as strict safety regulations to prevent accidents, restrictions on discharge, oil spill emergency plans, efficient surveillance, strict liability for pollution damage, new institutions (Ministry of Environment to have important responsibilities), international cooperation, and long-term monitoring and research programs to assess environmental effects.

A research program is at present under preparation by a ministerial committee. So far the following research areas have been defined: currents and transport phenomena in the North Sea and along the Norwegian coast; baseline biological studies in the area; biological and chemical degradation of petroleum and petroleum products in seawater; effects of petroleum components on marine organisms, and the combined effect of oil and other pollutants.

The organization of such a program is a problem in itself. It will be necessary to establish a new organization for coordination and resource allocation in this field.

CONCLUSION

In conclusion, there is the question whether petroleum activity in the North Sea so far has had any environmental effects. After all, there have been 10 years of activity—10 very risky years, based on a start with no experience in the field. The deep sea and rough weather conditions were new challenges to technology. It is not surprising that most projects met with technical problems and were delayed. In spite of this, however, there has been no major accident in the area. No important oil spill has taken place. Monitoring of oil content in seawater and fish shows insignificant presence of oil.

Personally, I think this apparent success is due in part to clever engineering, partly to strict agency control, and somewhat to luck — in that order.

For the future, we should definitely eliminate the third of these factors — the luck factor. We cannot adopt as a permanent solution the old way of keeping our fingers crossed.

The first major oil pollution in the North Sea, no matter what the short-term and long-term effects may be, will have unpredictable political implications in Norway. Such an event might drastically alter all present plans for further development on the continental shelf of the Norwegian Sea and adjacent areas.

REFERENCE

NORWEGIAN GOVERNMENT

1974 Petroleum industry in Norwegian society. Parliamentary Report No. 25, 1973-1074.

CHAPTER **6**

Recent Soviet research in the Arctic

A. F. TRESHNIKOV[1]

Abstract

Under the direction of the Main Administration of the Hydrometeorological Service of the USSR, the Arctic and Antarctic Research Institute (AARI) is responsible for complex scientific research in the polar regions. In the Arctic, its principal research task is to study the hydrometeorological conditions of the Arctic Ocean and adjacent seas which wash the coasts of the Soviet Union. Vital to this study is the continuing development and refinement of methods for forecasting weather and ice conditions with regard for long-range variations of global implication. Since the polar regions represent the earth's main sink for expenditure of solar energy which accumulates in the tropics, a primary program objective of AARI is to assess quantitatively the role of the ocean and atmosphere in the formation of energy balance.

INTRODUCTION

Dating back to about 1920, or the first years of the Soviet State, numerous geological expeditions have been organized to the far northern regions of Siberia to explore mineral resources such as coal, oil, gas, polymetallic ores, diamonds, and gold. Another valuable resource known to abound in the remote areas of Siberia were large tracts of forests rich in pine and spruce wood.

But with this natural wealth came also the problem of how to get it. In order to exploit various mineral deposits, it was necessary to deliver equipment and construction materials to the far north and to build settlements there. Hence, the main problem in the early years was that of transportation. Although the trans-Siberian railroad between Moscow and Vladivostok

---

[1]*Arctic and Antarctic Research Institute (AARI), 34 Fontanka, Leningrad 192104, USSR.*
(The author has worked in this Institute for 37 years, serving as director since 1960).

traversed the southern regions of the land, the only access to the remote
north country was by water — via the arctic seas and great Siberian rivers.

ICE REGIME

The arctic seas are covered with pack ice year-round except for only some
inconsistent areas of open water in the summer. Because of this dominating
feature, development of a regular navigation system demanded a knowledge
of the peculiarities of ice distribution and answers to such questions as
when does the shore ice break up? — what is the strength of ice? - and
how is the ice distribution influenced by wind, by water and air temperatures, sea currents, and by fresh water from rivers flowing into the arctic
seas? So before the northern sea route could become a functioning shipping
lane, it was necessary to study a whole complex of natural phenomena
determining the ice regime over many years.

Forecasting operations

In the 1930s, a network of permanently operating weather stations, now a
part of the international hydrometeorological network, was established along
the coast and on the islands of the Arctic Ocean. The continuous flow of
hydrometeorological information received from these stations is used for
weather forecasting and is presently disseminated on a worldwide basis
through the meteorological centers of Moscow and Washington.

Data collection. In addition to aircraft observations of the distribution of
ice, man-made satellites have been used in recent years to obtain information
on weather and ice conditions. Since the arctic seas are merely bays in the
Arctic Ocean which are open toward the north, there is a need to study
the deep-sea regions of the central Arctic Ocean as well; both manned and
automatic stations were established for this purpose on drifting ice islands.
At present, the "North Pole 22" drifting station and 10 automatic drifting
stations are in operation in the Arctic Basin.

Patterns and trends. Multi-year observations of the state of the atmosphere,
ice regime, and water-mass circulation have revealed certain regularities in the
development of these processes. On the basis of a succession pattern in
large-scale atmospheric processes over the Arctic, methods have been developed for the long-range forecasting of wind and temperature regimes of the
atmosphere. An understanding of the ice regime has made it possible to
make certain predictions of change in ice conditions depending on a
complex of local geographic conditions. It was found that since ice has an
inertia greater than air or water, knowledge of the thickness and character
of the ice in winter could enable forecasting of the state of ice several
months in advance of summer navigation.

Preparation of ice forecasts. At the same time that all data are being collected on the ice state in the Arctic Ocean in winter, usually from January to

March, a forecast of the wind regime and temperature variation for the spring and summer months is prepared. Then, by various methods, a determination is made of those changes expected to take place in the ice distribution as a result of currents and wind drift, and it is estimated how the ice will disintegrate under the influence of air temperature. A preliminary forecast of the general character of ice conditions is given in January for the next period of navigation. The basic forecast is then prepared in March for the first half of the navigation period until August, at which time the forecast is completed to the end of the period. Each month these forecasts are detailed and specified for use by the Ministry of the Merchant Marine Fleet of the USSR in planning formations of ship caravans.

Methods of ice destruction

Icebreaker fleet. Even the best forecasting system for ice and weather conditions in the Arctic does not exclude the need for use of icebreakers, however, since ice also remains in the shipping lane of the northern sea route during summer months. The mechanical performance of the icebreaker in accomplishing the largest volume of ice destruction per unit energy expenditure than any other method ranks it as the most efficient means of getting through ice for many decades. Therefore, the icebreaker fleet has been well developed in the Soviet Union to pilot the way for nearly every caravan of ships. This is why the 44,000 hp nuclear-powered icebreaker *Lenin* was followed up with the even more powerful nuclear *Arctica* — and now yet a third nuclear icebreaker, the *Sibir*, is being built.

The nuclear-powered vessels or the diesel-electric icebreakers *Yermak, Moskva, Leningrad,* and *Vladivostok* are used early in the navigation season, when the ice is hardly broken. In the milder ice conditions of mid-summer, the caravans are convoyed by middle-class icebreakers such as the *Kapitan Belousov, Kapitan Voronin,* and *Kapitan Melekhov.* When freeze-up occurs in autumn, the powerful class icebreakers are put back to use.

The Arctic and Antarctic Research Institute has taken part in creating various icebreakers. Although AARI neither builds nor designs the vessels, the Institute does house a special laboratory for studies dealing with how the ship's hull interacts with the ice. Models of future ice-class ships are tested by scientists in a special experimental ice tank. After the icebreakers have been built, the same investigators then participate in the seagoing trials under natural conditions in the Arctic. As a result of such comparisons, formulas have been obtained for calculating the resistance of a ship during its movement in continuous ice and in broken ice under varying conditions of ship size, hull shape, speed, and ice parameters such as thickness, strength, and distribution. The results of such studies are then reported to the institutions which design the icebreakers.

Use of thermal energy. Studies of active methods of icebreaking are also conducted at AARI. Upon first consideration, it would seem that ice formation depends entirely upon the amount of heat contained in water and subsequently released by it; therefore, a natural assumption would be that thermal methods of icebreaking should be given priority. The fact remains, however, that there is at present no feasible, artificial source of heat energy at man's disposal for breaking or melting ice. For example, if all 40,000 hp of the nuclear icebreaker *Lenin* were put to the task, it could not melt more than 6 m^3 of ice per min nor cut a channel greater than 86.4 m long and 50 m wide in 24 hrs in 2 m thick ice. As demonstrated by this calculation, there is no feasible prospect for the direct application of heat energy for navigation in general, although this method is widely used to extend the winter navigation season in icebound harbors.

Solar radiation. Since ancient times, attempts have been made to accelerate the ice-melting process by darkening the surface of ice and snow. The study of solar radiation conditions in the Arctic has made it possible to develop and recommend a method of cutting channels in the ice by darkening the surface with sand, pulverized slag, or coal dust. This method has been used to cut channels up to 100 km long in ice. If such a channel is cut by the radiation method, only 500 tons of coal dust is required. The volume of fuel that would have to be burned to melt away the ice along this same stretch, however, would amount to 90,000 m^3 of wood or 18,000 tons of coal.

Other methods. Explosive means of breaking up single ice masses have proved to be of little efficiency and are used only in isolated instances of clearing small ice dams or to free a wedged ship. Also tested have been chemical methods of powdering the ice with salt, but experiments have shown that the amount of salt required to melt a given volume of ice is equivalent to about one-third the volume of the ice itself. Application of ultrasound oscillations has not been overlooked, but since ultrasound is intensively absorbed by ice, a great amount of energy would be required. Other means of ice destruction have also been tested, but all methods have proved generally ineffective except for use with small ice volumes.

AIR-SEA INTERACTION

The study of mechanisms of exchange between the air and ocean is today recognized as one of the most important approaches toward understanding the physical basis of atmospheric and hydrospheric co-behavior. These studies include investigation of the thermal, radiation, and mechanical processes which operate both in the surface layer of the ocean and in the lower layer of the atmosphere, as well as an examination of the problems of phase transitions and diffusion of impurities. Therefore, micro- and small-scale processes are the basis of interaction dynamics. All the peculiarities inherent in the coexistence of atmosphere and ocean are ultimately realized repeatedly in the mutual effects of micro- and small-scale processes.

From the point of view of polar hydrometeorologists, the main trend in the study of meso- and macro-scale interactions is the improvement and development of methods for forecasting weather and ice conditions and long-term variations of the basic background of hydrometeorological processes.

In contrast to the conditions of low latitude, the peculiarity of air-sea interaction in the polar regions is associated with the presence of an ice cover. Sea ice is a product of a specific form of interaction between the atmosphere and ocean.

Air-sea interaction is the overall subject of the research program POLEX (Polar Experiment). In view of the fact that the polar regions of the earth are the main areas of energy sink, the main objective of the POLEX program is to assess, on a quantitative basis, the role of the ocean and atmosphere in the formation of their energy balance. Moreover, this program provides for the study of the air-sea interaction responsible for large-scale, long-period changes in hydrometeorological processes in the Arctic and Antarctic. And this makes it possible to develop methods of forecasting for several years ahead and to determine what direction the process of glaciation is following.

CONCLUDING REMARKS

Research in the Arctic must be pursued not on a separate basis, but rather with due regard for the global processes. If the tropics are a zone where solar energy accumulates, then the polar regions represent the sink for expenditure of energy brought there by air flows and ocean currents.

Topics under study at the Arctic and Antarctic Research Institute are many and diverse (Treshnikov 1972). Solutions to problems of magnitude require continuous, detailed studies of the processes of interaction between the air and ocean. Specific forms of oxygen and carbon dioxide exchange are closely connected with the sustenance of life in the ocean. Conversely, the sea seems to favor the terrestrial environment by its cleansing effect of absorbing much of the extra carbon dioxide exhausted to the air by industrialized society. Total assessment of such problems, however, involves also the task of predicting the impact of man's activity on natural processes. In Alaska and other northern areas where extensive oil development has been started, there exists a real threat that the arctic environment could change as a result of oil spillage. Conceivably, part of the recovered oil could spill over water and ice to become incorporated into the gyral over the Canadian Basin, where it might accumulate for many years.

REFERENCE

TRESHNIKOV, A. F.

1972 Soviet research in the Arctic. *Geoforum* 12.

Assessment of the Arctic Marine Environment: Selected Topics
Copyright © 1976 by Institute of Marine Science, University of Alaska, Fairbanks

CHAPTER 7

A study plan for the Alaskan Continental Shelf

H. E. BRUCE [1]

Abstract

An environmental assessment program of the Outer Continental Shelf (OCS) of the northeastern Gulf of Alaska was initiated in July 1974 by the National Oceanic and Atmospheric Administration (NOAA), working through an interagency agreement with the Bureau of Land Management (BLM). In October 1974, NOAA was requested by BLM to expand the program to include seven additional areas of the Alaskan Continental Shelf, extending from Yakutat Bay in the northeastern Gulf of Alaska to the Beaufort Sea in the Arctic.

The objectives of the Alaskan OCS studies are to develop an information and data base on the biological, physical, chemical, and geological processes of the Alaskan Outer Continental Shelf which will improve our ability to assess and predict the impact of Outer Continental Shelf oil and gas developments on the marine environment and resources of the area. The scope, technical guidelines, objectives, and management of this program are described in this chapter.

INTRODUCTION

Over the past few years, the Outer Continental Shelf (OCS) has taken on increasing importance in the United States movement toward energy self-sufficiency. This importance was underscored on 23 January 1974, when the federal government directed a three-fold expansion of the OCS lease program, increasing the annual lease rate to 10 million acres by 1975. Subsequent program evaluation has led to some modifications in the scope

---

[1] *U.S. Department of Commerce (NOAA), Outer Continental Shelf Energy Program, Juneau Project Office, Box 1808, Juneau, Alaska 99802.*

and schedule formerly proposed. However, Outer Continental Shelf oil and gas development still retains a high priority.

Because of concern that legal problems might delay development along the Atlantic coast, interest was centered on the Alaskan OCS. Interest was focused further in early 1974 on an area of the northeast Gulf of Alaska between Prince William Sound and Yakutat Bay (Fig. 7.1), a region believed to contain substantial petroleum reserves. The region, however, offers wider-ranging resources for the human economy and well-being beyond those of petroleum. A pristine area of unsurpassed natural beauty and grandeur, the waters of the northeastern Gulf of Alaska abound in oceanic life against the backdrop of a rocky, glaciated coastline populated by a variety of migratory birds and marine mammals.

Formidable environmental issues surrounding petroleum development in this region were cited by the Council on Environmental Quality (CEQ) following a study of the impact of oil and gas development along the OCS, undertaken in 1973 at the request of the President. In its report of 18 April 1974, CEQ indicated that the northeastern Gulf of Alaska ranked highest in environmental risk among the areas studied, noting, however, that "particularly in Alaska, we have little or no information on existing marine life." Expanding on its rationale for the "high-risk" ranking, CEQ indicated the following factors were considered: the relatively high probability of oil spills coming ashore from hypothetical production locations, especially those in the northeastern Gulf of Alaska (east of 150° W longitude); the relative slowness of crude oil weathering in this region due to temperature and sunlight considerations; the importance, again primarily in the northeastern Gulf of Alaska, of bird nesting and fish spawning areas; the frequency of storms; and the potential impact of earthquakes and tsunamis.

In May 1974, the Bureau of Land Management (BLM) requested that the National Oceanic and Atmospheric Administration (NOAA) initiate a program of environmental assessment in the northeastern Gulf of Alaska in anticipation of a possible oil and gas lease sale in the region early in 1976. These studies, outlined in the document, "Environmental Assessment of Northeastern Gulf of Alaska—First Year Program," were initiated in July 1974 and are now essentially completed.

In October 1974, a major expansion of the environmental assessment program was requested by BLM to encompass seven additional areas of the Continental Shelf of Alaska during the FY 1975 to FY 1976 period (Fig. 7.1).

GENERAL GUIDELINES FOR OCS ENVIRONMENT ASSESSMENT PROGRAMS

Three months prior to the requested expansion, NOAA had initiated a series of planning sessions for the purpose of defining the scientific concept which should underlie marine environmental assessment programs related to energy development. The document which resulted from this effort, "Report of NOAA Scientific and Technical Committee on Marine Environmental Assess-

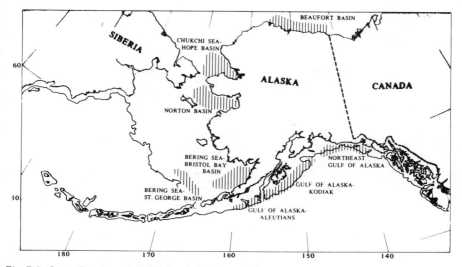

Fig. 7.1 Outer Continental Shelf areas of Alaska under consideration for funding.

ment," issued in November 1974, formed the basis for subsequent planning for studies of the Alaskan shelf and ultimately for the structure and content of the program proposal, "Environmental Assessment of the Alaskan Continental Shelf."

The 1974 report of the NOAA Scientific and Technical Committee on Marine Environmental Assessment (U.S. Department of Commerce) defined environmental assessment study concepts, including those for offshore nuclear power plants, superports, and ocean dumping, as well as oil and gas development. A subsequent report, which represents a refined version of the 1974 report and is aimed specifically to environmental assessment of OCS oil and gas development, is being developed by NOAA in response to a request from the U.S. Department of Interior, OCS Research Management Advisory Board. The material presented here is derived largely from the two above reports and the program proposal, "Environmental Assessment of the Alaskan Continental Shelf—First 18-Month Program.

The term "baseline study" is generally used to characterize the type of investigation required in connection with OCS developments. Baseline studies usually refer to those studies which provide a data and information base against which changes can be measured. However, the term used in this context is much too restrictive to describe the overall environmental studies program required for an OCS oil and gas area. A complete environmental studies program for an OCS oil and gas area should consist of a continuing sequence of information acquisition, compilation, and analysis and interpretation extending from the initial area selection through actual field abandonment. The overall environmental program then will involve a mix of baseline studies, experimental and special studies, and monitoring studies. Figure 7.2 illustrates an idealized OCS environmental studies program. The three types of studies are briefly described below:

64

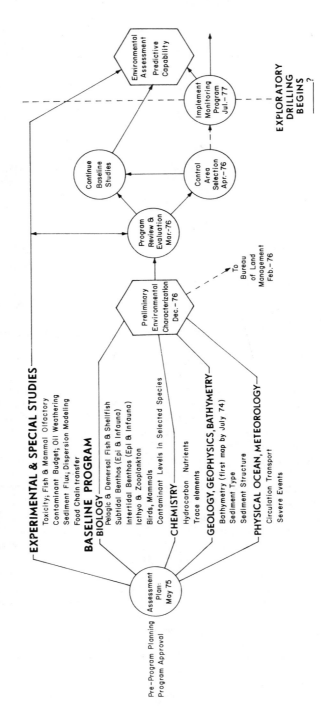

Fig. 7.2 Alaska-OCS development.

Baseline studies. Investigative activities conducted between the identification of a potential development area and initiation of major field development in order to provide comprehensive information needed for: evaluation and prediction of environmental consequences; planning or refinement of additional baseline, monitoring and special environmental studies; and documentation of the state of the OCS environment in its existing form.

Monitoring studies. Systematic sampling of both control and impacted sites subsequent to baseline studies to develop time series data for detecting impact, trends, and changes in the baseline.

Experimental and other special studies. Studies or experiments on dynamics, processes, and causal relationships necessary to understand and predict environmental effects and to improve the sophistication of baseline and monitoring programs.

Marine environmental assessment programs tend to be organized into discipline-oriented subprograms: biology, chemistry, marine geology, geophysics and bathymetry, physical oceanography, and meteorology. For purposes of developing a study program, it is probably appropriate that study elements be discussed by discipline. However, the fully developed study plan should reflect a single integrated program. Priorities should be established on the basis of relevance to important environmental issues in the area concerned and tailored to specific circumstances and needs. This will be discussed in more detail under program management and coordination and advisory functions.

In the process of developing the study plans for the Alaska OCS program, the following questions were addressed: Is it relevant? — to what extent does the proposed measurement or experiment contribute to the assessment or to prediction of environment impact, or post hoc assessment? — can it be achieved within time and resource constraints? — is it scientifically sound? — and, is it necessary in relation to information already available?

CRITERIA FOR EVALUATING "STUDY ADEQUACY"

The extreme complexity of the marine environment, the many little-understood relationships, and the difficulty of predicting significant long-range or low-level effects preclude an objective definition of an "adequate" study. This is particularly true when coupled with limited time and resource constraints. However, in designing a study of any particular area, the study plan or proposal should clearly identify both those questions that can be answered within time and resource constraints and those that are important but cannot be answered within the constraints.

GENERAL OBJECTIVES

Within the 5-year period expected to be available for studies of the Alaskan OCS oil and gas development area, attainment of the following general objectives is anticipated:

| | Improve the assessment of the relative hazards of proposed development in various locations, in terms of probability of accidents (e.g., from earthquakes and storms) and likely paths of dispersal of pollutants |

2   Establish baseline levels of key contaminants in the natural environment (sediments, water, organisms)

3   Describe distribution and general abundance of major biological components as a basis for qualitative predictions of possible impacts of major accidents

4   Synthesize information on systemic effects of target pollutants on selected organisms and conduct new laboratory studies where important gaps occur, and

5   Provide improved circulation models for the continental shelf and for nearshore and estuarine areas.

Within these constraints, we cannot expect to:

| | Provide the complete basic understanding of how natural environmental factors affect the marine ecosystem, and thus distinguish man-induced low-level effects with assurance |

2   Provide a complete understanding of low-level chronic effects of pollutants in critical life processes of many important marine organisms or

3   Completely understand the extremely complex food-webs through which contaminants pass and by means of which some may be accumulated and concentrated.

However, we should attempt to proceed as far as possible toward understanding how natural environmental factors affect the marine ecosystem. Although it is obvious that a complete understanding cannot be reached, as much as possible should be learned in order to provide some capability to detect man-induced, low-level effects.

The fact that some studies will be difficult to carry out within study time constraints should not preclude their being undertaken. Without them monitoring programs will only be able to detect gross changes after they have occurred.

PROGRAM STRUCTURE

We have identified five general questions which we feel address the primary physical and biological environmental issue associated with the oil and gas developments of the Alaska OCS.

The program is structured to respond to these questions through a series of tasks which address the various aspects of each question. The questions are listed below along with a summary description of our approach to producing answers.

:: What are the major biological populations and habitats subject to potential impact by petroleum exploration and development; what is the existing distribution and concentration of potential contaminants commonly associated with petroleum development?

Answers to these questions are intended in their simplest terms to provide a knowledge of the size and productivity of the biological resources which may be threatened by development, and an information base against which subsequent changes in populations or contaminant levels may be measured. Equally important, however, is the use of this information in formulating a basic understanding of the interrelationships among elements of the ecological system associated with the Alaskan shelf. A better understanding of these relationships is essential in predicting the effects of petroleum development.

The approach taken in this program to the above question initially involves a basic inventory of biological resources of the shelf area. In later years, the dynamics of the ecosystem will be more thoroughly explored in order to understand the mechanisms underlying natural population variability and to provide a picture of the dependencies and flow of energy and materials through the ecosystem.

Designing a program to answer these questions adequately within the limits of time and resources available is a challenging problem, not only because of the sheer variety of measurements possible, but also because of special difficulties relating to the statistical definition of natural variability.

In deciding which populations, habitats, and trophic interrelationships should receive priority, consideration must be given to the importance of each element to the ecosystem and to man, the probability of adverse impact, and the feasibility of the measurement program.

Importance. This consideration includes such issues as food-web relationships, the probability of impact on a particular population, the economic value of the resource, and its value to native subsistence.

Probability of adverse impact. The study program will stress those components of the ecosystem in which early adverse impact is most likely to be detectable. The likelihood of exposure to oil spills will be considered as well as the sensitivity of particular organisms to petroleum or heavy metal pollution.

Feasibility. The issues of cost, logistics, technology, and the general problem of natural variability must all be considered. The most pervasive problem to be dealt with in the proposed research program is the statistical definition of natural variation, both spatial and temporal, against which monitoring in later years can be employed to determine trends or deviations from a norm. Particularly in the case of a few biological indicators, this variation may be so great that meaningful characterization is impossible within time and manpower constraints. In these cases we recognize the fact that only rough characterization of populations will be possible, against which only major or catastrophic changes can be detected. This limitation can be offset to some extent through careful statistical design, as well as the establishment of representative experimental control sites, located outside potential impact areas, against which change within the impact area can be correlated.

:: What is the nature and effectiveness of physical, chemical and biological processes which transport pollutants?

In principle, the movement and dispersion of all physical and chemical properties associated with the environment can be considered to be related to this question. In practice, however, this program will focus on obtaining answers to direct questions regarding the movement of both oil spills and low concentrations of contaminants, particularly to and in critical areas identified in the biological programs discussed earlier. The program outlined in this section is thus specifically designed to:

1   Provide impact data needed in the planning stages of offshore petroleum development to minimize the potential risk to environmentally sensitive areas

2   Provide, in the event of an oil spill or the introduction of other contaminants, trajectories, coastal landfall and impact predictions required for containment and cleanup operations, and

3   Assist in planning the location of long-term environmental monitoring stations in the study area.

Oil or other contaminants introduced into the Alaskan environment will be transported by means of the atmosphere, water column, and sea ice acting as an intercoupled system. Winds will disperse and transport pollutants reaching the atmosphere. Currents, especially those acting on or near the surface, may transport contaminants over great distances, while on a smaller scale, diffusion and turbulence processes result in dispersion. Various constituents in the water mass, including suspended sediments, floating ice and free-floating biological elements such as phytoplankton, zooplankton, larvae and fish eggs, tend to concentrate contaminants. During the transport process, oil and other contaminants can undergo continual physical and chemical changes due to such processes as evaporation, flocculation, emulsification, weathering, biodegradation, and chemical decomposition.

Spills occurring on or adjacent to sea ice can be entrained, transported, and forced by the ice edge to concentrate in narrow open leads of water.

Each of these transport agents to some extent drives the other two; however, the energy transfer interrelationship coupling one medium to the other and the effectiveness of each in transporting contaminants are not presently understood in any quantitative sense.

:: What are the effects of hydrocarbon and trace metal contaminants on arctic and sub-arctic biota; what is the likelihood and timing of recovery of population from the effects of development?

Petroleum exploration and development may impact the diverse biological resources of the Alaskan marine environment in several ways; however, the most critical concern centers on the possible release of hydrocarbons or trace metal pollutants. The biological effects of such contamination may be categorized as follows: acute or chronic toxicity to marine, estuarine, or anadromous organisms; sublethal physiological effects of chronic exposure; effects of oil coating, such as impairment of the thermoregulatory function of marine birds and mammals; alteration of habitat; and elimination or contamination of food sources.

Although there is some literature on the toxicity of crude oil and its components, very little is applicable to arctic or subarctic conditions. The biological impact of petroleum hydrocarbons in this region, however, can be speculated to be more long-lasting and serious than in the more temperate regions of the earth because of the slower rate of evaporation of the more toxic aromatic fraction of oil, the possible slower rate of weathering, and the lower reproductive potentials of arctic and subarctic organisms. The absence of previous exposure is also a cause for concern, since arctic and subarctic organisms may lack defense mechanisms which have evolved in animals living in highly variable environments.

Effects on the physiology of Alaskan species may be singled out for special emphasis. Petroleum hydrocarbons may be damaging to the physiological systems vital for metabolism, migration, and successful reproduction.

Reduced photosynthesis, reduced fertility, and abnormal development are among other effects to be studied.

Coating by oil represents a major hazard, especially to marine birds and mammals. Oil removes the air-holding capacity and the insulating quality of fur and feathers, with the result that excessive body heat is lost to the environment and a critical physiological defense is breached. Young seals are especially vulnerable in this regard, because they are completely dependent for thermal insulation on a fine layer of hair until they have developed a thick layer of fat. Further biological damage is possible as the animal attempts to remove the coating through preening or other grooming activities.

Irritation of sensitive tissues such as skin, eyes, mucous membranes, lungs, and digestive tract can be anticipated as another direct effect of oiling. Any alteration or disruption of epithelial and endothelial tissues (the sites of initial interactions between aquatic animals and their immediate environments) will have serious survival consequences; therefore, it is particularly important to determine effects of oil on skin, gill, and gastrointestinal epithelial or endothelial tissues and cells and cellular products such as mucus.

Given the severe limitations imposed in the area of contaminant effects by a lack of existing literature and a shortage of skilled scientific manpower, even the few problem areas outlined above must be approached modestly. Tasks have been carefully selected, giving particular emphasis to those investigations which will aid in the interpretation of changes observed in the ecological baselines described earlier. In many instances, the establishment of causal relationships may provide the key to differentiating between natural and man-induced changes in the ecosystem.

:: What hazards does the environment pose to the safety of petroleum exploration and development activities?

The environment of the arctic and subarctic region contains a distinctive set of natural hazards to development which must be taken into account. A knowledge of the nature, frequency, and intensity of severe environmental events is essential, since the prime hazards to production-related structures and activities, as well as the greatest effect on the environment, will generally occur in conjunction with these extremes.

Geological hazards alone pose a gamut of complex problems which must be considered, including the instability of sea-floor substrate. Instability in the form of permafrost degradation, substrate fracturing, and resultant hydro-carbon leakage, substrate failure in the areas of rapid sediment accumulation, and sea-floor dislocations resulting from faulting can be anticipated in certain places.

Evidence suggests that extreme instability is likely along the outer continental shelf edge of the southern Bering Sea because of active faulting, potential slumping, and accumulation of subsurface gases. There is evidence of increasing rates of tectonic activity from the Bering Strait southward to the Alaskan Peninsula, while in the southern Bering Sea and Gulf of Alaska, tectonic activity poses a definite hazard to coastal zone and offshore development. Investigations of sea-floor instability resulting from these causes will focus on these areas.

These investigations will be designed to meet the following requirements. First, a study of sea-floor faulting and other tectonic activity within lease areas must provide information for use in the determination of which tracts might be unsuitable for development. Second, a regional study of faulting beyond the lease areas must be undertaken because pipeline and shore facilities will be distributed beyond the lease areas, and energy and vibrations originating from regional geological hazards will affect the lease tracts and coastal areas.

In the Beaufort and Chukchi seas, there is little evidence of recent tectonic activity. The existence of sub-sea permafrost indicates that potential instability should be anticipated along the inner shelves of these waters.

Here, sub-bottom sediments and the water within them remain below the freezing point year-round. It is expected that drilling, construction, and pipeline activities will cause a partial thawing and resultant loss of underlying support for structures. For this reason, siting of these activities will require information on the occurrence, distribution, and properties of sub-sea permafrost in both its vertical and lateral dimensions.

Onshore petroleum developments in Alaska indicate that permafrost is a hazard imposing serious constraints on the rate at which exploration and development can occur. This problem will exist on the OCS as well. Designing to overcome the structural and operational hazards which may be posed by sub-sea permafrost will require data on its distribution and engineering properties.

Offshore drilling and related support activities will directly and indirectly modify the rates and patterns of coastal erosion and deposition. Under most circumstances, design and location of these facilities and operations require adequate information on nearshore currents, transport processes, sea-floor morphology, climate, sediment sources, and location and dynamics of offshore bars and islands. Unique to coastal areas dominated by onshore permafrost is the rapid erosion of shorelines, particularly where ice-rich sediments are exposed. This process of thermal erosion by wave action and summer thawing can result in coastal erosion of 2 to 5 m per year and annual migration of barrier islands in the Beaufort Sea as much as 6 to 25 m. Facilities should be located in areas least subject to these rapid erosional processes, or else remedial action should be planned to counteract potential hazards due to erosion. The borrowing of fill material from beaches, offshore bars, and islands will modify coastal morphology, and this in turn can modify ice distribution and bottom gouging.

Another hazard peculiar to the arctic and subarctic areas is sea ice, which covers the waters of the Beaufort, Chukchi and Bering seas for a large portion of the year. Any structure for offshore petroleum development deployed in those areas must be designed to withstand the massive stresses of moving, drifting ice. Sea-floor completion of well heads and laying of connector pipelines are further jeopardized by the highly irregular subsurface relief of moving ice inasmuch as the keels of pressure ridges commonly come aground on the arctic shelves and extensively scour the sediments. The severity of the problem depends upon how frequently ice scouring occurs, how deeply the sea floor is penetrated, in what depth of water ice scouring occurs, and the magnitude of the plowing forces. This information is needed in order to allow for construction of well-head completions and associated pipelines below the maximum scour depth.

For offshore operations to proceed safely in a given geographical area, the type of ice conditions likely to be encountered throughout the year — and the worst conditions that can be anticipated — should be known. The ice stresses to be expected during the normal operation life of any offshore structures or marine transportation systems greatly exceed any stresses exerted by wind and waves, and a knowledge of their expected values is essential.

The dynamics of ice movement and its relationship to ice stresses upon offshore operations must also be understood so that a forecasting system can be developed to ward off severe ice conditions and thus prevent major catastrophes.

Severe oceanic events of concern relate to storms and seismic sea waves. Storm waves threaten offshore structures and shipping while storm surges may inundate coastal facilities, especially in the Beaufort Sea where the coast gradients are shallow and the normal tidal range is small. Since tectonic activity is high in the Gulf of Alaska and the southern Bering Sea, seismic sea waves (tsunamis) are also a hazard to be considered.

:: What conclusions may be drawn regarding the impact of OCS petroleum development on the Alaskan marine ecosystem?

Although data gathered for environmental assessment may be extensive, their full utility cannot be realized unless they are integrated into suitable systems that model both the static and dynamic aspects of the environment. The environmental systems must be examined from the standpoint of possible intensity and duration of perturbations that may be caused by impact of oil and gas development. Such an examination should include a description of the principal biotic systems, a knowledge of potential perturbing influences, and the availability of a coupling system of information on toxicities, physiology, and behavior, whereby influences are translated into effects. When these data are available, first-order effects of factors introduced by oil and gas development, such as mortalities of life stages of biota, may be reasonably easy to predict. The immediate consequences of an oil accident might be spectacular; yet the first-order results may be of considerably less consequence in the long term than less conspicuous but longer acting lower-order ones, such as those arising from chronic low-level release of contaminants. Although they may be considerably more difficult to predict and evaluate, possible second and lower-order effects such as decreased sizes of future reproductive populations, shifts of dominance within species guilds, transport of contaminants through the food-web, and destabilization of the ecosystem will also be assessed. Such releases may be especially harmful because of cumulative feedback of contaminants, their products, or their effects in the ecosystem. The two tasks described below are associated with the above problem.

Task 1. Upon completion of suitable background studies, develop conceptual and numerical models to describe the ecosystems of principal marine habitats that may be impacted by oil and gas development (e.g., littoral, benthic, open-water pelagic, seasonal ice, and multi-year ice).

Task 2. Use the information developed in the ecosystem models Task 1 above and the information developed in other environmental studies (e.g., circulation models and data on toxicities, physiology, behavior, and sediment transport) to develop systems that will predict possible short and long-term environmental effects of expected components of oil and gas development (e.g., physical alterations to the shoreline and sea floor, possible releases of contaminants, and harassment of biota), singly or in combination, on ecosystems of the regions.

PROGRAM BUDGET

Tables 7.1 and 7.2 give budget summaries showing the allocations of funds according to major objectives and discipline. Table 7.3 shows funding estimates of other agency programs that directly relate to the work to be undertaken by the BLM program being conducted by NOAA. The amounts indicated should be viewed as accurate estimates. Some adjustments have been made since these tables were developed, and other adjustments will undoubtedly occur. The overall scope of funding is not expected to change significantly.

DATA MANAGEMENT

A far-reaching goal of the marine environmental assessment program is the systematic collection and storage with specific retrieval capability of a core of marine environmental and ecological data and information for current and future use. Specialized investigations, coupled with organized collections of historical and monitoring data, must be readily accessible — not only to the scientific and engineering research communities — but to federal, state, and local coastal management planners, as well as to other interested individuals or groups. The primary objective of data management will be to ensure that information and services are responsive to this need.

The specific objective of the data management program is to ensure the smooth flow of field and laboratory data collected by Principal Investigators within the Alaskan project to other data users. The program is intended to satisfy the needs of future users of data through inclusion of data-retrieval systems as an integral part of the program. The main functions of the program, therefore, are to assure compatibility of data submission formats with retrieval system formats, scheduled flow of data, and adequate quality control.

A data management plan for the Alaskan program is being developed which will specify minimum requirements on data handling, information exchange, and archiving such that a melding of historical, research and monitoring data will be available in the future for operational forecasts and warnings of severe or catastrophic events. It will establish guidelines for the development of an operational data base which will interface with, and may become part of, future operational monitoring and forecast systems.

Contractual agreements for the accomplishment of program tasks will require specifics of data form, documentation and data handling procedures, including media, format, and schedules of availability to other investigators and the archive data base.

74

TABLE 7.1 Budget summary by major objective

	Beaufort Sea FY75	Beaufort Sea FY76	Bering Sea FY75	Bering Sea FY76	Gulf of Alaska FY75	Gulf of Alaska FY76	Total FY75	Total FY76
Section A, Baseline characterization	426.4	1494.4	413.4	2708.8		2648.7	839.8	6851.9
Section B, Transport processes	745.3	742.2	668.3	664.2		1337.2	1413.6	2743.6
Section C, Effects	15.2	136.5	24.2	328.1		268.2	39.4	732.8
Section D, Hazards	487.5	484.1	56.0	218.3		673.9	545.5	1376.3
Section E, Summary	–	–	–	–		–	–	–
Logistics	306.0	711.8	191.5	1170.6		1266.4	497.5	3148.8
Data management	16.6	55.6	16.6	55.6		55.6	33.2	166.8
Program and project management	151.3	536.8	88.7	330.9		877.1	240.0	1744.8
Contingency	158.0	499.3	160.0	535.6		573.1	318.0	1608.0
TOTAL PROPOSAL	2308.3	4660.7	1618.7	6012.1		7700.2	3927.0	18373.0

TABLE 7.2 Budget summary by discipline

	Beaufort Sea		Bering Sea		Gulf of Alaska		Total	
	FY75	FY76	FY75	FY76	FY75	FY76	FY75	FY76
Physical oceanography	249.2	397.9	620.9	494.4		1099.2	870.1	1991.5
Chemistry and microbiology	155.1	604.9	130.9	409.1		642.1	246.4	1394.3
Fish, plankton, benthos	202.7	658.5	113.1	1372.1		1379.1	268.2	3356.1
Effects	15.1	136.5	24.2	328.1		268.2	39.4	732.8
Birds	55.9	217.1	114.5	523.9		390.7	170.4	1131.7
Mammals	7.2	135.3	43.6	357.5		199.7	50.8	692.5
Geology	458.8	359.9	97.1	372.9		911.9	555.9	1644.7
Ice	617.3	662.5	17.6	61.4		37.1	656.7	761.0
Logistics	306.0	711.8	191.5	1170.6		1266.4	497.5	3148.8
Data Management	16.6	55.6	16.6	55.6		55.6	33.2	166.8
Program and project management	151.3	536.8	88.7	330.9		877.1	240.0	1744.8
Contingency	158.0	499.3	160.0	535.6		573.1	318.0	1608.0
TOTAL PROPOSAL	2308.3	4660.7	1618.7	6012.1		7700.2	3927.0	18373.0

TABLE 7.3 Agency matching funds*

	BLM	NOAA	USGS	USF&WS	CRREL	ADF&G
			(dollars in thousands)			
A	7691	1650	–	977	–	4470
B	4157	100	100	–	–	–
C	772	–	–	500	–	–
D	1922	50	900	–	100	–
Logistics	3647	14,140	500	–	100	–
Data management	200	–	–	–	–	–
Program management	1985	200	–	–	–	–
Contingency	1926	–	–	–	–	–
TOTALS	22,300	16,140	1500	1477	200	4470

*Several government agencies are supporting the program presented in the proposal to BLM by programs that they are carrying on in-house. Presented here are agency estimates of the level of these programs that directly relate to the work to be undertaken for BLM.

All environmental data acquired under NOAA sponsorship will be archived within the Environmental Data Service (EDS). Within EDS, the National Oceanographic Data Center (NODC) has been designated as the lead center for data archiving and services for all marine environmental assessment programs undertaken by NOAA. The data base will initially operate from the facilities of NODC and will be structured to interface with developing monitoring and forecast capabilities and plans. All data obtained during the course of the project is public property and available to the public from EDS in accordance with NOAA Circular 71-106.

The Project Data Manager will plan, schedule, and monitor the flow of data to archives; provide assistance to the Principal Investigators wherever possible and necessary to facilitate meeting project objectives in data processing and flow; inform the Project Manager of any constraints to scheduled data flow; and recommend to the Project Manager changes necessary to achieve objectives. He will coordinate data archive handling media and format with the Program Office to ensure compatibility with other projects and with the NODC to ensure compatibility between data sets.

PROGRAM MANAGEMENT FUNCTION AND ORGANIZATION

The responsibility for maintaining an integrated, directed program rests with the Program Manager and his staff. They are responsible for assuring that project work is integrated and coordinated both conceptually and logistically. They are responsible for providing overall leadership and direction in assuring that program goals are met and for managing the program resources.

A program management staff has been established in Boulder, Colorado, within the administrative framework of the Environmental Research Laboratories, to coordinate all marine environmental assessment activities related to energy development which are undertaken by NOAA (Fig. 7.3).

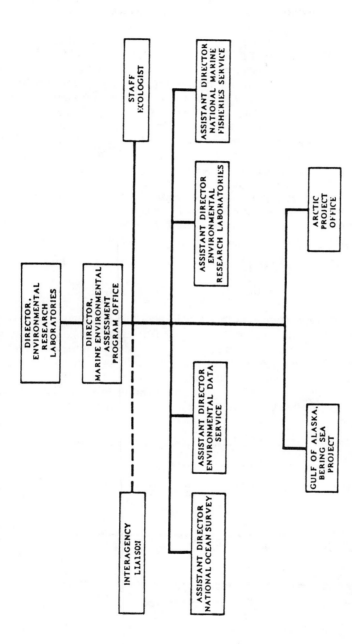

Fig. 7.3 Marine Environmental Assessment Program Office.

This office will manage all BLM-sponsored programs in NOAA, supervise field project offices in Alaska, and maintain active liaison with the State of Alaska and federal agencies involved in environmental research related to petroleum development. Liaison staff members representing these agencies will be assigned to the Program Office.

Field offices have been established in Juneau and Fairbanks (Figs. 7.4 and 7.5). The Juneau office will direct project implementation for the Bering Sea and Gulf of Alaska, serve as the center for all data management functions and provide a field point of contact with the State of Alaska. An Arctic Project Office has been established in Fairbanks, under contract to the University of Alaska, for the purpose of directing research and coordinating logistics support for field operations in the Beaufort and Chukchi seas.

Fulfillment of the program purposes and objectives can be accomplished only through a highly integrated and directed effort. Direct support of meritorious, but essentially unrelated, environmental studies may be included only when they result in direct spin-off benefits and can be carried out without impacting the primary mission of the program.

COORDINATION AND ADVISORY FUNCTION

In addition to NOAA, other federal agencies, the State of Alaska, and several academic institutions are involved in this program as contractors to NOAA. To improve communications and coordination among these efforts and to assure that the primary elements of relevancy and scientific soundness are properly considered, we plan to utilize appropriate advisory and coordinating groups or panels. Many of these will serve on an ad hoc basis, addressing specific technical or management problems.

Primary consideration will be given to establishing a basis and procedure for identifying user needs. A basic purpose of the Alaskan Outer Continental

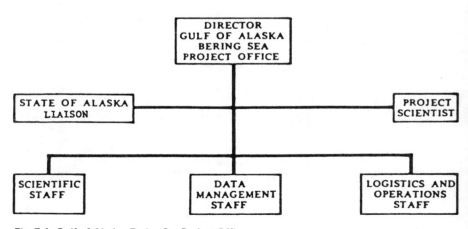

Fig. 7.4 Gulf of Alaska, Bering Sea Project Office.

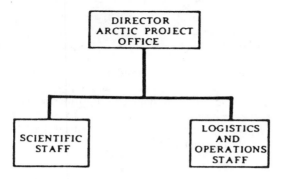

Fig. 7.5 Arctic Project Office.

Shelf environmental studies is to provide information which will guide future decisions relating to managing the resources of the Alaska OCS. A wide spectrum of management agencies and decision makers will be involved. These include federal regulatory agencies deciding on the stipulations required for permits, groups concerned with coastal zone planning, and citizen groups concerned with important public views. The determination of what decisions and issues will be confronting managers and the public, and what information they will need to make those decisions, will be at least as essential as a sound scientific program. We are conducting our studies with the above needs in mind and feel we can achieve reasonable success in providing a sound data and information base for managing and protecting the resources on the area.

Conference Discussion

Tywoniuk — Can you provide some examples of changes you made in your program as a result of the recommendation to put greater emphasis on dynamic processes as compared to conducting baseline studies?

Bruce — We carefully evaluated this recommendation in view of the studies related to processes that were already included in the first 18-month program. Our conclusions were that proper balance existed between baseline studies and studies oriented toward "processes;" we needed better definition of what were the most relevant and important process studies that should be included in an OCS environmental assessment program; and, before we could define special experimental studies focused on understanding critical environmental processes, we needed a better data base than currently existed and hence should not reduce the planned effort on baseline studies in favor of adding poorly defined "process-oriented" studies to the program during the first 18 months. We expect to shift emphasis strongly toward these studies in the following years of the Alaska OCS program.

Lutz — Has provision been made for the use of existing data along with new data and for the retrieval of data for other agencies, and for state and public use?

Bruce — Yes, we have given careful consideration to compiling and synthesizing existing data and information and to developing a data archiving and retrieval system which will make all data and information easily accessible in a usable form to the scientific community and to state, federal and local management and regulatory agencies. A data management plan to be followed is a part of each research contract.

CHAPTER **8**

PROBES: A study of processes and resources on the Bering Sea shelf

R. D. Muench, R. Elsner, *and* C. P. McRoy [1]

The problem

Existing information about the Bering Sea shelf ecosystem is sufficiently appealing to justify an intensive study program of one of the most productive shelf regions in the world. The probable dependence of this high productivity upon regionally unique oceanographic features suggests the likelihood of promising return — both for management of those resources and for contributions to the understanding of structure and function of marine ecosystems in general. Physical features on the shelf which we expect to have major influence upon its productivity are a continuous northerly flow of water and its drainage through the Bering Strait, the broad areal extent and shallow depth of the area, and a seasonal sea-ice cover that coincides roughly with the boundaries of the shelf.

The Bering shelf fishery, heavily exploited principally by Japan and the Soviet Union, less so by South Korea, Canada and the U. S., is concentrated in the so-called Golden Triangle region, extending roughly from Unimak Pass to the shelf extension north of the Pribilof Island. PROBES will be a study aimed at understanding the basis for the high biological productivity of the southern Bering Sea shelf. Specifically, it will address the factors controlling mass and energy transfer between primary producers and exploitable fisheries stocks; geographically, it will focus on the Golden Triangle region.

Ironically, the Bering Sea today is already the site of more scientific research than at any time in its history. Yet it is the very presence of the baseline studies conducted by the Outer Continental Shelf Environmental

[1] *Institute of Marine Science, University of Alaska, Fairbanks, Alaska 99701.*

Assessment Program of the Bureau of Land Management that makes PROBES a timely endeavor. PROBES will definitely benefit from the use of these data, and the return will be an understanding of ecosystem function to the ecosystem components description that is the usual result of baseline work. Similarly, the specific focus in PROBES, the use of Alaska pollock as a biological tracer, is possible because of the background data that have been gathered on the adult fishery by management agencies, especially the National Marine Fisheries Service. Without these two background data sets, PROBES would require an order of magnitude greater effort to pursue the same questions.

PROBES is organized into five component or disciplinary approaches to the ecosystem: physical oceanography, nutrient regeneration, primary production, consumer level trophic dynamics, and systems modeling. The research activities within these components are continually and stringently integrated by the ecosystems modeling effort. Through this approach PROBES becomes the integrated study of the ecosystem processes that support the entire trophic structure of the Bering Sea. The commercial fishery for pollock, through the data of the National Marine Fisheries Service, may be viewed as an extensive sampling program for adult fish that will be invaluable to the verification of the ecosystem models. The models will function to stimulate and direct questions concerning the structural, functional, and controlling aspects of the ecosystem. The model thus becomes a dynamic, steadily changing investigative tool.

The hypothesis

The research approach of PROBES is based on an hypothesis that was generated during a series of interdisciplinary meetings of interested scientists. We suspect that a combination of oceanographic factors, particularly the wide shallow shelf and its circulation pattern, plus the timing of crucial events in primary and higher-level production throughout the food-web, result in the highly efficient transfer of energy in the ecosystem. To test this hypothesis, we will examine one species — the pollock — as representative of a general pattern (Fig. 8.1). Confirmation of the pattern should then make it applicable to many other species, although details of life history would of course be different. Pollock was selected as the biological tracer in this system because it is presently the major fishery species in the Bering Sea. Also, during early life history, pollock larvae are not ecologically very different from the larvae of other fishes or from the larvae of crabs. In all cases, to survive, these planktonic creatures must find suitable food early in life. Thus the populations of all are affected by the same oceanic and ecological processes whose investigation is the major thrust of the PROBES study. PROBES will concern itself with the fish during their first year, since at this time recruitment levels are set for their subsequent entry into the fishery.

The hypothetical basis of pollock productivity relates to several features — physical, chemical, and biological — of the shelf. It is helpful, conceptually, to think of this region within the context of a spectrum of ecosystem types. Low-latitude coastal upwelling regimes fall at one end of that spec-

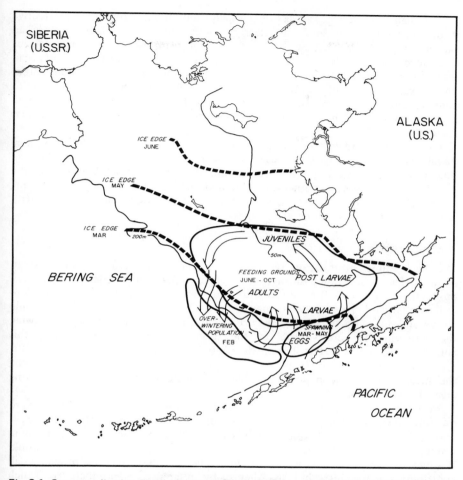

Fig. 8.1 Conceptualization illustrating the basic hypothesis relative to high upper trophic level productivity in the southeastern Bering Sea. Arrows represent proposed sense of migration of Alaska pollock during their life cycle.

trum, in that they are (theoretically or ideally) active all the time, as exemplified by continuous primary production. Arctic regimes, where the modifier refers to a physical process rather than a geographical location, fall at the other end of the spectrum in systems that are maintained annually by one or more pulses of primary productivity. The Bering Sea system pivots on a well-defined pulse event, the spring diatom bloom, which is in turn triggered by the equally well-defined physical processes characterizing the high-latitude spring. PROBES will address itself specifically to trophic studies, with the intent of quantifying upper-level energy transfer processes and their efficiency.

Although the Bering Sea is conceptually continuous with other productive marine ecosystems, it also has clearly unique aspects. Its waters overlie an exceptionally broad, shallow shelf and are in turn overlain by seasonal ice that normally extends to the shelf break. Both the ice and the water can act as conveyor belts for materials, and thus the problem is among other things one of horizontal advection. Also unique is the highly focused initiating bloom, a spring feature triggered by surface water temperature, stability, and available light, that follows both the thermocline and the calendar north.

Phenological effects are expected to be very important. While observational data show that there can be as much as a six-week variation in timing of the primary production bloom, total primary production for the year can be the same whether the bloom occurs in May or July. The timing of the diatom bloom is the first variable in an equation that concludes with the number of fish entering the fishery.

The approach

The background information on circulation and productivity is sufficient to allow formulation of a basic hypothesis concerning a high regional secondary productivity. Current flow information suggests that water circulates cyclonically in Bristol Bay. Pollock spawn in the region of Unimak Pass, while larvae and juveniles occur in sequence cyclonically around the Bay, following the retreating ice edge north in the spring as they develop. During this period, larval pollock distribution is controlled primarily by water currents, the fishes moving downstream to the north, west, and east. After reaching the juvenile stage, the pollock are able to migrate back to the over-wintering area along the shelf break.

Both the winter pack ice, which moves southward, and a net northward on-shelf water circulation tend to concentrate nutrient materials in the shelf area. Migration of adult fish to the shelf break underlying the region where water flows onto the shelf would act to recycle these materials, as would southward movement of ice onto the region. These factors probably all contribute to the high biomass, hence by inference — productivity, which characterizes the region.

The first phase of the PROBES program is directed towards confirmation of the general conditions on which the hypothesis is based. Primarily this will involve a series of physical and biological observations in the field during the spawning period of spring to early summer. The objectives of these first studies all seek to gain better resolution of the system in time and space. The physical oceanography work is designed to measure the general current pattern as well as the smaller-scale eddies in the circulation. The latter are particularly important to the biological studies in defining functional ecological units and will determine the design of the biological sampling program. The investigation of nutrient regeneration will contribute to the location and movement of water as well as examine the sources of nutrients and processes that contribute to the timing, maintenance, and distribution of the phytoplankton bloom. The productivity research is directed towards understanding the mechanisms regulating primary production, particularly the timing and distribution of the bloom. The primary production work also integrates with studies of higher levels by identifying the ecologically important phytoplankton species. The studies of upper trophic levels will determine the spatial distribution of pollock eggs and larvae, their food-web relationships, and trophic dynamics. The modeling work in PROBES has two functions central to the whole program. One is the construction of specific models as a tool to guide the field efforts. Second is the integration and synthesis of field observations into an ecosystem model with predictive capabilities. The overall integration of PROBES is maintained by the interdependence of all the individual research studies as they relate to the PROBES hypothesis. The future course of PROBES would not be radically affected were our hypothesis in error. Some combination of mechanisms must be responsible for the high regional productivity. Any test sufficient to disprove our basic hypothesis would provide sufficient input for new hypotheses based on more sound evidence. As a study dealing with a complex dynamic system, PROBES will retain the flexibility to accommodate changing ideas during the course of the work.

Acknowledgment

Contribution no. 295, Institute of Marine Science, University of Alaska.

CHAPTER 9

Canadian environmental studies of the southern Beaufort Sea

A. R. MILNE[1]

INTRODUCTION

Study site

The Beaufort Sea off northern Canada is dominated by the Mackenzie River, which through geologic time has been responsible for the deltaic formations of the continental shelf. These formations, similar in nature to those in the Gulf of Mexico, could yield 10 billion barrels of oil equivalent, according to an estimate made by the Geological Survey of Canada (1973). Between the Alaskan-Yukon border in the west to Cape Bathurst in the east, 20 million acres of the Beaufort Sea had been leased to the petroleum industry by 1972, mostly inshore of the 600-m isobath. These are shown in Figure 9.1. Since that time, artificial islands have provided the platforms for exploratory drilling in the shallow waters of the Beaufort Sea — drilling from these islands having been regarded as similar to land operations for regulatory purposes.

By early 1972, approval was sought to drill in deeper waters using drill-ships during the short arctic summer in the open waters northward of Tuktoyaktuk Peninsula and Richards Island. Although marine drilling tech-nology and experience in drilling offshore wells are well advanced, it was clear that the Beaufort Sea is a relatively unknown frontier which could be seriously damaged by an oil-well blowout and by new industrial activities. More recently, the Canadian Marine Drilling Company proposed to drill from ice-strengthened drill ships in the summer of 1976. The company must have a drilling authority, and the constraints on this authority are subject to

[1] *Environment Canada, Institute of Ocean Sciences, 512 Federal Building, Victoria, B.C. V8W 1Y4 (Canada).*

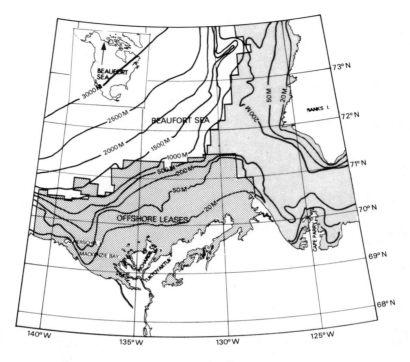

Fig. 9.1 Offshore lease areas in the Beaufort Sea.

recommendations based on a set of environmental studies called the Beaufort Sea Project.

Project history

The Beaufort Sea Project began as a collection of 29 studies of the environment of the Beaufort Sea. It is funded jointly by the petroleum industry and by agencies of the Canadian federal government, with 4.1 million dollars coming from industry and 7.5 million dollars worth of support and new funds from the government. Direct management of the project is through the Department of the Environment in Victoria, B. C., while the petroleum industry, through an industry project manager, monitors the progress and conduct of the 21 studies they are supporting.

PROGRAM OBJECTIVES

The project must attempt to relate its recommendations to the following questions where the prime concern is damage which could result from an oil-well blowout.

What is at threat from offshore exploratory drilling? — what threats does the marine environment pose to offshore exploratory drilling? — how are pollutants transported from drilling sites, and to where? — how can changes in the marine environment be identified and monitored? — and, how can offshore drilling be conducted with a minimum but acceptable risk to the environment?

Scope

The short duration of the project — two years to December 1975 — has produced concern regarding the completeness of studies; hence clarification of the purpose of the project and how its output will be used is necessary. The purpose of the project is to speed up the pace of Beaufort Sea investigations, some of which have been underway for years, and to collect all information relative to the questions mentioned earlier. The output of the project, in addition to its technical reports will have a summary with recommendations. This latter document will be a preliminary environmental assessment on the impact of exploratory drilling in the Beaufort Sea. The recommendations will relate to the threat potential of each proposed drilling site for 1976 and the requirements, or otherwise, of a formal impact assessment for these sites.

Since the project's inception, the number of studies has grown from the original 29 to 38. These are listed in Table 9.1. In addition, 6 special "overview reports," spanning the subject material of the 38 studies, are being produced. Their titles are listed in Table 9.2. The intention in this short report is not to describe all the studies in detail but to cover their scope and relevance to offshore petroleum exploration.

Baseline studies

The project includes baseline studies of phytoplankton and benthic communities, of fish, seabirds and mammals, as well as related biological research on marine ecosystems. These latter studies include the effects of crude oil on ringed seals, on invertebrates, and on the bottom organisms responsible for nitrogen fixing. The possibilities of biodegradation of oil are also being examined. Baseline studies on existing pollutants include beach walking to determine the status of shoreline debris, tar balls, and oil seeps, and include sampling of the sea for dissolved petroleum hydrocarbons. Also important to consider are the needs of the indigenous human predators regarding the cropping of marine wildlife.

Physical oceanographic studies include those on surface and bottom currents, Mackenzie River flow effects on the Beaufort Sea, identification of water masses and their movements and storm surges.

Climatology studies include those on the atmosphere, wind-waves, and ice. A major meteorological study is concerned with the predictability of weather, wind-waves, and ice motion based on the utilization of a small-area computerized prediction system.

The sea-bottom is being selectively examined for sub-bottom frozen materials and for scouring by sea ice. Two oil-related geological studies are on the dispersal of Mackenzie River sediments and on coastal sedimentary processes.

TABLE 9.1 Studies in the Beaufort Sea Project

Mammals and birds

    Ringed and bearded seals
    Polar bears
    Seabirds
    Bowhead and beluga whales
    Effects of crude oil on ringed seals

Fishes, invertebrates and plants

    Fishes of the Yukon coast
    Fishes of offshore waters and Tuktoyaktuk vicinity
    Fishes of the outer Mackenzie delta
    Nitrogen fixation in arctic marine sediments
    Biodegradation of crude petroleum by indigenous microbial flora
    Effects of crude oils on arctic marine invertebrates
    Biological productivity

Baseline pollutants

    Coastal distribution of tar and debris
    Chemical oceanography and the measurement of petroleum-based hydrocarbons

Oceanography

    Mackenzie River input to the Beaufort Sea
    Bottom currents in open water
    Surface currents in open water
    Physical oceanography
    Storm surges

Geology

    Frozen seabed materials
    Bottom scour by sea ice
    Sediments and sedimentary processes
    Sediment dispersal

Climatology and climate prediction

    A real-time prediction system
    Meteorology and wave climatology
    Ice climatology
    Distribution of ice thickness
    Satellite observations of the Beaufort Sea
    Sea ice morphology
    Movement of shorefast ice in Mackenzie Bay

Oil-in-ice and countermeasures

    The spread of oil in the Beaufort Sea
    Entrainment of crude oil in sea ice
    Oil, ice and climate
    Physics of an underwater blowout
    Oil-spill countermeasure
    Effects of crude oil on Balaena Bay

TABLE 9.2 Overview reports of the Beaufort Sea Project

Birds and marine mammals
Fishes, invertebrates and marine plants
Pathways and fate of pollutants
Geology, climatology and prediction
Oil in sea ice
Oil spill countermeasures

Finally, there are major studies on interactions of oil and sea ice and on oil countermeasures.

What is the southern part of the Beaufort Sea like? It is usually covered by ice for up to nine months of the year, but during the short three or four months of summer, drilling from floating vessels is proposed. With the onset of winter in early October, the extent of shorefast ice progresses seaward with relatively smooth ice reaching up to 30 km northward of the shore to the 20-m isobath. Multi-year ice and ice island fragments of glacial origin are sometimes incorporated in the fast ice.

Well offshore and beyond the continental shelf, the polar pack ice circulates slowly clockwise with the Beaufort Sea Gyre. Between the shorefast ice and the ice of the Beaufort Sea Gyre is a shear zone where the ice moves sporadically.

In shallow coastal waters, the bottom is saturated with scours caused by moving ice. In waters of 15 to 50 m depth, deeper scours are caused by pressure-ridge keels, and the number of scours per unit area decreases with depth. Relic permafrost is a feature of some shallow-water areas.

Important Beaufort Sea wildlife are the many species of fish, ringed seals, beluga and bowhead whales, seabirds, white foxes and polar bears, all of which are used extensively by northerners in communities on the rim of the Beaufort Sea. In the spring and summer, seabirds and mammals are dependent on leads and open water as wildlife highways — which unfortunately would be where oil would collect if a blowout remained free-running throughout the winter.

PROJECT ACTIVITIES

Research craft

During the summer of 1974, the new 191-ft government vessel *Pandora II*, carrying the submersible *Pisces IV*, was used to support marine geophysical studies and sea-bottom investigations. The charter vessel *Theta* of similar size was used to support oceanographic, fisheries, and biological studies. These vessels spent only one month in the Beaufort Sea due to adverse ice conditions but returned again in summer 1975. Fixed-wing aircraft and helicopters provided indispensible research platforms and transportation. The ships and aircraft utilized a DECCA positioning system operated by the Polar Continental Shelf Support base at Tuktoyaktuk. Based on aerial surveys covering almost 30,000 mi^2, the population of ringed and bearded seals is estimated at between 1.0 and 1.5 million. Three hundred polar bears were tagged, of which 35 were recaptured, demonstrating that some bears moved between Alaska and Canada. The area north of Tuktoyaktuk Peninsula was found to be an important polar bear feeding area. It is of interest that in years of low lemming populations, white foxes move offshore in spring to prey on the seal pups which remain helpless in their nursery lairs.

Biological surveys

The studies of seabirds commenced in the spring of 1974, but reconnaissance surveys were hindered by fog and storm conditions and by lack of open water. Reproduction for some species was very poor due to a late spring and a short cold season. In the case of snow geese, there were almost no young to migrate. Further study during spring 1975 using DEW-Line radar observations has provided useful offshore migration information.

The beluga whales which migrate to the Mackenzie delta to calve are estimated to number 5000. The population of bowhead whales is much smaller, estimated at a little over 100 animals.

Several studies were made to examine the movements, distribution, populations, and food habits of coastal and anadromous fish. A total of 5000 fish of 23 species were caught, and 1500 were tagged by Freshwater Institute, Winnipeg and Yukon Fisheries. Findings indicate that anadromous fish are concentrated near coastlines. There is a justifiable concern about the possible effects of an oil spill in the lagoon areas and bays, which are not only important feeding areas for fish fry but which also provide shelter for molting seabirds. No tar balls, oil spills, or natural oil seeps have been discovered along Canada's Beaufort Sea coastline. From water samples taken from *Theta*, the water was found to be free of particulate pollutants, and the level of dissolved petroleum hydrocarbons was equivalent to that in the cleanest of seas.

Physical studies. Surface currents were measured by dropping and tracking disposable drift drogues containing radio beacons from a Twin-Otter aircraft. Although severe ice conditions last year limited their deployment to near-shore, the results showed that the north winds tend to cause eastward currents in Mackenzie Bay and along the Tuk Peninsula, whereas eastern winds cause westerly movements. Current speeds were generally less than 0.5 knots but reached 1.3 knots.

Physical oceanographers have monitored the water column for temperature, salinity, turbidity, and water mass movements in the summer of 1974 from *Theta* and in spring 1975 from fixed camps on the sea ice by using helicopters. Of particular interest are the interactions of the Mackenzie River with the Beaufort Sea. Less open water than usual in 1974 resulted in a thicker (7 m) and colder (2° C) freshwater layer than usual (2.5 m and 6° C, respectively) north of Richards Island, apparently caused by the damming effect of the sea ice.

An environmental prediction system has been designed for rapidly updating forecasts of weather, winds, waves, and ice movement. Designed by Atmospheric Environmental Services, Toronto, parts of the system are under test this year. In the future, offshore ice-mounted weather buoys will be used to enhance the forecasting ability of the system. Design and installation costs are estimated at almost 2 million dollars. A computer model for predicting the magnitude of storm surges has also been prepared for incorporation into the environmental prediction system.

In order to determine the ages and frequencies of sea-bottom scours, the Geological Survey of Canada has used side-scan sonar to survey the bottom and has constructed sea-floor mosaics. It has been suggested that the majority of scouring occurs in the wintertime.

Ice climatology

Aerial ice reconnaissance observations made during the 20 years from 1953 to 1973 have been re-examined by computer methods on a 15 x 20-mile grid scale. Work is also underway on the floe size distribution. Another interesting aspect of this work deals with the long-range predictability of good or bad ice years. A pulse-radar method of determining ice thickness from aircraft is being developed. The distribution of ice types and thicknesses is being obtained from the analysis of microwave radiometric satellite images and from microwave imagery obtained from aircraft flights in cooperation with the AIDJEX project. The ice studies are of importance regarding "threats from the environment" mentioned earlier.

Oil in ice

Research concerning the behavior of oil in ice, on the other hand, is attempting to assess a major threat to the environment from offshore drilling. It is important to understand the spreading characteristics of crude oil and its ultimate destiny in sea ice. Complementary to this understanding is the possible effect on wildlife mentioned earlier and the effects on climate of oil accumulations. It is estimated that a single blowout free-running for a year could contaminate 9 km^2 in ice-covered seas, assuming a 1-cm average thickness of oil. The effects on climate from a single blowout are likely insignificant. An oil countermeasure study is concerned with the possibilities of oil containment, clean-up, and disposal at a variety of sites and seasons in the Beaufort Sea. No testing of equipment under actual operating conditions is presently funded; at this stage, the concern is to estimate the probable effectiveness of available equipment and techniques.

CONCLUSIONS

There will remain some significant gaps in knowledge after the project has been completed. We will fall far short of having good estimates of the biological productivity of the southern Beaufort Sea; in particular, estimates of fish production will be poor. It will be possible to delineate biologically sensitive areas upon which part of the productivity depends, but the impact of an oil inundation will be difficult to estimate. The heavy ice in the summer of 1974 appeared to have nearly catastrophic effects on the higher life forms in the Beaufort Sea, such as the snow geese which failed to reproduce successfully, the slower growth rates evident in ringed seals, and the incidence of starving bears. Little enough is understood regarding the stresses that natural catastrophes impose on the more primitive elements in the food-chain that the higher life forms depend on. Even less will be known on the significance of man's intrusions into the marine ecosystem. On the positive side, the Beaufort Sea Project will considerably advance our knowledge on the limits of environmental predictability, particularly with regard to ocean currents, sea-ice movements, bottom scouring, and storm surges.

REFERENCE

GEOLOGICAL SURVEY OF CANADA

    1973   An energy policy for Canada. Information Canada, Montreal, p. 36.

Assessment of the Arctic Marine Environment: Selected Topics
Copyright © 1976 by Institute of Marine Science, University of Alaska, Fairbanks

CHAPTER **10**

Marine terminus of the trans-Alaska pipeline

D. W. Hood [1]

Abstract

Expected to begin operations in mid-1977, the trans-Alaska pipeline is scheduled to carry oil at a rate of up to 2 million barrels per day from the arctic shore to a tanker terminal on the southern coast. Treatment will be required for as much as 1 million barrels of ballast water discharged by incoming tankers. The terminal facility, loading dock, and the effluent disperser for the ballast treatment plant are located in Port Valdez. In 1971-1972, a 15-month concentrated study was made to obtain background information on the baseline condition of the Port, including its physical circulation and dispersion, flushing rates, processes of primary productivity, the biology of the benthos, effects of crude oil on productivity, *in-situ* biodegradation rates of hydrocarbons, and sedimentary geology. Additional studies of a broad scope were implemented to carefully monitor the system through start-up of terminal operations and for at least a year after operations begin. After start-up, direct measurements of dilution rates will be made using in-place diffusers and dye-labeled ballast water. The purpose of this chapter is to summarize some of the findings and conclusions of the preliminary baseline studies as they relate to assessment of this type of arctic environment.

INTRODUCTION

Geography of Port Valdez

Trending east and west as a northwest extension of Prince William Sound, Port Valdez is approximately 21 km in length and 4.5 km wide. The fjord

---

[1] *Institute of Marine Science, University of Alaska, Fairbanks, Alaska 99701.*

of Port Valdez, carved from troughs of precipitous mountains, reaches a maximum depth of 240 km. At its entrance, the fjord bends at nearly right angles into Valdez Narrows and Valdez Arm. At this point, a minimum sill depth of 110 m occurs and represents the narrowest passage for the projected voyage of potential tankers plying the North Pacific waters from Puget Sound to the Alyeska pipeline terminal. Prince William Sound contains about 3000 miles of mostly wilderness coastline, encompassing many islands and fjords. Though comparable in size to Puget Sound, this region has only about 5000 inhabitants, engaged primarily in the fishing or lumbering industries.

The town of Old Valdez, the only settlement in Port Valdez prior to oil development, was destroyed by the Great Alaska Earthquake of 1964. The town was rebuilt on higher ground and restored its population of about 1000 prior to 1973. The Alyeska pipeline terminal construction camp at Dayville and the community of Valdez have increased in population manyfold during the construction phase of the pipeline, but numbers will probably reduce to around a total of 5000 people during the operational phase of the project. Recreational opportunities are plentiful in the exotic surroundings of Port Valdez, and there is an abundance of fish and wildlife. Heavy runs of silver, pink, and chum salmon occur during the late summer months; bear and deer are hunted at various times of the year; and there are over 150 avian species, about 70 of which contain thousands of individual birds.

Fjordal characterization of Port Valdez

The Alaskan fjords typified by Port Valdez are positive estuaries, since they receive more water by runoff and precipitation than is lost by evaporation (Pritchard 1952). The classical circulation pattern prevails, involving a seaward movement of brackish surface layer and a landward movement of the deeper waters. Alaskan fjord dynamics have been studied by Trites 1955; Tully 1958; Rattray 1967; Pickard 1967; Matthews and Rosenberg 1969; Quinlan 1970; Nebert 1972; and Matthews 1972. These investigations have found that circulation in Alaskan fjords is nearly always sufficient for continuous, or at least frequent, renewal of water within the fjord. The bottom water renewal shows a cyclic variation. The freezing air temperatures and negligible freshwater runoff in winter causes vertical circulation tending to renew the waters of the fjord. Complete renewal at least once a year is assured by a combination of thermal-haline convection and by dense water inflow from outside. The density flow can occur anytime during the year depending upon the character of the outside water, which is controlled by the dynamic processes on the continental shelf.

The climate is maritime but altered by the surrounding mountains and glaciers. Mean temperatures range from 7 to 13° C during the summer months and from -10 to -4° C in the winter. The maximum daily temperature exceeds 20° C on the average of only 4 days per yr, and a minimum temperature lower than -18° C occurs only 4 days per yr. The record minimum is -34° C. The mean annual precipitation at Valdez is 158 cm.

Daily precipitation of greater than 0.25 cm occurs on the average of 120 days per yr. Much of the precipitation occurs as moist dense snow resulting in a mean snowfall of 245 inches.

Environmental criteria for waste disposal

The discharge of foreign materials into the balanced ecosystem of a near-pristine fjord system without significant disruption requires a finely tuned disposal system coupled with a full knowledge of the stress load involved and the environmental forces controlling the system. Identification of the kinds and amounts of waste materials, the rate and geographic extents of dispersal and dilution, an understanding of the chemical and biological transfer and sensitivities of critical forms of biota, the physical transport, transfer and exchange of the fjord waters with larger systems, the influence of outside factors on the water quality of the stressed system, and a knowledge of the degradation rate and ultimate fate of the contaminants are all needed to effect a satisfactory predictive model of what to expect from given waste loadings. This complex problem does not lend itself to an exact solution but requires a combination of experimentation, observation, and computation by interdisciplinary scientists who are willing to estimate their conclusions while continuously re-evaluating and testing previous hypotheses. In general, the tendency is to be on the ultra-safe or overly cautious side, a situation which tends to reduce the level of flexibility and increases cost. The desirable level of waste disposal allows for a condition of discharge which accommodates that amount of material not harmful to the indigenous or acceptable ecosystem. To reach this goal requires an ultimate knowledge of the system far beyond that usually acquired in short-term limited studies. Short-term studies, such as the initial one in Port Valdez, introduce a baseline perspective on which future testing and monitoring programs can be based in developing the capability for predicting and regulating the environmental effects imposed by a waste-discharge level. As the discharge system is refined, an understanding can be developed, and the optimal level of waste discharge can gradually be established.

The facility for storage and loading of oil and the ballast treatment plant for the trans-Alaska pipeline terminal will be located on the south side of Port Valdez near Jackson Point (Fig. 10.1). The facility will have a capacity for a daily loading of 2 million barrels of crude oil into tankers and processing in excess of 1 million barrels of tanker ballast water received from incoming tankers. The ballast water will be treated by skimming, aeration and aluminum hydroxide flocculation to give a daily average concentration of 8 mg/liter and a daily maximum of 10 mg/liter of oil and grease. The treated water will then be discharged through an elaborate diffuser system into Port Valdez. The State of Alaska discharge permit under which this facility operates defines a mixing zone in the Port for discharge that extends 150 m from the diffuser center line and 150 m beyond each end of the diffuser horizontally and 5 m from the surface and 0.5 m from the bottom vertically. Concentrations of 0.1 mg/liter oil and grease may not be exceeded in the horizontal direction and 0.05 mg/liter in the vertical direc-

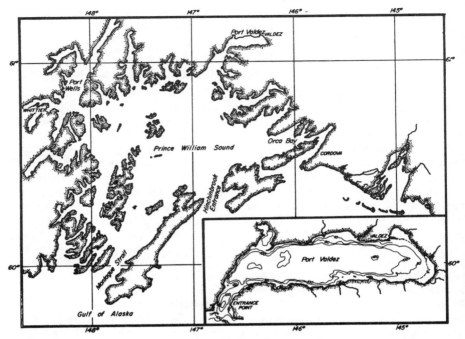

Fig. 10.1 Prince William Sound region with index map showing Port Valdez bathymetry in inset (Sharma and Burbank 1973).

tion outside the limits of the mixing zone. The above permit was granted after a 15-mo intensive study of the Port area (Hood et al. 1973).

The design and construction of a diffuser system which would meet these limits obviously required a full knowledge of the physical characteristics of the Port and advance design features of the treatment plant and diffuser system. Studies are now underway which will carefully monitor the system through start-up of terminal operations. This phase of the study is very broad in scope and includes further work on circulation, present hydrocarbon content and biodegradation rates, biological beach surveys, establishment of distribution of benthic organisms, selection of monitoring organisms, trace metal distribution, kinds of plant and animal plankton, and toxicity effects of effluents on photosynthesizing organisms. This study, beginning approximately 18 months prior to start-up of the facility, will continue until at least one year after operations begin. After start-up, direct measurements of dilution rates will be made using the in-place diffusers and ballast water labeled with dye as a tracer. The beginning effort of what will be a rather long-term study of the Port Valdez ecosystem and the impact of oil development on this system has been completely documented (Hood et al. 1973). Selected aspects discussed below are especially relevant to assessment of one type of arctic marine environment.

PHYSICAL BASELINE STUDIES

Bathymetry

The bathymetry features of the basin into which a diffuser outfall is to be placed has important relations to circulation, sedimentation, biological communities, water quality, and the flushing characteristics of the system. A very significant feature of Port Valdez (Fig. 10.1) is that it is deep (230-250 m), which allows for choice of depth for the diffuser to take full advantage of jet dispersion in the vertical direction, followed by horizontal eddy diffusion to dilute the wastes. The bottom topography is generally flat, featuring deep rocky shores with deltaic deposits and river flats on the east. There are two submarine fans about 3 km south of Valdez and two more near Shoup Bay, which is fed by Shoup Glacier, at the western end. The relief of these features is only 10-20 m, and they probably do not significantly affect the bottom circulation.

At the entrance to Valdez Narrows, there are two sills. The most southern is 110 m, and the northern one is 128 m. The entrance sill depth is an important limitation to deepwater exchange between outside deep water and fjords. In Alaska fjord systems, there appear to be no cases in which fjord exchange does not occur at least yearly. Part of this is due to convective cooling and partly to other factors. In some systems, renewal is more frequent, resulting from the processes of upwelling of dense deep ocean water along the continental shelf in the summer which in turn flows into the fjords and replaces less dense bottom water. The upwelled water frequently does not reach the surface, and whether it freely enters the fjords clearly depends upon the sill depth and the depth of upwelling. The relatively deep sills of Port Valdez apparently do not limit deepwater exchange significantly.

Hydrography and circulation

A strong stratification of the water column occurs in the Port during the months of May to October (Fig. 10.2). During the months of December to April, vertical homogeneity occurs. In April, some surface stratification is initiated, leading quickly to distinct stratification associated with the period of highest primary productivity of the year in April and May. The circulation of the Port is generally counterclockwise, and, except for the top 20 m or less, the typical circulation expected in estuaries does not apply. The reason for this apparent anomaly is that the freshwater input to the Port seldom exceeds 5 percent of the tidal volume (tidal range is 5-6 m). Wind and tidal currents dominate the flow, leading to horizontally uniform temperatures and salinities throughout the Port. Currents near Jackson Point and the outfall site reached a maximum observed velocity of 40 cm/sec with a preferred velocity of 7 cm/sec. Transport away from the outfall site occurs through these small currents, which are variable in both direction and velocity of flow.

The rates of dispersion of water at potential outfall sites near the facility were measured on each of the six R/V *Acona* cruises by dye dispersion and

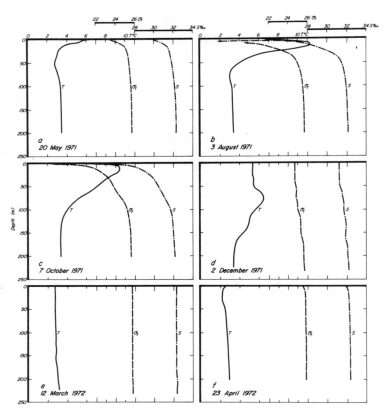

Fig. 10.2 Vertical profiles of temperature, salinity, and density (σt) at one station in Port Valdez for six different times during period May 1971 to April 1972 (Muench and Nebert 1973).

detection techniques. After the initial studies, it became apparent that the site between Jackson Point and Rocky Island (Location No. 1, Fig. 10.3) had the best water transport characteristics, and most of the study time was spent at this location. Data at this site were obtained at three depths of dye input (3, 10, 25 m) under hydrographic conditions representing most of those expected in a normal year. Quantitative determination of the dye concentration was made on three sample streams pumped continuously from the depth of dye input and at 2 m above and below this depth with continuous-flow fluorometers.

The dye dispersion fields obtained at Location No. 1 during early December 1971 for an injection depth of 75 feet (23 m) are shown in Figure 10.3. A plot of the dye concentration data against distance from the injection site is shown in Figure 10.4; in this figure it can be observed that the 10-fold dilution distance for the 23-m (75 ft) injection is 0.17 n. miles. In all studies made, the average 10-fold dilution rate for the 15 and 23-m

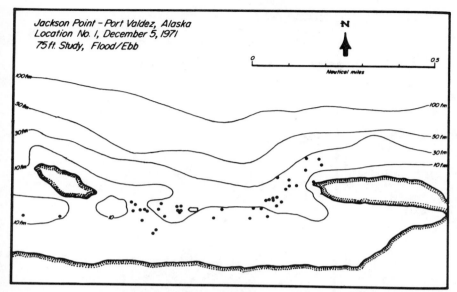

Fig. 10.3 Positions where dye was found near location 1 on 5 December 1971 at depth of 75 ft (23 m) (Nebert et al. 1973).

injection depth was 0.36 n. miles. This computation is very heavily biased toward the longer distances for dilution, because only peak-height concentrations obtained when crossing the dye plume were plotted, and then only the highest concentrations of these were used to plot the dilution curve. The rationale for this treatment is that in waste disposal considerations, the real concern is the maximum concentration to which the organisms will be exposed. Other concentrations are really of little importance unless a requirement for the actual concentration field is established. From the data obtained, it appears that during winter months, dilution rates are more rapid — due probably to the homogeneity of the water and therefore a heavy contribution to dilution coming from vertical turbulence. When the water density is stratified, as occurs in the Port between April and December, little vertical mixing occurs. In fact, the sampling streams taking water only 2 m above and below the injection depth often show zero concentration of dye at the same time the center stream is showing peak concentration for that respective distance. During periods of strong vertical stratification, the maximum distance required for 10-fold dilution was about 0.75 n. miles. These findings impose a heavy requirement on the diffuser system of the terminal, since only two- to threefold dilution can be expected in the mixing zone after discharge.

After the effluent waste has once left the mixing zone, transport in Port Valdez is by weak and variable currents heavily dependent on wind and tides. Many current measurements were made with the use of parachute drogues, but the results only showed trends, because of the changing currents and the heavy influence of wind on the drogues. Direct current

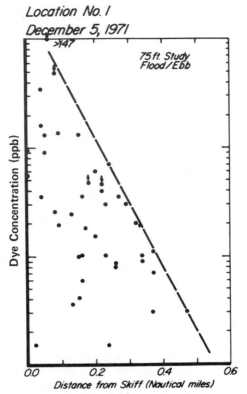

Fig. 10.4 Dye concentration (ppb) versus distance (n. miles) for location 1 on 5 December 1971 (Nebert et al. 1973).

measurements were also made near Jackson Point and in Port Valdez Narrows. The conclusion was drawn that currents are generally counterclockwise in the Port and would tend to transport the effluent in the general direction of the Narrows without excessive accumulation at any one location.

The flushing rate of the entire bay was computed by the classical estuarine circulation entrainment model and by direct current measurements in Valdez Narrows. Assuming cross-channel homogeneity for the current measurements made in a 10-day period in the winter of 1971, a flushing rate of 40 days was calculated for the winter period. The estuarine circulation model gave longer periods but is not considered reliable, because the freshwater signal is too small to be an effective tracer for water turnover in the bay.

Sedimentology

The nature of sediments, their transport and distribution in an estuary have an important influence on the kind of organisms, the chemistry of sediment-water interaction, the capacity of the estuary to handle wastes, and the ultimate fate of many kinds of chemical compounds. In the glacier-fed fjords, the suspended load, composed largely of ground rocks from the glacier valley, is available for interaction with dissolved or suspended material and may thereby remove it from liquid phase components to solid phase components subject to deposition in areas where transport energy permits. Whether the suspended sediments react with the "dissolved" hydrocarbons of the water column depends on their chemical character. The clay fraction or the charged edges of ground rock are most active in sorption-desorption reactions with charge particles. However, direct reaction with the hydrophobic hydrocarbon molecules from petroleum would not be expected on pure clay surfaces. Sorption of hydrophobic molecules to clay already coated with organic matter is more likely. This phenomenon would place considerable significance on the history of the suspended sediments as to their sorptive capacity for hydrocarbon material.

Three major units of surficial sediments are found in Port Valdez. Mixtures of silt, sand, and gravel are found at the mouth of the Lowe River, Valdez glacier stream, Sawmill Creek, and Valdez Narrows. Fine to near fine silt occurs in an elongated narrow area extending west-northwest and east-southeast from the head of Port Valdez to Mineral Creek and Gold Creek. The predominant unit covering the rest of the Port consists of coarse clay. The distribution of suspended sediment in Port Valdez on 4 September 1972 is shown in Figure 10.5.

ASSESSMENT OF THE CHEMICAL BASELINE

To gain an understanding of the baseline chemistry of an impact area in conducting an environmental assessment study would appear to be self-evident; however, adequate chemical studies are too seldom undertaken.

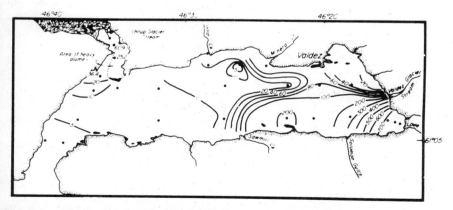

Fig. 10.5 Distribution of the surface suspended sediment load (mg/liter) in Port Valdez on 4 September 1972 during flood tide: low tide 1.1 m (3.5 ft) at 1647 ADST, high tide 3.7 m (12.0 ft) at 2056 (Sharma and Burbank 1973).

Without careful chemical studies, some of the most sensitive environmental signals of damage may be overlooked.

By tradition, workers in the marine area measure salinity and temperature in order to obtain density contours and use dissolved oxygen values to estimate the margin of safety in loading of organic matter to the system. Oxygen variability is imposed by the dynamics of physical and biological stresses and is probably the most used single parameter for evaluating water quality. Of probably greater importance toward a basic understanding of the chemistry of the impact area in northern latitudes, however, is a knowledge of the pH, alkalinity, and carbon dioxide system.

The pH, alkalinity and CO_2 system

Hydrogen ion in the near-neutral solutions of the environment controls many chemical reactions which regulate ion exchange, chelation of trace metals, sorption and desorption, membrane transport, and many other important phenomena occurring in the sea. The pH is largely controlled by component concentration of the carbon dioxide system, which in turn is affected by photosynthesis, respiration, circulation of water, and air-sea exchange. A full understanding of this system, particularly of its subtle interactions, is by no means complete at this time. In coastal environments, the changes in pH can be between 7.2 and 9.0, there can be an almost 100-fold change in concentration of hydrogen ion, and the partial pressure of CO_2 has been found over as wide a range as 50 ppm in highly productive coastal inlet areas to over 600 ppm in upwelled areas of coastal Alaska (J. J. Kelley, personal communication). In contrast, oxygen values of less than 5.0 ppm are seldom found even in deep water of the fjords — and that occurs only when the continental shelf source water is low in oxygen as a result of intrusion from the oxygen minimum zone of the Pacific.

There are advantages to carefully watching the carbon dioxide system in conjunction with certain other studies if the objective is to evaluate changes in the ecosystem. In the first place, it is possible to measure precisely the components of this system. This is in contrast to the imprecision experienced in sampling and measuring biological populations and the very difficult statistical problem involved in detecting significant changes. Secondly, the CO_2 system is very sensitive to both photosynthesis and respiration and tends to integrate major changes in the activity of organisms. In principle, at least, it is possible to use the distribution of components of the CO_2 system as an aid in estimating primary and secondary productivity under environmental conditions. Thirdly, continuous measurements of the partial pressure of carbon dioxide and pH of surface water are easily made from a ship underway, thus giving many data points and a general synopsis of the area studied.

Trace metal dynamics

Trace-metal contamination can be the most critical of environmental impacts by some type developments. An assessment of the level of critical metals including Cu, Zn, Hg, Pb, Se, V, and Cr contained in the sediments, biota, and water is needed to establish the natural concentrations and their distribution. Present knowledge is lacking on the process dynamics of most trace metals in the water column which would lead to an understanding of the effect of shifts in such parameters as pH and kinds of organic matter on toxicity to biota. However, the analytical techniques for determination and sampling methods to avoid contamination have been highly perfected.

Inorganic nutrient studies

Inorganic nutrient data give considerable insight into physical processes occurring within an impact area that would influence the distribution both horizontally and vertically of added contaminants. These ions are very sensitive indicators of vertical transport because of their tendency to become depleted at the surface in warm seasons due to utilization by plants and their increase in concentration with depth due to decomposition of sinking organisms. In cold seasons in high latitudes, the surface waters down to several hundred meters may become homogeneous with respect to these parameters. The analysis is easily automated and gives many high quality data points from samples collected by conventional oceanographic techniques.

In the first 15-month study of Port Valdez, emphasis was placed on dissolved oxygen and inorganic nutrient distribution. Oxygen values were uniformly high in the surface throughout the year. During periods of high plant productivity, oxygen reached values as high as 8.0 ml/liter. In winter the surface values were about 6.5 ml/liter. Bottom water reached its minimal annual oxygen concentration in the winter; however, values of less than 5.0 ml/liter were never found (Fig. 10.6). The near-surface water reached a minimum pH value of 8.10 in December and a maximum value of 8.86 in July. The bottom water reached values as low as 7.96 during the winter.

The inorganic nutrients, including nitrate, nitrite, ammonia, phosphate, and soluble silica, showed a maximum, except for ammonia, during the month of March when the Port waters were homogeneously mixed vertically and before phytoplankton growth had initiated. In April, when the waters began to stratify, the spring phytoplankton bloom occurred. By early May the surface waters were depleted of nutrients to very low values (0.02 μg-atoms nitrate-N/liter), except near areas of freshwater input, and then concentrations remained low until fall mixing occurred in October (Fig. 10.7).

THE BIOLOGICAL CONDITION

Primary productivity

The most fundamental biological process in the oceans is the fixation of light energy, nutrients and carbon dioxide by phytoplankton — providing the base of the food pyramid for all life in the sea. A detailed understanding of the plant species composition that are responsible for primary production and of the factors which affect their growth is probably the most significant biological information that can be obtained for an impact area. Until recently such data were difficult to obtain. Not only was there the inherent difficulty of measuring gross primary productivity with enough frequency to

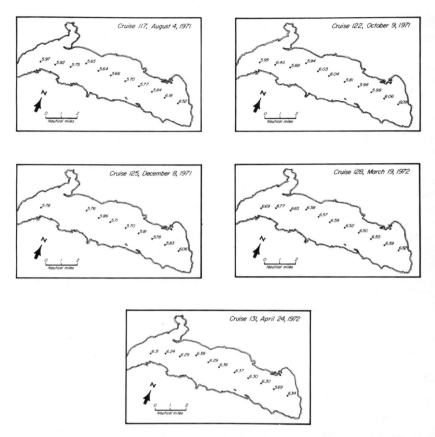

Fig. 10.6 Oxygen values at 1 m above the bottom in Port Valdez during five cruises of the R/V Acona from August 1971 to April 1972 (Hood and Patton 1973).

be meaningful, entailing the tedious work of identifying the many species present — but there was the problem of obtaining an index of the contribution of photosynthetic activity of each species present. Now, however, the innovative techniques of radio-autography (V. Alexander, personal communication) allow for an approximation of the relative importance of each species to total photosynthesis. The radio-autography method has the added advantage in toxicity measurements of determining the sensitivity of each organism to a toxic substance in a natural mixed population community.

The 1971-1972 studies of Port Valdez were not able to use the radio-autographic technique, since it had not yet been perfected; however, the net productivity was measured. Primary production reached its maximum value in late April and early May (Fig. 10.8), at which time values as high as 175 mg C/m^2-hr were measured. Productivity during the rest of the year was low, with winter values of 0.2 mg C/m^2-hr observed. The average annual production in Port Valdez was about 150 g C/m^2-yr, with a range of

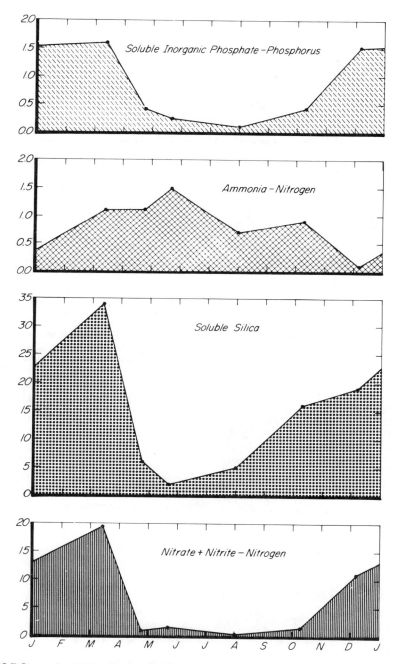

Fig. 10.7 Seasonal variations in the distribution of nutrients at the 50% light depth at station 142 in Valdez Arm (Goering, Patton and Shiels 1973).

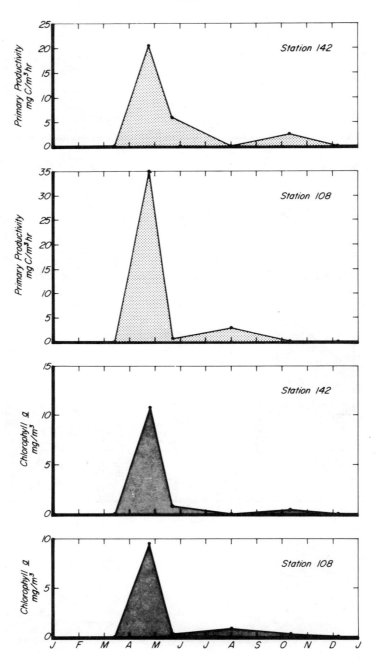

Fig. 10.8 Seasonal cycles of surface carbon-14 uptake and chlorophyll a concentration at station 142 in Port Valdez and station 108 in Valdez Arm (Goering, Shiels and Patton 1973).

111-113 C/m^2-yr. Daily rates during spring bloom of up to 4 g C/m^2-day approached the daily rates reported for the most productive marine phytoplankton crops ever observed.

Benthic plants are of minor importance (estimated less than 1 percent contribution) to productivity of Port Valdez; however, large stands do occur in nearby bays where they would be expected to be of major importance to the marine communities of these areas.

Secondary production

The primary trophic level represented by phytoplankton is very sensitive to environmental change, and the techniques for measuring the effect of contaminants at the primary level are perhaps more satisfactory, in general, than those directed at higher levels of the food-chain. The secondary productivity grouping, however, includes all the animals of the sea — i.e., fish, crabs, seals, etc., — and is of greater interest because of its direct link to food items and products of commerce. The manner in which contaminants brought by developmental activities of any kind, including structures, sea traffic noise, and atmospheric changes, may influence populations used by man is certainly of greatest concern and usually receives the majority of the attention of most environmental studies. Because of the number of organisms and the impracticality of obtaining detailed data of a quantitative nature on more than a few types of organisms, studies usually concentrate their efforts on sessile forms that remain exposed to the conditions because they do not have movement capability, or focus on exotic species of particular interest or on sensitive forms of commercially important species. Since environmental changes cause natural fluctuations among the animals of the sea, a longer period of time is usually needed to establish a norm. Also, because the numbers and the species compositions in this group often fluctuate so widely both in space and time, the sampling and statistical analysis of the data are very important in this area of investigation.

In the Port Valdez studies, our emphasis on the animals present centered on the benthic organisms. The National Marine Fisheries Service had a program involving beach surveys and the migration of salmon, particularly the fingerlings as they leave the hatchery streams.

Assessment of benthic infauna

The primary purpose of the benthic survey was to provide a framework for future monitoring of the benthic infauna, in order to determine the quantitative effect of the addition of petroleum fractions to waters of Port Valdez and restricted portions of the adjacent Valdez Arm. Benthic infauna consists of those organisms that were quantitatively collected by the bottom sampler used in these studies and included a few species of slow-moving epifauna. The specific objectives were to identify and locate the major infaunal associations of the benthos of Port Valdez and adjacent waters and to examine the population density and biomass of selected species in relation to various biological and physical parameters, specifically the adjacent biological associations and sediment properties.

Fifty-nine stations were occupied four times during the one-year study. Most sampling was done by five replicate grabs of a Van Veen grab. Gleason's index of diversity was calculated by station and dendograms resulting from weighted and unweighted cluster analyses of benthic station data. Maps of the clustering of stations based on population densities of the ubiquitously distributed Biologically Important Species are shown in Figure 10.9. The complete report of this study is given in Feder et al. (1973).

OIL IMPACT STUDIES

Hydrocarbon levels and biodegradation rates

Hydrocarbon levels of water, sediments, and biota are usually carefully measured in any oil impact studies. The quality of these analyses and the sophistication of the analytical equipment is continuously increasing. In the preliminary experiments, largely liquid-solid column chromatography for separation of hydrocarbon types and gas chromatography for analysis for specific hydrocarbons were used. The results showed concentrations of normal paraffin hydrocarbons to be <0.1 ppb in the water column, 0.5 to 2.5 ppm in recent sediment, and 0.5 to 1.9 ppm in the biota. No evidence was found to indicate that the kinds of hydrocarbons present were derived

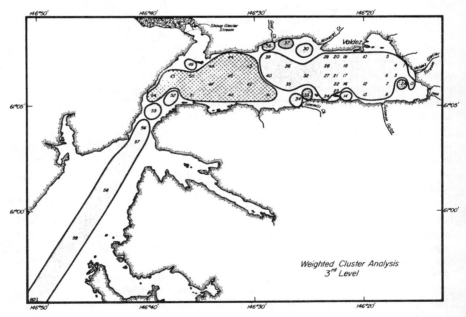

Fig. 10.9 Map showing clustering of stations based on population densities of the ubiquitously distributed "Biologically Important Species." The groups were determined by means of a weighted cluster analysis interpreted at the third level of a printed dendrogram (Feder et al. 1973).

from petroleum origins, except at sites known to be so contaminated. Continuing studies will utilize the more sophisticated techniques of mass spectrometry and improved gas chromatography in the hydrocarbon analysis. Emphasis will be placed on the aromatic and other petroleum origin fractions.

The biodegradation studies (Robertson et al. 1973) were made with radio-active carbon-labeled dodecane as tracer incubated *in situ* in the natural waters of Port Valdez, in order to estimate the rate of decomposition of hydrocarbons in this ecosystem. An average rate of 1 μg-atom C/liter-day was observed at a depth of 10 m over a period of 35 days. These experiments, indicating relatively rapid decomposition rates for hydrocarbons in Port Valdez, will be continued in the follow-up study but with more emphasis on other types of hydrocarbons.

Effect of crude oil and its components on the growth of phytoplankton

Phytoplankton growth is very sensitive to the quality of water in which it occurs. Not only do the relative concentrations of available forms of nutrient nitrogen, phosphorus and silica influence the growth of specific species — but trace metal concentrations, the kinds of organic matter present, and presence of contaminants also greatly affect the ability of these organisms to grow normally and to photosynthesize primary energy in the form of organic matter.

A variety of techniques are available to test the physiological capacity of phytoplankton populations as influenced by contaminants that reach the environment. Many laboratory investigations have also been conducted on the concentration of toxic substances that limit growth as measured by photosynthetic activity of single species. The most commonly used methods (Shiels et al. 1973) involve a measurement of O_2 production or radioactive C-14 uptake under conditions similar to those used for measuring primary productivity. The comparison is made of growth occurring under natural conditions with and without the addition of gradient concentrations of the test substances. Other methods entail the effect of short-term exposure on the growth of natural populations of phytoplankton as an indication of species differentiation. The recent use of auto-radiography has greatly enhanced these types of measurements.

Most of the tests now used are for short-term exposure and fall short in projecting long-term effects. Large-scale experiments over long periods are underway, however, in many parts of the world, where various methods are used to determine the effect of toxic substances on community metabolism. Although these methods are giving very interesting results, they are usually laborious, costly to run and not yet perfected for routine use. Perhaps the most notable of these experiments is CYPEX, conducted at Sidney, Vancouver, Canada.

During the preliminary investigations in the Valdez study, effects of crude oil both on photosynthesis and on the diversity of species were considered. The summary results of the effect of Prudhoe crude oil on phytoplankton

growing under natural conditions in Port Valdez is presented in Figure
10.10. These data indicate that at low concentrations (0 to 0.017 ppm) of
crude oil, growth is actually stimulated by addition of the oil. At about
0.05 ppm, there is some evidence of inhibition, and 50 percent inhibition
occurred at 20 ppm. It is interesting that the treated ballast water with 5.5
ppm crude oil content gave only limited inhibition in these experiments. It
is apparent that a significant reduction of the volatile aliphatic and aromatic
compounds occurs during the ballast treatment process, which effectively
reduces the toxicity of the crude oil residue. Further investigations of this
type showed the effect of crude oil to be a function of a combination of
factors, including water temperature, season, phytoplankton population,
nutrient level, and possibly light level.

Preliminary evidence of the effect of crude oil on the relative growth of
phytoplankton species in a natural population was obtained in the Port
Valdez study. Seawater was collected from 5 m depth near Jackson Point,
and aliquots were dispensed into four 1-liter bottles. The contents of one
bottle were preserved immediately for species identification and counting,

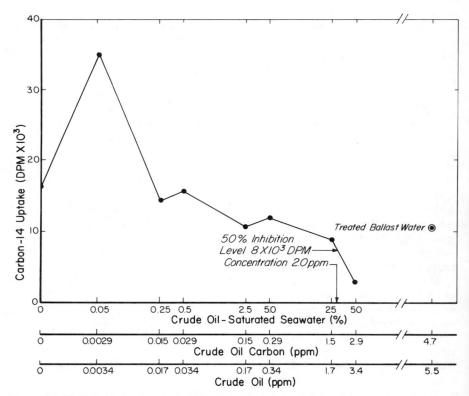

Fig. 10.10 Photosynthesis by phytoplankton incubated for 4 hrs under natural conditions of
light and temperature in relation to dilutions of seawater saturated with Prudhoe Bay crude oil
(Shiels et al. 1973).

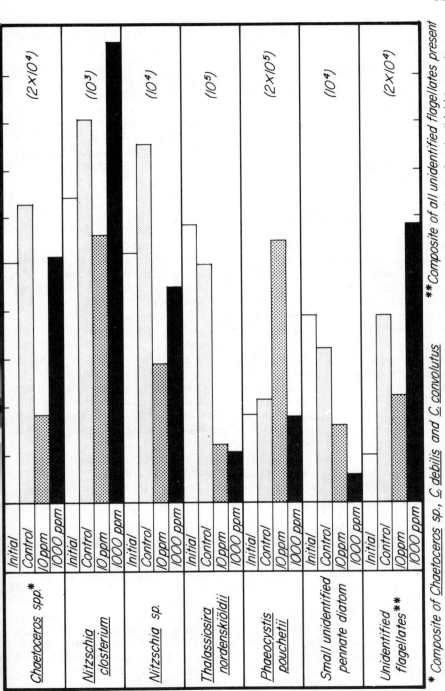

* *Composite of Chaetoceros sp., C. debilis and C. convolutus*   ** *Composite of all unidentified flagellates present*

Fig. 10.11 Relative abundance of seven major phytoplankton species, or species groups, in relation to additions of crude oil (v/v) in experiments incubated for 48 hrs under natural conditions of light and temperature during May 1972 in Port Valdez. Standing stocks (cells/liter) are computed by multiplying the cell number by the factor given in each row (Shiels et al. 1973).

and the other three samples were exposed to 0, 10, and 1000 ppm Prudhoe Bay crude oil. After 48 hrs incubation at ambient light and temperature conditions, the samples were again analyzed for species and numbers. The results of this experiment are shown in Figure 10.11. The possible effect of crude oil on the growth rates of different species of phytoplankton is evident in these data. The reason for the higher oil concentration stimulating the production of higher numbers of cells of some species is not clear; however, the results do contribute to the concept that the change of phytoplankton species may result in an area as a result of exposure to oil. A better understanding of how species succession might be altered by manmade and natural influences is essential if reliable predictions are to be made of the ecological consequences of oil pollution.

SUMMARY AND CONCLUSIONS

The mixing, flushing and biodegradation processes which were measured for Port Valdez indicate that a maximum of 5 ppb concentration of hydrocarbons will accumulate in the Port as a result of ballast treatment at the terminal facility. It was found that the Port has a sluggish circulation, largely wind-driven, but does have a slow movement in a counterclockwise direction. Surface currents of greater than 1 knot were not found in the main portion of the Port, and rates were slower at depth. Near the planned diffuser pipe location for the ballast treatment plant effluent, extensive dye measurements showed a minimum 10-fold dilution rate within 0.6 n. miles. The diffuser design will have to accommodate nearly a 100-fold dilution to take 10 ppm or less hydrocarbons in the treatment of plant effluent down to 0.1 ppm within a 150-m horizontal mixing zone as required by the permit for effluent discharge from the State of Alaska and by regulation of the Environmental Protection Agency. Even more stringent limits are required for vertical dispersion.

From the toxicity data obtained in these studies, there is no evidence that concentrations below 0.05 ppm of Prudhoe Bay crude oil would affect the biota of Port Valdez. This value was used as the target concentration for outfall design. It is apparent that an outfall can be designed that will reduce the effluent concentration of hydrocarbon in the immediate mixing zone to at least 0.5 ppm. The mixing processes of the Port, which were measured extensively by dye dispersion techniques, will reduce this concentration by an order of magnitude within less than 0.6 n. miles from the point of injection under the least advantageous mixing conditions.

It was also computed that the maximum concentration the crude oil could reach in the main body of Port Valdez would not exceed 0.5 ppm. This estimate does not include enhanced flushing through tidal entrainment, mixing of surface with bottom water and neglects biodegradation, adsorption, volatilization, and other factors which would reduce the concentration of oil. Biodegradation alone in Port Valdez has the capacity to decompose more oil than would be discharged with the ballast water. Since this process is not light-dependent and is slowed in high nutrient waters only by lower temperatures, this process will continue to function at about the same rate throughout the year. It is not expected that the hydrocarbon concentrations in the Port will exceed a few parts per billion as a result of ballast treatment discharge.

Studies will continue over the next several years in Port Valdez in order to establish more firmly the baseline conditions under which this ecosystem operates. It is hoped that by understanding the ecosystem in detail, so that cause and effect relationships can be established, industrial development might proceed together with environmental maintenance in this most beautiful and biologically rich area.

Acknowledgments

The study here reviewed was conducted under the major sponsorship of the Alyeska Pipeline Service Company. The National Science Foundation provided part of the ship time through Grant No. OCE72-02718 and the Sea Grant program helped with the costs of publication. The individuals who contributed to the original study as presented in environmental studies of Port Valdez Occasional Publication No. 3 of the Institute of Marine Science are gratefully acknowledged. Contribution no. 289, Institute of Marine Science, University of Alaska.

REFERENCES

FEDER, H. M., G. J. MUELLER, M. H. DICK, and D. B. HAWKINS

    1973   Preliminary benthos survey. In *Environmental studies of Port Valdez*, edited by D. W. Hood, W. E. Shiels, and E. J. Kelley. Occas. Publ. No. 3, Inst. Mar. Sci., Univ. Alaska, Fairbanks, pp. 305-391.

GOERING, J. J., C. J. PATTON, and W. E. SHIELS

    1973   Primary productivity. In *Environmental studies of Port Valdez*, edited by D. W. Hood, W. E. Shiels, and E. J. Kelley. Occas. Publ. No. 3, Inst. Mar. Sci., Univ. Alaska, Fairbanks, pp. 225-248.

GOERING, J. J., W. E. SHIELS, and C. J. PATTON

    1973   Nutrient cycles. In *Environmental studies of Port Valdez*, edited by D. W. Hood, W. E. Shiels, and E. J. Kelley. Occas. Publ. No. 3, Inst. Mar. Sci., Univ. Alaska, Fairbanks, pp. 253-279.

HOOD, D. W., and C. J. PATTON

    1973   Chemical oceanography. In *Environmental studies of Port Valdez*, edited by D. W. Hood, W. E. Shiels, and E. J. Kelley. Occas. Publ. No. 3, Inst. Mar. Sci., Univ. Alaska, Fairbanks, pp. 201-221.

HOOD, D. W., W. E. SHIELS, and E. J. KELLEY (Eds.)

    1973   *Environmental studies of Port Valdez*. Occas. Publ. No. 3 and 3A, Inst. Mar. Sci., Univ. Alaska, Fairbanks, 477 + 800 pp.

MATTHEWS, J. B.

    1972   Some aspects of the hydrography of Alaskan and Norwegian fjords. *In* Proc. 1st Int'l Conf. on Port and Ocean Engineering under Arctic Conditions, edited by S. Wetteland and P. Bruun. Vol. 1, pp. 829-834.

MATTHEWS, J. B., and D. L. NEBERT

    1969   Numeric modeling of a fjord estuary. Tech. Report R69-4, Inst. Mar. Sci., Univ. Alaska, Fairbanks.

MUENCH, R. D., and D. L. NEBERT

    1973   Physical oceanography. In *Environmental studies of Port Valdez*, edited by D. W. Hood, W. E. Shiels, and E. J. Kelley. Occas. Publ. No. 3, Inst. Mar. Sci., Univ. Alaska, Fairbanks, pp. 103-149.

NEBERT, D. L.

    1972   A proposed circulation in Endicott Arm, an Alaskan fjord. M.S. Thesis, Univ. Alaska, Fairbanks, 90 pp.

NEBERT, D. L., R. D. MUENCH, and D. W. HOOD

    1973   Dye dispersion studies. In *Environmental studies of Port Valdez*, edited by D. W. Hood, W. E. Shiels, and E. J. Kelley. Occas. Publ. No. 3, Inst. Mar. Sci., Univ. Alaska, Fairbanks, pp. 153-198.

PICKARD, G. L.

    1967   Some oceanographic characteristics of the larger inlets of Southeast Alaska. *J. Fish. Res. Bd. Canada* 24: 1475-1505.

PRITCHARD, D. W.

    1952   Estuarine hydrography. In *Advances in geophysics*, Vol. 1, edited by H. E. Landsberg. Academic Press, New York, pp. 243-280.

QUINLAN, A. V.

1970 Seasonal and spatial variations in the water mass characteristics of Muir Inlet, Glacier Bay, Alaska. M.S. Thesis, Univ. Alaska, 145 pp.

RATTRAY, M., JR.

1967 Some aspects of the dynamics of circulation in fjords. In *Estuaries*, edited by G. H. Lauff. Publ. 83, Amer. Ass. Adv. Sci., Washington, D. C., pp. 52-62.

ROBERTSON, B. R., S. D. ARHELGER, R. A. T. LAW, and D. K. BUTTON

1973 Hydrocarbon biodegradation. In *Environmental studies of Port Valdez*, edited by D. W. Hood, W. E. Shiels, and E. J. Kelley. Occas. Publ. No. 3, Inst. Mar. Sci., Univ. Alaska, Fairbanks, pp. 447-479.

SHARMA, G. D., and D. C. BURBANK

1973 Geological oceanography. In *Environmental studies of Port Valdez*, edited by D. W. Hood, W. E. Shiels, and E. J. Kelley. Occas. Publ. No. 3, Inst. Mar. Sci., Univ. Alaska, Fairbanks, pp. 15-100.

SHIELS, W. E., J. J. GOERING, and D. W. HOOD

1973 Crude oil phytotoxicity. In *Environmental studies of Port Valdez*, edited by D. W. Hood, W. E. Shiels, and E. J. Kelley. Occas. Publ. No. 3, Inst. Mar. Sci., Univ. Alaska, Fairbanks, pp. 413-446.

TRITES, R. W.

1955 A study of the oceanographic structure in British Columbia inlets and some of the determining factors. Ph.D. Thesis, Univ. British Columbia, Vancouver, B. C., Canada, 123 pp.

TULLY, J. P.

1958 On structure, entrainment and transport in estuarine embayments. *J. Mar. Res.* 17: 523-535.

Part II

CONTEMPORARY TOPICS

section 3

the northern seabed condition

Assessment of the Arctic Marine Environment: Selected Topics
Copyright © 1976 by Institute of Marine Science, University of Alaska, Fairbanks

CHAPTER **11**

Chemistry of deep-sea sediments in the Canada Basin, west Arctic Ocean

A. S. NAIDU, C. J. LEE[1], *and* T. C. MOWATT[2]

Abstract

Twenty-four surficial (top 3 cm) sediments from the Canada Basin were analyzed for a suite of alkali and alkaline earth elements, first group of trace transition heavy metals, organic carbon, and carbonate contents. Average concentration (dry wt percentages) of the analyzed components in the arctic sediments are presented and compared with those found in deep-sea sediments of tropical-temperate oceans. Relative metal paucity in arctic deposits is discussed.

Based on the elements analyzed, deep-sea arctic ocean sediments are chemically quite similar to common igneous rocks. Therefore, extreme caution must be exercised by geologists while applying geochemical criteria in distinguishing between para- and ortho-gneisses and schists.

INTRODUCTION

The polar oceans in their distinctive depositional setting (Naidu and Hood 1972; Naidu et al. 1974) offer a unique opportunity to study geochemical processes in an ice-stressed oceanic regime. However, study of the published English literature clearly demonstrates the existence of only minimal chemical data on contemporary deep-sea sediments of the polar areas relative to the more temperate regions of the world. Limited geochemical studies of the deep Arctic Ocean have been made by Belov and Lapina (1966), Li et al. (1969), Bostrom (1971), Spiro and Zelenova (1971), Naidu and Hood

---

[1] *Institute of Marine Science, University of Alaska, Fairbanks, Alaska 99701.*
[2] *U.S. Bureau of Mines, Anchorage, Alaska 99510.*

(1972), and Naidu et al. (1974). The investigation of Angino (1966) is a noteworthy contribution to knowledge of Antarctic Ocean sediments. The purpose of this chapter is to present some new data on the concentrations of a suite of major, minor, and trace elements, as well as carbonate and organic carbon contents, in some deep-sea deposits of the Canada Basin, west Arctic Ocean.

In this study, the deep-sea area refers arbitrarily to extensive marine basins which have depths in excess of 100 m. Generally, this marine regime forms either a part of the continental rise or is an integral part of the abyssal oceanic basin. This is particularly true of the Canada Basin, which is situated north and northwest of arctic Alaska and Canada, respectively (Carsola et al. 1961).

SAMPLES AND ANALYTICAL TECHNIQUES

For purposes of this study, only surficial bottom sediments retrieved from either short gravity cores or van Veen grabs were considered. A suite of 12 samples (T3 series in Table 11.1) were collected under Fletcher's Ice Island T3. The other 12 sediment samples were obtained from U. S. Coast Guard ice breakers *Staten Island* and *Glacier* between 1969 and 1971. All samples were preserved in a frozen state prior to analysis. Locations of the sediment samples are shown elsewhere by Naidu et al. (1974). They were collected from the west Arctic Ocean situated north of Alaska and northwest of arctic Canada. This area is encompassed roughly by north latitudes 71 and 76°, and west longitudes 142 and 158°.

The elemental analyses were carried out by atomic absorption spectrophotometry, while the clay mineral analyses were accomplished by X-ray diffraction. Carbonate in the sediments was analyzed manometrically, and total carbon was determined by gravimetric method. The difference between these latter two values was taken as the organic carbon content in each sediment sample. Analytical procedures have been described at length elsewhere (Naidu et al. 1972; Naidu and Hood 1972; Mowatt et al. 1974, and Naidu 1976). The significance in the differences in the means of various elemental abundances between the arctic and tropical-temperate deep-sea sediments was statistically checked by conducting either the student-t or z tests.

RESULTS AND DISCUSSION

Deep-sea sediments of the Canada Basin were analyzed for chemical abundances of major, minor and trace element, organic carbon, carbonate, and clay mineral types (Table 11.1). Average abundances of the above elements were compared with those of the same elements analyzed from deep-sea sediments of the tropical-temperate regions (Table 11.2). Statistical analysis indicates that there are significant regional differences in the average abundances of all elements except Fe, Mn, and Zn, at least at the 95 percent confidence level. The relative difference for Zn is discernible at the 90 percent confidence level. Of note is a relative deficiency of all elements except Cr in the Canada Basin sediments. In the case of Cr there is a significant enrichment (Table 11.2). The significance of the regional difference in the mean concentrations of Li has not been determined because of the very limited number of analyses (a total of 2) available for that element in deep-sea sediments of the tropical-temperate regions. It is quite apparent from an examination of the data in Table 11.1 that there is a slight enrichment of almost all metals in the T3 series of samples. These samples were collected from the abyssal regime of the Canada Basin situated farther away from the Alaskan-Canadian continent than the rest of the samples. There appears to be, however, a relatively low concentration of organic carbon in the suite of T3 series of samples.

The relative deficiency of all the metals except Fe, Mn, and Cr in the Canada Basin sediments, as compared to nonpolar sediments (Table 11.2), may be attributed to the possible effects of several physicochemical factors. However, it would seem that the most influential factors which might be related to such regional differences in the chemistry of surface marine sediments are variations in the concentrations of elements in terrigenous detrital grains; the rates of sedimentation of terrigenous detritus; the amounts of volcanogenous, biogenous, and chemogenous influxes; the mineralogy; the size distribution of the sediment grains; and the extent to which sediment chemistries from various regions are modified by post-depositional changes. After careful consideration of all the above possible factors, as elaborated by Naidu et al. (1974), it is concluded that the observed differences between the deep-sea sediment chemistries of the Canada Basin and tropical-temperate regions (Table 11.2) are most likely attributable to the regional differences in the volume of volcanogenous influxes and are due indirectly to some extent to regional differences in climate. At the present time, the Canada Basin floor of the Arctic Ocean is tectonically quite inactive as compared to the seabeds of the Atlantic, Pacific, and Indian oceans. By implication, it is believed that the influx of volcanogenous material into the Canada Basin sediments would be minimal. In view of the fact that copious quantities of a variety of several heavy metals are commonly associated with submarine volcanic emissions, the overall metal deficiency in the Canada Basin sediments can be considered quite usual.

The frigid climate of the Arctic can conceivably bring about an overall metal deficiency in the Canada Basin sediments in several ways. Biologists have generally observed a net yearly low organic productivity in the surficial water columns of the Canada Basin. This low productivity has been ascribed to the near-freezing temperatures of the surface water, as well as to a euphotic zone limited by the pack ice cover. It is contended that, as a result of low biological density, contribution of several biophile elements in the Canada Basin sediments via organic detritus deposition is quite restricted. Further, after deposition on the sea bottom, organic matter tends to facilitate the immobilization of many metals, by means of adsorption or metal-organic complex formation. However, in the Canada Basin, the sediments are not very enriched in organic matter, as suggested by the presence there of low organic carbon contents (Table 11.1).

TABLE 11.1 Chemical and clay mineral compositions of deep-sea sediments, Canada Basin, west Arctic Ocean. Chemical abundances are in weight percentages. The clay mineral abundances are expressed as weighted peak area percentages (Biscayne 1965).

Sample no.	Water depth (m)	C_{org}	CO_3^{-2}	Fe	Mn	Na	K	Ca	Mg	Li
T3-1	3637	0.74	4.2	5.15	0.48	2.03	2.37	1.99	1.15	0.0065
T3-2	3650	0.71	3.1	4.90	0.40	2.59	2.19	1.68	1.20	0.0063
T3-3	3795	0.73	0.6	5.45	0.32	2.84	2.12	0.96	1.30	0.0073
T3-4	3792	0.74	2.9	5.25	0.48	4.13	2.39	1.29	1.23	0.0070
T3-5	3761	0.75	1.2	5.20	0.36	2.74	2.26	0.94	1.28	0.0068
T3-6	3827	0.79	1.1	5.40	0.33	3.46	2.25	0.82	1.35	0.0073
T3-7	3830	0.77	0.9	5.25	0.38	3.74	2.49	0.88	1.38	0.0068
T3-8	3833	1.16	1.2	5.10	0.49	3.43	2.27	1.11	1.28	0.0073
T3-9	2860	0.39	5.3	6.05	0.58	4.32	2.87	6.52	2.13	0.0074
T3-10	1705	0.39	8.0	5.10	0.37	2.97	1.82	5.92	1.15	0.0050
T3-11	1160	0.47	4.1	4.60	0.50	3.82	1.97	5.12	1.38	0.0055
T3-12	3835	0.79	0.8	5.10	0.50	3.27	2.27	1.14	1.15	0.0075
BSS 51	2477	0.91	1.25	4.55	0.86	2.85	2.38	0.50	0.88	0.0047
BSS 54	3458	0.88	2.65	4.88	0.33	2.56	2.54	1.20	1.03	0.0053
BSS 97	1500	1.34	3.15	4.06	0.03	3.04	2.27	1.80	2.03	0.0052
BSS 92	1600	1.11	1.75	3.65	0.03	1.74	2.38	0.70	0.75	0.0063
BSS 91	1800	1.41	0.15	4.06	0.08	2.38	2.22	1.00	0.80	0.0042
GLA7122	1006	1.35	2.61	5.10	0.26	2.40	1.71	1.15	1.13	0.0059
GLA7158	1011	1.03	1.87	5.00	0.09	2.45	1.71	1.10	1.10	0.0059
GLA7185	1053	0.98	1.98	5.80	0.09	1.90	1.86	1.00	1.03	0.0060
GLA7157	1829	1.20	1.41	5.20	0.53	1.90	1.91	0.85	0.98	0.0057
GLA7120	2000	1.07	2.36	4.80	2.58	3.15	1.71	0.75	1.05	0.0052
GLA7186	2150	0.73	3.61	5.20	1.86	3.00	1.56	0.70	0.95	0.0052
GLA7121	2200	1.35	1.12	5.20	1.20	2.30	1.71	0.60	0.70	0.0050

TABLE 11.1 (*continued*)

Rb	Cu	Ni	Zn	Co	Cr	SMT*	ILT*	KLT*	CLT*
0.0090	0.0060	0.0090	0.0120	0.0055	0.0130	11	62	12	15
0.0083	0.0052	0.0090	0.0093	0.0053	0.0110	7	57	13	23
0.0115	0.0125	0.0080	0.0094	0.0041	0.0130	14	57	11	18
0.0100	0.0082	0.0100	0.0110	0.0055	0.0145	13	56	13	18
0.0112	0.0052	0.0060	0.0104	0.0044	0.0140	14	57	13	16
0.0112	0.0055	0.0060	0.0095	0.0041	0.0145	8	62	12	18
0.0098	0.0050	0.0070	0.0084	0.0052	0.0145	15	55	13	17
0.0112	0.0049	0.0090	0.0094	0.0074	0.0145	12	52	20	16
0.0125	0.0085	0.0110	0.0130	0.0121	0.0160	6	58	14	22
0.0075	0.0048	0.0070	0.0102	0.0074	0.0100	3	65	16	16
0.0085	0.0046	0.0080	0.0104	0.0087	0.0090	6	64	11	19
0.0110	0.0052	0.0090	0.0088	0.0078	0.0120	5	62	16	17
0.0083	0.0080	0.0065	0.0083	0.0034	0.0080	7	64	11	18
0.0063	0.0061	0.0058	0.0103	0.0062	0.0135	4	59	16	21
0.0160	0.0060	0.0041	0.0083	0.0035	0.0135	11	55	11	23
0.0092	0.0092	0.0065	0.0122	0.0019	0.0090	5	60	10	25
0.0089	0.0053	0.0050	0.0065	0.0030	0.0100	5	60	10	25
0.0128	0.0032	0.0052	0.0129	0.0046	0.0090	20	53	9	18
0.0130	0.0030	0.0050	0.0090	0.0044	0.0080	NA	NA	NA	NA
0.0132	0.0037	0.0055	0.0111	0.0042	0.0090	18	50	11	21
0.0120	0.0036	0.0054	0.0111	0.0046	0.0085	25	49	10	16
0.0116	0.0070	0.0050	0.0103	0.0047	0.0080	12	55	11	22
0.0104	0.0040	0.0052	0.0106	0.0044	0.0090	10	64	6	20
0.0098	0.0030	0.0054	0.0156	0.0044	0.0085	19	48	10	23

*SMT: smectite; ILT: illite; KLT: kaolinite; CLT: chlorite.

Unlike the case in the low-latitudes, the prevalence of cold climate in the arctic hinterland may well result in relatively lower concentrations of heavy metals, alkali, and alkaline earth elements in the weathered terrigenous detritus that is eventually supplied to, and deposited in, the Canada Basin. This can apparently result from decreased leaching of silica and alumina from the parent rock source of the terrigenous debris under the relatively lower intensity of chemical weathering conditions prevailing in the Arctic. But, such a contention has yet to be resolved firmly. However, it is of interest to note that the contemporary sediments of the Alaskan-Canadian arctic continental margin generally do have relatively low contents of the above suite of elements (Naidu and Mowatt 1974, 1975). These sediments are to be considered as potential source material of considerable significance for the Canada Basin sediments, in terms of volume.

An important result of the present study is the finding of no great enrichment in most of the elements analyzed in the deep-sea sediments of the Canada Basin in comparison to concentrations of the same elements in common igneous rocks (Turekian and Wedepohl 1961). This was somewhat unexpected in view of the common observation that there is some degree of fractionation of most elements in contemporary deep-sea sediments in nonpolar areas relative to continental igneous rocks. In fact, these differences in chemistries of primary rocks and marine sediments are frequently used by

TABLE 11.2 Differences in the average abundances of some elements in the deep-sea sediments of the Canada Basin (Arctic Ocean) and tropical-temperate oceans. The abundances are in percentages of dry sediment weights.

Element	Canada Basin (Arctic Ocean)	Tropical-temperate oceans	't' or 'z' Test-results
Fe	5.00	5.16[abd]	Insignificant
Mn	0.55	0.66[abd]	Insignificant
Na	2.88	3.35[b]	Significant*
K	2.13	2.39[b]	Significant*
Ca	1.66	2.84[b]	Significant*
Mg	1.18	1.85[b]	Significant*
Li	0.0061	0.0057[c]	Not calculated
Rb	0.0105	0.0119[cd]	Significant*
Cu	0.0057	0.0338[abd]	Significant*
Ni	0.0068	0.0200[abd]	Significant*
Zn	0.0103	0.0125[d]	Significant**
Co	0.0053	0.0101[abd]	Significant*
Cr	0.0113	0.0074[abd]	Significant*

*Differences significant at least at 95% confidence level
**Difference significant at 90% confidence level
[a]Cronan (1969)
[b]Goldberg and Arrhenius (1958)
[c]Horstman (1957)
[d]Wedepohl (1956).

geologists as useful criteria to differentiate between para- and ortho-gneisses and schists. However, based on the results of this study, it is clear that the application of geochemical criteria to infer the origin of metamorphosed rocks must be made with caution.

Acknowledgments

The authors wish to thank Dr. K. Hunkins of the Lamont-Doherty Geological Observatory of Columbia University for providing the sediment samples collected by him from Fletcher's Ice Island T3. Dr. Hunkins' sampling program was supported by the Office of Naval Research under Contract no. NOOO14-67-A-0108-0016. The other sediment samples were collected either by Dr. D. C. Burrell or by the senior author (A.S.N.) from U. S. Coast Guard ice breakers. The help of Dr. Peter W. Barnes of the U. S. Geological Survey (Menlo Park), Dr. David Mountain, and the crews of the USCGC *Glacier* and *Staten Island* in the collection of samples is gratefully acknowledged. This work was supported by the Office of Marine Geology, U.S. Geological Survey, Menlo Park, under Contract 14-09-001-12559. Contribution no. 280, Institute of Marine Science, University of Alaska.

REFERENCES

ANGINO, E. E.

1966 Geochemistry of Antarctic pelagic sediments. *Geochim. Cosmochim. Acta* 30: 939-961.

BELOV, N. A., and N. N. LAPINA

1966 Geological studies of the floor of the Arctic Ocean during the last twenty-five years. In *Problems of the Arctic and Antarctic: Collection of articles*, No. 1, edited by N. A. Ostenso. Arctic Inst. North Amer., Washington, D. C. (original in Russian) 1(1): 1-15.

BISCAYE, P. E.

1965 Mineralogy and sedimentation of recent deep-sea clays in the Atlantic Ocean and adjacent seas and oceans. *Geol. Soc. Amer. Bull.* 76: 803-832.

BOSTROM, K.

1971 Origin of manganese-rich layers in Arctic sediments (*Abstract*). Proc. 2nd Int'l Symposium Arctic Geol., San Francisco, p. 9.

CARSOLA, A. J., R. L. FISHER, C. J. SHIPEK, and G. SHUMWAY

1961 Bathymetry of the Beaufort Sea. In *Geology of the Arctic*, Vol. 1, edited by G. O. Raasch. Univ. Toronto Press, pp. 678-689.

CRONAN, D. S.
　　1969　Average abundance of Mn, Fe, Ni, Co, Cu, Pb, Mo, V, Cr, Ti
　　　　　　and P in Pacific pelagic clays. *Geochim. Cosmochim. Acta* 33:
　　　　　　1562-1565.

GOLDBERG, E. D., and G. O. S. ARRHENIUS
　　1958　Chemistry of Pacific pelagic sediments. *Geochim. Cosmochim.
　　　　　　Acta* 13: 153-212.

HORSTMAN, E. L.
　　1957　The distribution of lithium, rubidium, and cesium in igneous
　　　　　　and sedimentary rocks. *Geochim. Cosmochim. Acta* 12: 1-28.

LI, Y. H., J. BISCHOFF, and G. M. MATHIEU
　　1969　The migration of manganese in the Arctic Basin sediment.
　　　　　　Earth and Planetary Science Letter 7: 265-270.

MOWATT, T. C., A. S. NAIDU, and N. VEACH
　　1974　Clay mineralogy of the lower Colville River Delta, north arctic
　　　　　　Alaska. State of Alaska, Div. Geol. and Geophys. Survey,
　　　　　　Open File Report 45: 1-39.

NAIDU, A. S.
　　1976　Clay minerals and chemical stratigraphy of unconsolidated sedi-
　　　　　　ments, Beaufort Sea, Arctic Ocean, Alaska. Final technical
　　　　　　report submitted to the U. S. Geological Survey. Inst. Mar.
　　　　　　Sci., Univ. Alaska, pp. 1-22.

NAIDU, A. S., D. C. BURRELL, and D. W. HOOD
　　1972　Clay mineral composition and geologic significance of some
　　　　　　Beaufort Sea sediments. *Jour. Sedimentary Petrology* 41:
　　　　　　691-694.

NAIDU, A. S., and D. W. HOOD
　　1971　Chemical composition of bottom sediments of the Beaufort
　　　　　　Sea, Arctic Ocean. Proc. 24th Int'l Geol. Congress, Montreal,
　　　　　　Canada, Section 10, pp. 307-317.

NAIDU, A. S., and T. C. MOWATT
　　1974　Clay mineralogy and geochemistry of continental shelf sedi-
　　　　　　ments of the Beaufort Sea. In *The coast and shelf of the
　　　　　　Beaufort Sea*, edited by J. C. Reed and J. E. Sater. Arctic
　　　　　　Inst. North America, Arlington, Virginia, pp. 493-510.

NAIDU, A. S., and T. C. MOWATT

    1975   Depositional environments and sediment characteristics of the Colville and adjacent deltas, northern arctic Alaska. In *Deltas, models for exploration*, edited by M. L. S. Broussard. Houston Geol. Soc., Houston, Texas, pp. 283-309.

NAIDU, A. S., T. C. MOWATT, D. B. HAWKINS, and D. W. HOOD

    1974   Clay mineralogy and geochemistry of some Arctic Ocean sediments: Significance of Paleoclimate interpretation. In *Climate of the Arctic*, edited by G. Weller and S. A. Bowling. Geophysical Inst., Univ. Alaska, Fairbanks, pp. 59-67.

SPIRO, N. S., and A. F. ZELENOVA

    1971   Effect of climatic changes on chemical content of deep-water muds exemplified by deposits of Arctic and Atlantic Oceans (*Abstract*). Proc. 2nd Int'l Symposium Arctic Geol., San Francisco, pp. 49.

TUREKIAN, K. K., and K. H. WEDEPOHL

    1961   Distribution of the elements in some major units of the earth's crust. *Bull. Geol. Soc. Amer.* 72: 175-192.

WEDEPOHL, K. H.

    1956   Spurenanalytische Untersuchungen an Tiefseetonen aus dem Atlantik. *Geochim. Cosmochim. Acta* 18: 200-231.

Assessment of the Arctic Marine Environment: Selected Topics

CHAPTER **12**

Sedimentation in a "half-tide" harbor

Part 1. Sedimentation under ice-free conditions

C. H. EVERTS [1]

Abstract

Harbor shoaling becomes a significant problem when coastal waters are laden with suspended solids and the tidal range is high. Because of currents resulting from the tides and severe ice conditions which exist along much of the Alaskan coast, some small-craft harbors are constructed as enclosed basins sited adjacent to, rather than within, navigable estuaries. Determination of the shoaling rate and relative importance of factors involved in the shoaling process is essential to the effective design of these harbors.

The unique feature of the "half-tide" harbor is a sill placed in the navigation channel at an elevation greater than that of the basin. Flotation for vessels is provided during low-tide stages, but the sill restricts navigation into and out of the harbor to times of higher tidal elevations. This report describes the history and processes of shoaling observed during the ice-free season at an enclosed small-craft harbor at Dillingham, Alaska, since its construction in 1961. (Results of a study of the shoaling process during the winter ice-cover season at Dillingham Harbor are presented in Part 2 of this chapter).

[1] *Coastal Engineering Research Center, Kingman Building, Fort Belvoir, Virginia 22060.*

INTRODUCTION

An enclosed small-craft harbor at Dillingham, Alaska, has shoaled about 2 m/yr since it was constructed in 1960-1961. Without high-cost annual dredging, the usefulness of the harbor would be severely limited. At the time of construction, some data were available to indicate that estuarine waters at Dillingham carried 50 to 1000 mg/liter suspended solids. Because the tidal range in the estuary is high, it was also known that a large estuary-harbor exchange of water would occur with each tidal cycle. However, the magnitude of the shoaling problem was not anticipated.

Future decisions to construct harbors like that at Dillingham in similar high-tide range, sediment-laden estuaries may depend upon an accurate forecast of maintenance costs. Consequently, the shoaling rate must be predicted beforehand. Such a prediction, by whatever method used, requires that the shoaling process be known. In addition, design techniques for such harbors cannot be significantly improved unless a better understanding of the sedimentation process in and near the harbor is acquired.

The purpose of this paper is to present the results of a study of the shoaling process in Dillingham Harbor during the ice-free season. The shoaling history of the harbor is given, and a field study conducted to collect suspended sediment and current velocity data is detailed. Results of a study of the shoaling process during the winter ice-cover season at Dillingham Harbor are given in Part 2.

Harbor characteristics

Dillingham Harbor is a "half-tide" type harbor (Fig. 12.1). Because of currents resulting from the high tides, as well as severe winter ice conditions, "half-tide" harbors are constructed as enclosed basins sited adjacent to, rather than within, navigable estuaries. Harbor depths are generally specified near mean lower low water (MLLW) to reduce initial excavation costs. The unique feature of these land-contained harbors is a sill at an elevation above the harbor bottom and across the navigation channel where it enters the harbor. The sill provides flotation for vessels during low tide stages while it restricts navigation into or out of the harbor to times of higher tidal elevations.

Dillingham is located near the confluence of the Nushagak and Wood rivers at the upper end of Nushagak Bay (Fig. 12.1). The source of fine-grained suspended material in the bay is primarily the surrounding eroding bluffs composed of glacial till. Lesser sources of sediment are the Nushagak River and eroding tidal flats in the bay. The tide at Dillingham is semi-diurnal with a mean period of 12.4 hr and a mean higher high water (MHHW) elevation of 6 m above mean lower low water.

Project design

Dillingham Harbor is located about one-half mile west of Dillingham (Fig. 12.1). It serves local and transient fishing boats to a length of 15 m as well as commercial barges carrying commodities to outlying villages in the region.

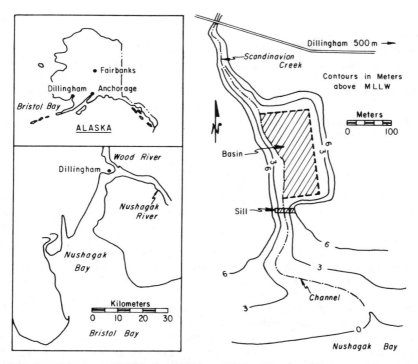

Fig. 12.1 Location map of Dillingham Harbor, Alaska, showing the three elements which comprise a half-tide harbor: basin, sill, and navigation channel.

Harbor construction began in September 1960 using a 30-cm pipeline dredge. Because of the autumn freeze-up, the dredge was shut down in October, but work continued throughout the winter using a dragline and bulldozer. In April 1961 dredging resumed, and excavation of the basin was completed in late September of that year.

Project elevation of the basin was initially 0.6 m (MLLW) with sidewall slopes of 1:5. At project elevation, the basin area was 21,500 m^2, providing moorage for about 140 boats. Auger holes drilled prior to construction in the basin area and adjacent to the Scandinavian Creek channel indicated the area was at one time a deltaic tidal flat composed of glacial or fluvio-glacial silt and minor amounts of sand overlain by peat.

The natural course of Scandinavian Creek was incorporated into the design of the harbor to serve as the navigation channel. Since construction, the channel has been self-maintaining and apparently stable in location and in plan geometry. Because the volume of the harbor greatly increased the tidal prism through the channel, the channel cross-section likely increased, then stabilized, after construction. From the sill location to Nushagak Bay, the channel is 400 m long with a gradient of 1:40. It has a V-shaped cross-section with sidewall slopes of 1:4 to 1:6. Sidewall slopes are steepest near the sill, becoming more gradual toward the bay. The tidal flat through which the channel is incised is 180 m wide. Summer discharge from Scandinavian Creek is usually less than 0.15 m³/sec.

The rock sill was designed with a top elevation of 2.1 m (MLLW) to maintain a water depth in the basin of 1.5 m at lower tide stages. The sill is located approximately 30 m from the basin in the channel. Rocks comprising the sill are angular to sub-rounded cobbles and boulders 15 to 50 cm in diameter. The sill sidewalls are 1:3 and the sill width is 22 m In some instances, because of leakage through the sill at low tide, the basin water level drops 0.1 to 0.3 m below the sill.

SHOALING HISTORY

In late May 1962, eight months after the harbor was constructed, soundings indicated 1.4 m of material had been uniformly deposited in the basin. Redredging was completed in late June 1962, but one year later the basin had again filled, in this case to 2.1 m (MLLW). For two years the harbor was not dredged. By June 1965, soundings showed that the basin had filled to 4 m above MLLW. Annual maintenance during the ice free season began in 1969 and is continuing to the present using an 8-inch cutterhead hydraulic dredge. At least two months of double-shift dredging are presently required per year to remove the winter accumulation and the accumulation which occurs during the dredging season.

Table 12.1 presents sediment accumulation data obtained from pre- and post-dredging surveys in the basin. On the average, 24,000 m³ of deposited sediment are required to uniformly fill the basin one meter at the sill elevation. Winter accumulation (October-May) averaged about 3100 m³/mo, while summer filling (June-September) was about 5400 m³/mo. The average yearly accumulation was 42,500 m³, or 57 m³ per tidal cycle (0.0024 m elevation increase per tidal cycle).

TABLE 12.1 Sediment accumulation data from Dillingham Harbor, Alaska.

Sedimentation period	Pre-winter basin elevation (MLLW datum)	Winter accumulation (m³)	Pre-dredge elevation (MLLW datum)	Summer accumulation (m³)
9/61-5/62	0	35,000		
7/62-7/63	+0.6	30,500		
6/26/69-10/21/69				
10/21/69-6/10/70	+3.0	8,600		
6/10/70-10/2/70			+3.4	10,800
7/2/70-6/10/71	+1.2	25,200		
6/10/71-10/2/71			+2.5	18,200
10/2/71-6/19/72	+0.3	36,700		
6/19/72-9/20/72			+1.2	23,700
9/20/72-6/8/73	+0.6	29,100		

During periods when sediment was deposited only below the elevation of the sill, as occurred after annual dredging commenced in 1969, the accumulations were nearly uniform over the entire basin except in the immediate region of the sill. Figure 12.2a shows a typical winter accumulation from September 1972, when the bottom was near MLLW, to May 1973. Bottom accretion was obtained by subtracting the autumn sounding record from the spring record. The accumulation gradient at the boundaries of the basin reflects less deposition on the sidewalls. All autumn to spring accumulations were similar to that shown in Fig. 12.2a, with the exception of the 1 8-m contour, which generally extended nearer the south end of the basin.

The deviation of the sediment surface from a level within the basin after a winter's accumulation (September 1972–May 1973) is shown in Figure 12.2b. The western boundary of the basin was maintained slightly below the level surface by the low discharge of Scandinavian Creek and the tidal prism in the creek. The trough near the sill resulted from scour by a fluid jet

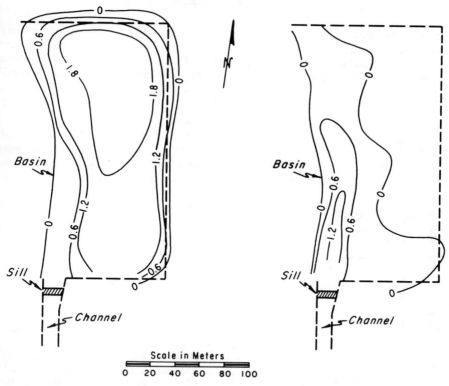

Scale in Meters

0   20   40   60   80   100

a) Basin Depth Changes
(September 1972 to May 1973)

b) Deviation of Basin Bottom
from Level (May 1973)

Fig. 12.2 Topographic changes in Dillingham Harbor during the period September 1972 to May 1973, showing nearly uniform accretion throughout the basin. Contours are in meters.

created during the interval of high inflow velocities just after the flooding tide rose above sill elevation.

Sounding records obtained after several years' accumulation above the sill elevation (1963 to 1965 and 1965 to 1969) indicate the basin then shoaled more near its center and landward portions. A slight depression of about 1 m in the total accumulation of 3.5 to 5 m was also measured along the east and south boundaries of the basin. During these periods, Scandinavian Creek was naturally maintained at a higher elevation along the western boundary of the basin. Figure 12.3 illustrates the thalweg profile as measured in June 1965 after the basin had silted 4.0 to 4.5 m (MLLW). The thalweg shoaled to sill elevation because the sill acted as a control point below which the creek could not be naturally maintained. Sixty meters seaward of the sill, the channel resumed its natural gradient of about 1:500.

DEPOSITIONAL PROCESSES

During the ice-free seasons of 1969, 1973, and 1974, currents were measured and suspended sediment samples were collected in the basin over the sill and in Nushagak Bay. The objective of the measurement and sampling program was to detail the magnitude and mechanism of the depositional process in the basin.

Currents

Flow into and out of the basin is almost entirely tide-dominated, with negligible influence from Scandinavian Creek. Current measurements were made at the sill to establish the vertical velocity gradient in and out of the harbor, and to obtain the mean current velocity throughout a tidal cycle.

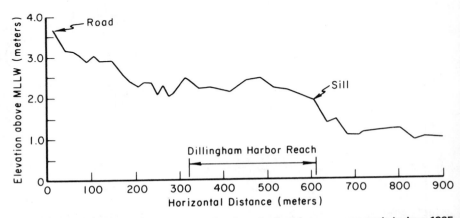

Fig. 12.3 Thalweg Profile of Scandinavian Creek as obtained from a survey made in June 1965, after the basin had silted 4 m above MLLW. The profile begins upstream at the Dillingham road and extends to an elevation of + 1.0 m (MLLW) near Nushagak Bay.

Such data are useful in determining the time and elevation-dependent water and sediment discharge into the basin. Current measurements were also made in the basin to establish horizontal as well as vertical current patterns. A Gurley Model 667 current speed indicator suspended from an anchored skiff was used to obtain velocity data.

Current measurements as shown in Figure 12.4 were made over the sill during the night of 20-21 September 1973. The predicted tidal amplitude was 6.5 m, with a beginning low water elevation of -0.6 m and an ending low water elevation of +1.5 m. The tidal period was about 13 hr. Wind was 16 km/hr from the southwest. Air temperature was 10° C, and the water temperature was 6.7° C.

Current velocities at the sill were nearly uniform through the water column, except when the water surface elevation was within 1 m of the sill. At that low tidal elevation, the mean velocity was high and the near-surface velocity was greatest. Current velocities obtained during another tidal cycle on 21 September 1973 were similar in magnitude and vertical gradient to those measured on 20-21 September. Discharge into or out of the harbor was always unidirectional.

Water and suspended sediment was sometimes discharged into the basin in the form of a jet when the tidal elevation was less than 1 m above the sill. Within a radius of 15 m of the sill, the highest current velocities were directed to the bottom, because the basin-side face of the sill acted as a spillway. The currents then surfaced and were dissipated within an additional 15-m radius.

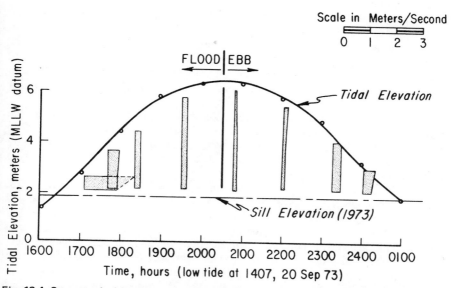

Fig. 12.4 Current velocities above the sill as measured during a tidal cycle on 20-21 September 1973. Note the uniform velocity distribution with depth, except when the water depth over the sill was less than 1 m.

Current measurements were also made in the basin at various times in the tidal cycle and at various locations on 21 September 1973. In all cases, the current speeds were less than 0.07 m/sec. In the basin away from the sill region, currents were less than 0.015 m/sec when tidal elevations were above 3 m (1 m above the sill elevation). Below the elevation of the sill, current speeds were near zero. From the sill elevation to the water surface, measured current speeds were uniform. Currents in the basin were generally within 10 degrees of a line directed toward or away from the sill.

Suspended sediment concentrations

Sediment in and near Dillingham Harbor predominantly moves in suspension. No evidence of significant bedload movement into or out of the harbor was observed. The channel and harbor bottom at all times was composed of cohesive silt and mud-sized material. Time and depth-dependent suspended sediment concentration is of interest, because it influences the rate of basin infilling.

Suspended sediment samples were collected over the sill through 14 tidal cycles in 1969 (2), 1973 (9), and 1974 (3). Samples were collected at various depths from a skiff anchored in the center of the channel. A pump sampler was used, and sample sets through the water column were obtained at approximately 1-hr intervals. Samples from five tidal cycles were analyzed for concentration by measuring the liquid mass and the dried solids mass which had been separated using a filter.

Figure 12.5 shows the suspended solids concentration of samples collected on 20 September 1973, 21 June 1974, and on 24 June 1974. Suspended concentrations at the sill were greatest as the rising tide initially passed the sill. Concentrations then declined rapidly to the time of high water. Concentrations during the ebbing portion of the tide were uniformally low. The vertical concentration was nearly uniform, although there was a slight trend to higher concentrations nearest the sill. Figure 12.6 illustrates the mean concentration of suspended material passing over the sill as a function of time in the tidal cycle.

On 24 May 1967 two samples were collected at different locations in the basin (unpublished report, U. S. Army Corps of Engineers, 1968). The samples, taken at mid-depth during a rising stage of the tide, are the only ones available from within the basin. The suspended concentrations were similar at 234 and 278 mg/liter.

In conjunction with sampling at the sill on 20-21 September 1973, samples were collected in Nushagak Bay 400 m seaward (south) of the sill (Fig. 12.1). This was about 50 m seaward of the channel mouth at MLLW elevation. The purpose for collecting the samples was to determine the difference in suspended concentration at the sill and in the adjacent bay. Three samples collected from the water surface in Nushagak Bay were analyzed for sediment concentration. The bay concentrations from earliest to latest in the tidal cycle were respectively 43, 52, and 70 percent as large as the sill concentrations obtained at the same time. The earliest sample was taken at a tidal elevation of 3 m or 1.2 m above the elevation of the sill.

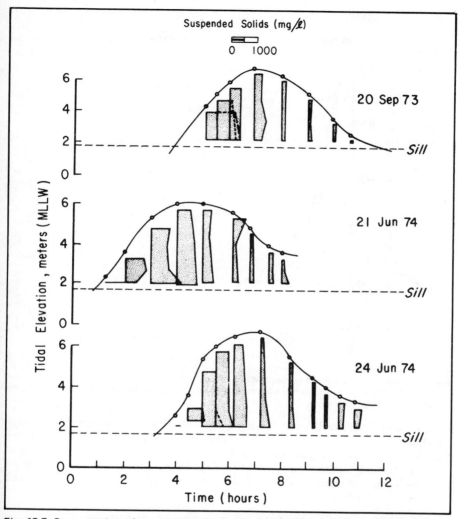

Suspended Solids (mg/ℓ)

0   1000

20 Sep 73

21 Jun 74

24 Jun 74

Sill

Tidal Elevation, meters (MLLW)

Time (hours)

Fig. 12.5 Concentration of suspended solids at the sill as measured during three tidal cycles. Note the approximately uniform concentrations through the water column at any given time, and the decrease in concentration of sediment leaving the harbor on the ebbing tide.

The last sample was taken at a tidal elevation of 6 m about 1 hr before the time of high water.

Sediment size distribution

Four-liter water samples were collected at the sill to determine the size distribution of material entering the basin. Figure 12.7 illustrates the mean particle size, as a function of time and elevation in the water column, for all samples in which the suspended concentration exceeded 250 mg/liter. For

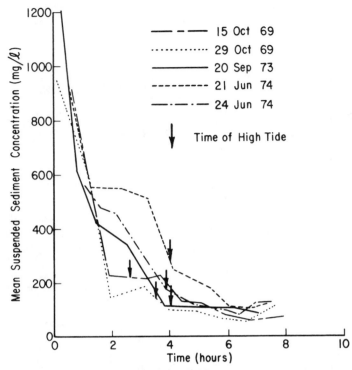

Fig. 12.6 Suspended sediment concentrations averaged through the water column as measured over the sill. Note the rapid decrease in concentration from the time the water level rises to the sill until the time of high water.

samples with suspended concentrations less than 250 mg/liter (including most of those collected during the ebb portion of the tidal cycle), the concentration for a pipette analysis was too low, when clear water was decanted, to form a 500-ml sample (Guy 1969). Sample size was limited because of transportation restrictions. The median particle diameter of 26 flood tide samples was .0059 mm.

DISCUSSION AND CONCLUSIONS

Data collected during the ice-free season in and adjacent to Dillingham Harbor provide information on shoaling rates and the shoaling process in an enclosed harbor where shoaling is primarily caused by fine-grained sediment that enters the basin during the flood tide and subsequently settles out of suspension before it is removed on the ebb tide.

Shoaling rates

The volume shoaling rate from June through September at Dillingham was 5400 m^3/mo, or 1.7 times the October-May rate of 3100 m^3/mo. The

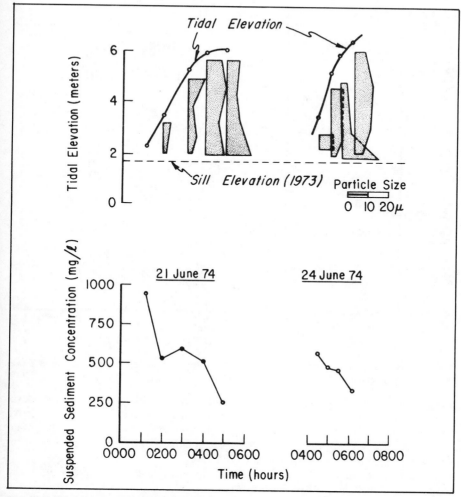

Fig. 12.7 Size distribution of suspended sediment collected at the sill on 21 and 24 June 1974. The median particle size of all samples was 0.0059 mm.

summer sedimentation rate averaged 0.007 m/day. In contrast, a summer rate, measured by Everts and Moore (in press) in a large sedimentation tank emplaced on a tidal flat near Anchorage, Alaska, was about 0.02 m/day. The tide range above the tank sill was similar to that at Dillingham, but the average concentration of suspended sediment in the estuary at Anchorage was about four times as large (1075 ppm) as that at Dillingham (280 ppm).

Sediment shoaling rates appear to be a function of the bottom elevation in the basin at the time of shoaling. Figure 12.8, based on dredging records and sounding data, indicates the shoaling rate decreased as the bottom shoaled above sill elevation. Although the data are limited and somewhat

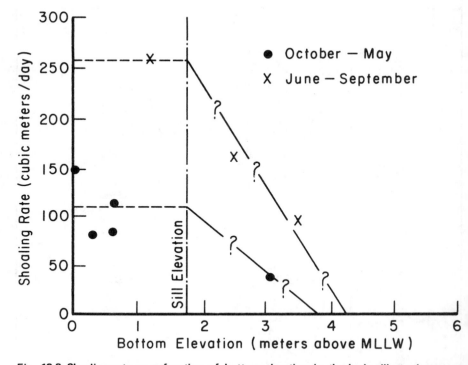

Fig. 12.8 Shoaling rate as a function of bottom elevation in the basin, illustrating a rate decrease as the bottom shoals above the sill elevation.

scattered, the shoaling rate does not appear to be related to bottom elevation when it is below the elevation of the sill. As the basin shoals above the sill, the total quantity of water and sediment discharged to the basin decreases. Therefore, the total mass of sediment available for deposition declines, resulting in a decreased shoaling rate.

A nearly constant 80 percent of the material that was carried into the harbor remained there. Based on data from five tidal cycles (Fig. 12.6), the average was 81.2 percent, with a range of 76.0 to 84.2 percent.

In contrast with the depositional percentage, the absolute magnitude of the sediment mass carried into the harbor was highly variable from tidal cycle to tidal cycle. The maximum variation for the five tidal cycles shown in Figure 12.9 was 70 percent. Values were computed by combining the mean suspended sediment concentration across the sill (Fig. 12.6) and the mean water discharge across the sill using the tidal hydrograph. The mass of sediment deposited in the basin varied from 47,600 to 80,700 kg/tidal cycle and averaged 64,400 kg/tidal cycle.

The summer depositional rate of 88.5 m³/tidal cycle with a bulk density of 1.5 g/cm³ is consistent with the mean 64,000 kg/tidal cycle depositional rate calculated from velocity and concentration measurements. Fine-grained material of a size similar to that at Dillingham (0.0059 mm), when deposit-

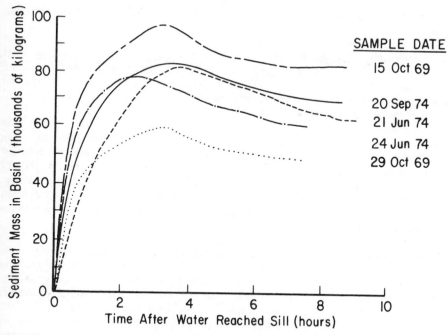

Fig. 12.9 Cumulative sediment mass budget for Dillingham Harbor for five selected tidal cycles when samples were collected. Note that most material enters during the first two hours after the water level reaches the sill.

ed in the sedimentation tank at Anchorage (Everts and Moore, in press) also exhibited a bulk density of about 1.5 g/cm^3.

Shoaling mechanism

Flow into the basin is dominated by tidal fluctuations in Nushagak Bay. The rise and fall of the tide in the basin is nearly in phase with, and equal in magnitude to, that in the bay. At Dillingham Harbor, for example, the Keulegan Repletion Coefficient (Keulegan 1967), used to predict the tide range in a restricted water body connected by a channel to a region that forces the tide in the water body, is about 25. When the coefficient exceeds 2.5, the tide ranges on either side of the channel are about equal.

Basin processes. Sediment infilling was nearly uniform throughout the basin when the bottom elevation was below the elevation of the sill (Fig. 12.2a). An exception was the region near the sill (about 10 to 15 percent of the total basin area), which was scoured by a jet produced during the lower stages of a flooding tide. Deposition did not result in a level bottom (Fig. 12.2b), but conformed to the previous dredged bathymetry. Sediment, once deposited on the bottom, is not resuspended, so it retains the bottom contours which existed prior to deposition. The uniformity of shoaling rates suggest the depositional mechanism is similar in type and magnitude throughout the basin.

Suspended sediment concentration and velocity gradient data are important in determining how much sediment enters and leaves the basin, and at what elevation it enters. At any given time in the tidal cycle, the suspended sediment concentration (Fig. 12.5) and current velocity (Fig. 12.4) at the sill are uniform with depth. This results because of mixing in the channel and mixing caused by the flow constriction at the sill.

Suspended sediment appears to be distributed equally throughout the basin at any given time. Currents moving radially away from the sill carry near-constant concentrations of material in all directions. The vertical velocity of the suspended particles (mean fall velocity of a 0.006-mm grain = 0.0129 cm/sec) is generally less than one or two percent of the horizontal velocity. Suspended sediment, therefore, appears to move to all locations in the basin before settling below sill elevation. Suspended material settles as it would in a nearly-still fluid.

Channel processes. Over 50 percent of the suspended material brought into Dillingham Harbor is brought in during the first hour of the three-hour long flood tide cycle at the sill (Figs. 12.6 and 12.9). By entering early in the flood cycle and at an elevation near that of the sill, the material is preferentially deposited. Two probable causes of the high concentrations at the sill early in the flooding tide are an increase in the concentration coming in from Nushagak Bay and an increase caused by the resuspension of sediment which had settled during the previous low tide in the channel between the sill and the estuary.

Extremely high concentrations of sediment from Nushagak Bay are probably not the main cause of harbor shoaling. Although they are highest early in a flooding tide, bay concentrations average about one-half the concentration reaching the harbor.

Current velocities in the channel are very low when the tidal elevation is below the sill; consequently, suspended material settles in the channel during those times. As the subsequent flood tide rises over the sill, channel currents increase rapidly to accommodate the increased water discharge required to fill the basin. It is at that time, when material first begins to enter the harbor and the submerged channel cross-sectional area is smallest, that velocities are greatest and previously deposited material is resuspended. The amount that is resuspended (Fig. 12.5) decreases during the remainder of the flood cycle as the current (Fig. 12.4) declines and the cross-sectional area of the channel increases. The amount of sediment entering the harbor from this channel source is about 40 percent of the total sediment that is brought into the harbor.

GENERAL IMPLICATIONS

Attention to the causes of shoaling may assist in reducing the harbor shoaling rate. Also, an understanding of the shoaling mechanism is a requirement in developing methods to predict the shoaling rate prior to construction of a harbor.

A portion of the material that enters the harbor is sediment that was deposited in the navigation channel at low tide stages and resuspended on the next rising tide. Economies can be achieved when the entrance channel is designed to reduce the accumulation of sediment there. In general, this will require that the channel be designed as short as possible.

Sedimentation in the half-tide harbor at Dillingham is the result of fine-grained sediment that settles from suspension after entering the harbor basin on a flood tide and fails to be removed during the ebb portion of the tide. The shoaling rate and sedimentation process are similar throughout the basin in the region away from the sill. This implies that a one-dimensional model of the process could be applied to predict the shoaling rate in a similarly designed harbor.

When predicting harbor shoaling rates where the tidal range (6 m), sill elevation (2 m), and sediment characteristics (median size = 0.006 mm of rock fragments, not clay minerals) are similar to those at Dillingham, about 80 percent of the incoming sediment might be expected to be deposited. The total amount that enters, however, will vary widely between tidal cycles, depending on conditions in the estuary. Consequently, a number of tidal cycles must be sampled in the estuary to obtain a representative tidal cycle sediment concentration.

Although agitation dredging is not recommended practice in most areas, because of environmental as well as engineering considerations, it might be feasible in a region such as Nushagak Bay. As a result of the high ambient suspended sediment concentrations and thorough mixing, direct discharge to the bay by resuspension in the basin during an ebbing tide may be an acceptable alternative to hydraulic dredging. Agitation dredging would neither increase bay concentrations nor be environmentally undesirable. Also, little of the discharged sediment would return to the harbor at a later time.

Acknowledgments

Field data were collected by the Alaska District, Corps of Engineers, under the supervision of Harlan Moore. Sediment samples were analyzed in the Alaska District Soils Laboratory by Clair Johnson. Billy Joe Adams and Dean Dewey provided support for the field effort at the harbor, and Ed Pohl coordinated the study with the Coastal Engineering Research Center which funded it. The findings of this paper are not to be construed as official Department of Army position unless so designated by other authorized documents.

REFERENCES

EVERTS, C. H., and H. E. MOORE

    (In press)   Shoaling rates and related data from Knik Arm near Anchorage, Alaska. U.S. Army Coastal Engineering Research Center, Fort Belvoir, Virginia 22060.

GUY, H. P.

1969 Laboratory theory and methods for sediment analysis. In *Techniques of water resources investigations of the United States Geological Survey*, Book 5, Chapter C1. U.S. Geol. Survey, Reston, Virginia 22092, 20 pp.

KEULEGAN, G. H.

1967 Tidal flow in entrances: Water-level fluctuations of basins in communication with seas. Tech. Bull. No. 14, Committee on Tidal Hydraulics, U. S. Army Corps of Engineers, Vicksburg, Mississippi.

Assessment of the Arctic Marine Environment: Selected Topics
Copyright © 1976 by Institute of Marine Science, University of Alaska, Fairbanks

Part 2. Sedimentation during periods of ice-cover

C. H. EVERTS [1]

Abstract

Seasonal ice cover often limits the use of small-craft harbors in the northern regions to a few summer months. The role that ice plays in sedimentation may be an important consideration in harbor design. A study was conducted during the ice season in 1973 and 1974 at a harbor near Dillingham, Alaska, to determine the winter shoaling rate and the mechanisms that contribute to winter sedimentation. The purpose of the study reported herein was to develop a method to quantitatively predict the difference between winter and summer shoaling rates and to evaluate possible solutions to reduce winter shoaling. (Sedimentation under ice-free conditions is discussed in Part 1 of this chapter).

INTRODUCTION

Small-craft harbor use in the north is generally limited to between three and six ice-free months. During the non-use months, ice plays a role in sedimentation which may be an important consideration in harbor design. A study during the ice season at a half-tide harbor near Dillingham, Alaska, was conducted to determine the mechanisms that contribute to winter sedimentation. The purposes of the study were to measure seasonal differences in elements responsible for the shoaling problem, to develop a method to quantitatively predict the difference between winter and summer shoaling rates, and to evaluate possible solutions to mitigate winter shoaling.

Results of a study of the shoaling process during the ice-free season at Dillingham Harbor are presented in Part 1 of this chapter. Also included in Part 1 are discussions of the physical setting of the harbor, the characteristics of a half-tide harbor, and the design of Dillingham Harbor.

---

[1] *Coastal Engineering Research Center, Kingman Building, Fort Belvoir, Virginia 22060.*

DATA COLLECTION AND RESULTS

A systematic series of routinely collected photographs, suspended sediment samples, ice thickness measurements, and meterological measurements were obtained at Dillingham Harbor in 1973 and 1974.

Ice conditions

Four photographs of the channel, sill, and basin of Dillingham Harbor were obtained daily (5 days/wk) at 1300 hr, a time of high sun angle, to provide a chronology of ice conditions in and adjacent to the harbor. Sites from which the photographs were taken are shown in Figure 12.10. Photos were obtained from 10 October 1973 to 16 May 1974, and between 21 October 1974 and 27 November 1974.

Ice thickness and ice movement measurements were made in conjunction with the daily photograph study. Ice thickness was measured near the sill. Horizontal ice movement was measured by placing metal drums on the ice near the sill and observing their horizontal movement. The submerged ice

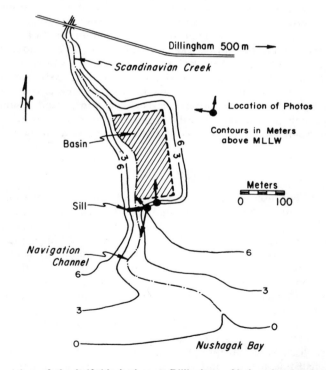

Fig. 12.10 Plan view of the half-tide harbor at Dillingham, Alaska, showing locations from which four ground photos were obtained on a near-daily basis. Photo orientation is illustrated by arrows. The three elements that comprise a half-tide harbor are the basin, sill, and navigation channel.

thickness (defined as the actual thickness x 0.9) is shown in Figure 12.11 with reference to the mean monthly tidal amplitude. Submerged thickness values are used, because it is the submerged volume which causes a winter decrease in the water and sediment volume that enters the basin during the flood tide.

Typical freeze-up conditions are shown in Figure 12.12. At the end of October, ice began to form to the harbor basin. At the same time, thin flow ice, carried by tidal currents, moved into the basin from Nushagak Bay. Until mid-November, ice that formed in the basin was broken up with the rise and fall of the tide. Also, mid-November ice carried in from the estuary had formed a 0.5 to 1.0-m thick jam near the basin entrance with the bay ice generally pushed under the basin-formed ice. The jam was lobate in shape, 40-m long, and oriented toward the north. Horizontal ice movement ended in the channel and basin by mid-November.

Typical winter conditions are shown in Figure 12.13. From mid-November until spring thaw, ice in the channel and basin moved vertically as a solid block with the rise and fall of the tide. The ice was broken only in the vicinity of the sidewalls. Ice thickness in the basin reached a maximum of about 2 m by mid-March.

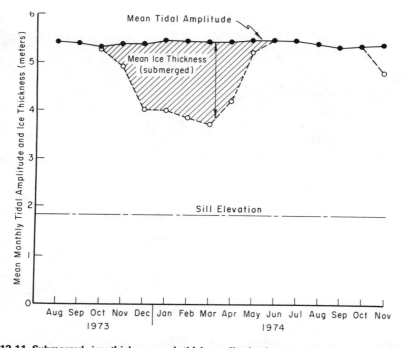

Fig. 12.11 Submerged ice thickness and tidal amplitude, by month, showing the relative volume of ice and bay water in the basin at the monthly average high tide. In the months December - March, the volume of water entering the harbor each flood tide is about one-half the summer volume.

In an observation and sampling program conducted on 16 January 1973, there was no indication that the harbor ice, when lowered on an ebbing tide to the basin bottom, picked up sedimentary material and incorporated it in the ice column. Bottom sediment appeared to be unfrozen in three holes drilled through the ice at various locations in the basin. When an ice-sediment contact was observed above the water surface elevation at low tide — i.e., on the sidewalls of the basin above sill elevation — the surface sediments were frozen into a mush but did not appear to adhere to the ice moving up and down with the tide.

At the beginning of the ice season, water froze from the surface down in a clear layer. Incoming water filled the basin beneath the ice. Later, during the winter, when the flood tide reached the sill, the incoming flow was partially blocked by the ice jam on the basin side of the sill. As the water surface elevation rose, flood water breached the ice jam and flowed on top of the basin ice. Later, it froze there. Flooding appeared to occur when the flood tide was between two-thirds and five-sixths the maximum tide elevation. The flood water contained sediment that settled out on the ice surface in thin bands which graded from near-zero sediment concentration to sediment layers of 1 or 2 cm. Sediment-ice units that formed on the surface were 1 to 10 cm thick.

Fig. 12.12 Views of the harbor and vicinity during freeze-up. Photos were taken during flood tide near the time of high water on 12 November 1973. Photograph orientations are as shown on Figure 12.10.

Fig. 12.13 Views of the harbor and vicinity on 19 February 1974 during flood tide when the water surface was near mid-tide level. Photograph orientations are as shown in Figure 12.10.

Ice began to thin by thawing in mid-March, as mid-day air temperatures rose above 0° C. Beginning in the second week of April, ice began to move out of the channel. By mid-April, the channel was clear of ice. At that time, ice in the basin near the channel began to break off the solid cake and drift into the bay (Fig. 12.14). Later in May, wind stress was an important factor in breaking up the basin ice. By mid-May, all the ice was gone from the harbor. Estuary ice had disappeared two weeks earlier.

Suspended sediment

A total of 298 water samples were obtained on a one-per-day basis. They were collected 1 hr before the time of high water in the period August 1973 to November 1974. Each sample of 1 liter was obtained between the water surface and a depth of 0.3 m at the sill. A collection time of 1 hr before high tide was selected for comparison purposes. It corresponded to the time during the flood cycle when the rate of concentration change with time varied least. The mean concentration of suspended sediment entering the harbor during the flood tide was about 2.5 times the concentration

Fig. 12.14 Views of the habor and vicinity on 1 May 1974 during flood tide and near the time of high water. Photograph orientations are as shown in Figure 12.10.

obtained 1 hr before high water (see Fig. 12.6, Part 1). The sill was the only collection site accessible year round.

Samples were analyzed to determine the absolute suspended sediment concentration and its seasonal variability. Figure 12.15 shows the mean monthly suspended sediment concentration and the range of concentrations for the sampling period. Sediment concentrations averaged about 250 mg/liter.

Approximately every fifth sample (54 total) was analyzed for grain size distribution using standard pipette techniques (Guy 1969). The samples were washed and settled in distilled water so that flocculation induced by solutes in the ambient bay water was eliminated. As shown in Figure 12.16, the mean of all the median sizes was 0.0081 mm. A mean 36 percent of the material was finer than 0.001 mm. On the average, less than 5 percent of each sample exceeded a grain size of 0.07 mm. From March through

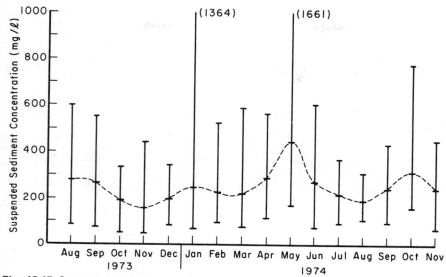

Fig. 12.15 Suspended sediment concentration as obtained one hour before high water at Dillingham Harbor. The average concentration is about 250 mg/liter with little seasonal variation. Concentrations entering the harbor are much greater earlier in the flood tide cycle.

September, the median particle size averaged 0.014 mm, while from October to March it was about 0.005 mm. However, a difference of only 4 percent for sediment less than 0.001 mm was found between winter (38 percent) and summer (34 percent) samples.

Atmospheric and hydrologic conditions

Air temperature and wind velocity measurements were made at 1300 hr at times when photographs were taken during the winter and near 1200 hr during the summer. Winter measurements were made near the sill at ground elevation (+8-m MLLW). Summer measurements were made either near the sill or from the deck of a dredge working in the basin (at an elevation of between +5 and +8-m MLLW). Meteorological data were obtained to determine their relationship to the sediment load carried into the harbor.

Figure 12.17 illustrates the mean monthly air temperature, wind speed, and wind direction. Wind directions are highly seasonal with winds from the north, northeast, and east during the ice season, and from the south and southeast in the summer. Winds from the south are oriented parallel to the axis of Nushagak Bay and toward the harbor.

Salinity, as measured in the water samples collected 1 hr before high tide at the sill, exhibited a marked seasonal variation (Fig. 12.18). Salinities were generally less than 1 ppt (part per thousand) between May and November, and nearly 2 ppt from December through April.

Water temperature measurements were not made on a routine basis. The winter water temperature in Nushagak Bay is about 0° C, while summer temperatures rarely exceed 10° C.

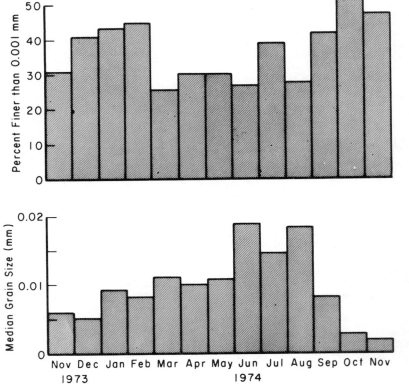

Fig. 12.16 Sediment size data obtained from samples collected at the sill of Dillingham Harbor.

DISCUSSION

Data collected in and adjacent to Dillingham Harbor provide information on the shoaling process in an enclosed harbor, where shoaling is primarily caused when fine-grained sediment settles from suspension. For the purposes of the following discussion, winter will be considered as the October–May shoaling period, and summer as the June–September period. These periods correspond to the dates when harbor surveys were made and roughly correspond to the season limits for harbor navigation.

During the ice-free season, sediment was transported into the basin in suspension during the flood portion of the tidal cycle (see Part 1, this chapter). Shoaling resulted when the fine-grained sediment settled from suspension before it was removed during the ebb portion of the tide. The shoaling rate and sedimentation process were similar throughout the basin. Over 50 percent of the suspended material was carried in during the first hour of the 3-hr long flood cycle at the sill. A near-constant 80 percent of the material that entered the harbor during the summer was deposited. The

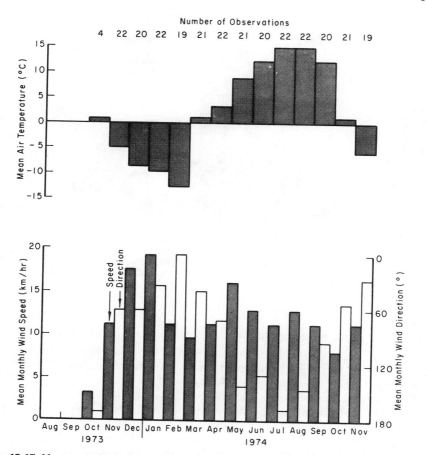

Fig. 12.17 Mean monthly air temperature, wind speed, and wind direction at Dillingham Harbor. The number of observations per month is shown at the top of the figure.

absolute mass of sediment that entered the harbor, however, varied by over 70 percent from tidal cycle to tidal cycle. The mean summer shoaling rate, as obtained from bathymetric surveys and dredge records, was 0.007 m/day (about 64,000 kg/per tidal cycle), or 1.7 times the winter rate of 0.004 m/day (38,000 kg/per tidal cycle).

The winter shoaling rate at Dillingham Harbor was 59 percent of the summer rate. It appears the winter rate can be estimated using the summer rate. This could result in a considerable cost savings when a model based on ice-free harbor processes is used to predict the shoaling rate prior to harbor construction. The objective of this section is to discuss the factors that appear to be important in causing the seasonal variation in the shoaling rate. Values placed on each factor in some instances are only approximate because of lack of available data.

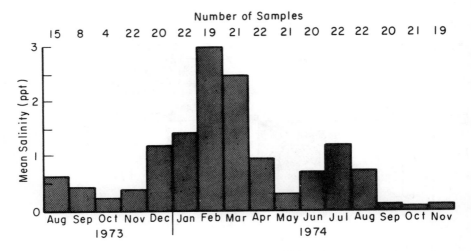

Fig. 12.18 Salinity measured one hour before high tide at Dillingham Harbor, showing seasonal variation.

Sediment mass entering harbor

Because of the submerged ice volume in the basin (average thickness about 1 m; Fig. 12.11), the discharge of water to Dillingham Harbor during the flood tide is 25 percent less per tidal cycle in the winter.

Two sources for sediment carried to the basin with the flood-tide discharge are the navigation channel leading to the basin, and Nushagak Bay. Bay concentrations in suspension exhibit only a slight seasonal variation at Dillingham (Fig. 12.15), probably because of the small contribution by rivers. Winter concentrations averaged 4 percent more than summer concentrations. These results, however, apply only to the Nushagak Bay area; elsewhere, significant seasonal variations in suspended sediment concentration may be expected, especially where river discharge accounts for much of the suspended sediment.

The average winter ice thickness in the channel is about 1.0 m. Sediment is deposited in the channel when tidal elevations are below the sill (+2.0 m MLLW). This channel deposition is dependent on the mass of suspended sediment available in the water column for deposition. Because of ice in the channel, this mass is reduced in the winter by 60 percent. Since 40 percent of the mass entering the harbor appears to be from the channel source (see Part 1, this chapter), the displacement of fluid by channel ice reduces the calculated total winter sediment mass entering the harbor by 24 percent. Considering the 25 percent decrease in water discharge to the basin in the winter, which carries 20 percent less sediment in suspension, the mass of sediment carried into the harbor per tidal cycle in the winter is 59 percent of the mass that enters each tidal cycle in the summer.

Settling velocity

Sediment size, water viscosity, and flocculation all affect the settling velocity of suspended material in the basin. The combination of these factors may reduce the fall velocity in the winter by 18 percent.

The effect of salt on the settling rate of clay minerals is distinguishable at seawater concentrations of about 1 to 2 ppt (Whitehouse et al. 1960). Flocculation, and hence the settling rate of the flocs, increase to seawater concentrations of 10 to 15 ppt. Suspended sediment in water samples collected in 0.5 ppt Nushagak Bay water in the summer did not appear to flocculate. Possibly when winter salinities reach 3 ppt (Fig. 12.18), flocculation would occur. Data obtained from winter water samples are not available to test this hypothesis.

Sediment size affects deposition, because the larger sizes settle at a higher rate. Median particle size varied from 0.014 mm in summer to 0.005 mm in winter (Fig. 12.16); however, the percent of material finer than 0.001 mm varied much less (34 and 38 percent, respectively, winter versus summer). Since about 80 percent of the sediment that entered the harbor in summer was deposited, only the finer sizes, on the average, were not deposited. A comparison of size density distributions for the summer and winter seasons indicates that only 3 percent less of the total sediment mass entering the basin is deposited in the winter because of size differences. This small percentage would not hold where sizes were much finer than those at Dillingham (median = 0.008 mm).

Water temperature variations at Dillingham (seasonal range is about 0° to 10° C) affect water viscosity. At 0° C winter water temperature, the fall velocity of an 0.008 mm particle is about 70 percent the velocity it would be at 10° C. Assuming a mean summer water temperature of 5° C, the winter fall velocity is reduced by about 15 percent. This value is approximate, because the settling velocity difference increases as particle size decreases.

Settling distance

To be deposited, a sediment particle must settle from the elevation at which it enters the basin to below the sill elevation. Summer studies indicate most of the basin is stagnant and not affected by tidal exchange below the sill.

Submerged ice causes the water and sediment to be discharged to the basin at a lower elevation relative to the sill than it would be if no ice were present. Thus, the distance a suspended particle must fall to pass below sill elevation is reduced in winter. This calculated reduction is 25 percent, suggesting that 25 percent more of the sediment entering the basin will be deposited as a result of a reduced settling distance in the winter than during summer.

Mass deposited

The mass of sediment carried into the basin each tidal cycle during the winter is 59 percent as great as the mass per tidal cycle discharged to the basin in the summer. As a result of an 18 percent lower settling velocity of the suspended mass in the winter — and of a 25 percent decreased average distance a particle must fall to be deposited in the winter — 7 percent more of the suspended material will be deposited in the winter. Thus, assuming that 80 percent of the mass carried into the harbor is deposited in the summer (87 percent in winter), the calculated winter shoaling rate is 64 percent of the summer rate, or within 5 percent of the measured winter rate. The close agreement, however, may be fortuitous rather than actual, because of the number of assumptions and estimates made in the calculations.

ENGINEERING IMPLICATIONS

Although winter shoaling rates were only about one-half those of the summer season, winter sedimentation accounts for about one-half of the total yearly shoaling problem. This is because the summer navigation season lasts only about four to five months. Since winter shoaling occurs when the harbor is not in use, methods different from those presently used in the summer — i.e, dredging — may be employed to decrease the shoaling problem.

Sediment is deposited because it is carried into the basin in suspension on the flooding tide and settles out before it can be removed, still in suspension, with the ebbing tide. Two ways to keep deposition from occurring are to stop sediment from entering the harbor entirely, or to maintain it in suspension once it is in the harbor. The first solution requires some sort of harbor closure structure. The second requires a system to keep the sediment in suspension. Each solution may be affected by ice.

Harbor closure

Harbor closure is probably the only practical method to keep sediment out of the harbor. No economical means is available to separate out the sediment while allowing the water to pass.

The best location for a closure structure is near the channel-basin intersection (Fig. 12.10), where the banks are sufficiently high to prevent overtopping during extreme tides. Maximum closure benefits will result if the harbor is closed by an impermeable barrier immediately after the navigation season has ended and kept closed until navigation use is required again in the spring.

Positive aspects of a harbor closure structure include the following considerations: closure will result in a total absence of winter sedimentation in the harbor basin. If the basin is drained, there would be little ice formation; consequently, the spring navigation season could open approximately two weeks earlier. Gravity drainage could be accomplished at low-tide stages through the closure structure. Also, before freeze-up of the exposed basin silts and muds in the autumn, maintainence procedures — i.e., dragline — could be employed to clear the harbor of summer-deposited material at less expense than dredging, which interferes slightly with navigation. Finally, there would probably be no need for annual removal of harbor structures such as piers and piles.

There are also some negative aspects of a harbor closure scheme: currents in the channel would be very low as a result of the structure, and considerable in-channel deposition from suspension could be anticipated. The sediment might be swept clear in the spring when the closure structure is removed, but much of the sediment might also be carried into the harbor and deposited there. In the absence of a large channel, of course, this would not be a problem. The expense of emplacing and removing a harbor closure structure might be excessive. And, if there is freshwater flow to the basin, a diversion or pumping scheme would be required to prevent the buildup of ice. At Dillingham, this would require not only a diversion of Scandinavian Creek — but possibly also the diversion of springs which discharge to the northeast portion of the basin.

Sediment in suspension

Maintaining sediment in suspension might be accomplished in a number of ways, possibly including the use of air or water jets on the basin bottom. Methods to prevent long-term deposition, such as bottom agitation, might also be employed. Each system would require periodic maintenance and would consume power. Ice would be a factor in the success of each method.

Acknowledgments

Field and laboratory work was supported by the Alaska District, U. S. Army Corps of Engineers, under the supervision of Ed Pohl (1973-1974) and Doug Staub (1975). The hospitality of Billy Joe Adams, the dredgemaster at Dillingham, was invaluable to field parties sampling at the harbor. Efforts by Harlan Moore and Dean Dewey greatly aided the field program. Samples and photos from the period during the ice season were obtained by Delpin Lopez, without whose able assistance the project could not have been completed. The findings of this paper are not to be construed as official Department of Army position unless so designated by other authorized documents.

REFERENCES

GUY, H. P.

    1969   Laboratory theory and methods for sediment analysis. In *Techniques of water resources investigations of the United States Geological Survey*, Book 5, Chapter C1. U. S. Geol. Survey, Reston, Virginia 22092, 20 pp.

WHITEHOUSE, U. G., L. M. JEFFERY, and D. B. DEBBRECHT

    1960 Differential settling tendenbies of clay minerals in saline waters. *In* Proc. 7th National Conference on Clays and Clay Minerals. Pergamon Press, New York.

Assessment of the Arctic Marine Environment: Selected Topics
Copyright © 1976 by Institute of Marine Science, University of Alaska, Fairbanks

CHAPTER **13**

Fill materials between Barrow and the Colville River, northern Alaska

J. C. LaBelle[1]

Abstract

Fill materials for foundations for engineering structures are scarce in most parts of the arctic coastal plain of Naval Petroleum Reserve No. 4. There has been an increasing demand for such materials as northern Alaska has begun to develop in the past 30 years, and it is anticipated that the need will continue to increase greatly during the foreseeable future, with continuing development of the petroleum reserve and with improvement in living conditions for the expanding population. Average demands in the past have been in the order of 10,000 cubic yards per year, but projected new programs may increase the requirement 10 to 20 times in some years. Reported herein are results of summer field studies conducted in 1972 and 1973.

INTRODUCTION

A program to identify and inventory fill material resources within 25 miles of Barrow and within 50 miles of Cape Halkett was contracted to the Arctic Institute of North America by the Office of Naval Petroleum and Oil Shale Reserves. Field work for the program was carried out during the summers of 1972 and 1973 (LaBelle 1973, 1974).

---

[1]*Arctic Environmental Information and Data Center, University of Alaska, 707 A Street, Anchorage, Alaska 99501.*

Regional setting

The arctic coastal plain of Naval Petroleum Reserve No. 4 (NPR4) in northern Alaska (Fig. 13.1) is characterized by low topographic relief, thousands of lakes and swamps, and numerous meandering streams. The coastal plain surface continues beneath the ocean and forms the shallow continental shelf.

The emergent region is typified by numerous thaw lake basins, some filled and others drained. It is estimated that 50 to 75 percent of the coastal plain near Barrow is covered by lakes or marshes that occupy low areas of former lake basins. The basins are elongated, and their long axes are parallel and oriented a few degrees west of north. They range from a few yards to nine miles long. The remaining 25 to 50 percent that is relatively higher ground includes the "initial surface residuals" — remnants of the original surface that have not undergone thaw depression (Hussey and Michelson 1966).

In the Barrow area, relief between the initial surface residuals and the basins is between 10 and 15 ft. Other features of approximately 10 to 20 ft of relief are superimposed on the initial surface and apparently are not associated with the lake basins. All relief features exhibit the polygonal ground pattern of permafrost regions. The pattern is produced by contraction of the ground during extreme low temperatures.

The shoreline bounding the northern side of the region is composed chiefly of low banks and bluffs, fronted by occasional narrow beaches of sand and gravel. Bluffs about 30 ft high, fronted by a narrow beach, extend along most of the coast for 40 miles southwest of Barrow. Gravel and sand beaches backed by low tundra lie between Barrow and the base of the Point Barrow spit. From there, sand and gravel beaches form the spit out to Point Barrow. Extending southeast from the northernmost point of the spit is another low, sandy feature called Eluitkak Spit, which terminates at Eluitkak Pass. Beyond the pass, a chain of barrier islands stretches southeastward. The head of a submarine canyon, the Barrow Sea Valley, lies offshore a short distance north of Point Barrow.

South of the barrier islands is a shallow lagoon known as Elson Lagoon, the south shore of which is receding rapidly. The shoreline there is characteristic of a submerging coast, with very narrow or no beaches, truncated lake basins, estuaries, and low but steep undercut shores slumping into the sea. Thermal niching of the frozen land is the predominant erosion factor, causing coastal recession at rates up to 33 ft/yr (Lewellen 1970).

Eastward from Cape Simpson, the shoreline consists of low bluffs fronted by occasional narrow sandy beaches. The shoreline is interrupted by Smith and Harrison bays and by the extensive Ikpikpuk and Colville river deltas. One section of coast, near Lonely DEW station and Pogik Point, includes a wide, gravelly beach and barrier island.

The southern two-thirds of the area is mantled with extensive sand and silt dunes. Active dunes and blowouts are notable along the western banks of most of the larger streams and rivers crossing the area, as well as along the western shores of several lakes. Between the lakes and streams, stabilized

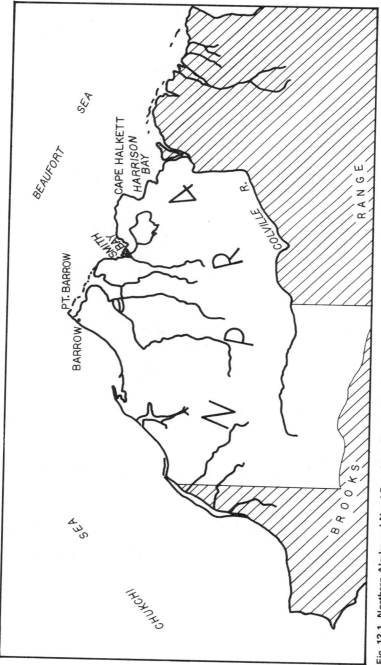

Fig. 13.1 Northern Alaska and Naval Petroleum Reserve No. 4.

dunes, vegetated principally with willows, are abundantly scattered over the surface to depths up to 20 ft (Black 1951).

The Colville River, the principal river of northern Alaska, terminates in a 200-mi^2 delta of extensive mud flats, sand bars, sand dunes, tundra, and lakes. The Colville is the only river on the eastern side of NPR4 that has its headwaters in the mountains. Several other streams cross the region on their way to the sea, but all have their beginnings in the foothills or in the coastal plain itself, and have beds of fine sand and silt with little coarse material.

North of Teshekpuk Lake, several sublinear relief features were earlier considered to be old sand and gravel beach ridges (Rex 1953). Recent investigators, however, describe these features as initial surface residuals.

Geologic setting

For the most part, the arctic coastal plain underlying NPR4 is underlain by Cretaceous strata mantled unconformably by a thin cover of dominantly marine Quaternary sediments called the Gubik Formation (Black 1964).

Some of the best exposures of the Gubik Formation are in the cliffs along the shore southwest of Barrow. The marine sediments indicate near-shore deposition. Abundant interfingered and crossbedded channel sands and gravels grading upward and laterally into shallow lagoonal sands and silts characterize the exposed portions of the formation (O'Sullivan 1961).

"A very generalized section of the Gubik would be: a basal gravelly sand, overlain by white laminated sand which may in some instances be missing (and replaced by a dark silt), overlain by orange sand or sand interlayered with a dark silt-sand which grades upward through a layered zone into the orange sand. The thin basal gravel may be absent, and the material under the main gravel may vary from reworked silty clay to sand with lenses of gravel and sand" (O'Sullivan 1961).

The above-mentioned basal gravels, noted especially along the Colville River bluffs, have been assigned by Hopkins (1967) to specific marine transgressions, each of which reached limited inland levels. "The Beringian (marine transgression) beds at Ocean Point constitute the basal part of the Gubik Formation in that area; they consist of richly fossiliferous crossbedded sand and sandy gravel that in some places is more than 15 m thick.... The Beringian beds rest on bedrock of Cretaceous age and are covered by about 10 m of gravel and sand that probably was deposited during the Anvilian transgression. The Beringian beds at Ocean Point wedge out 40 km inland from the present coast at an altitude of 30 m; farther inland, the deposits of the Anvilian transgression rest directly on bedrock."

DEPOSITIONAL PROCESSES

Shoreline processes

Winds near Barrow usually come from easterly or westerly directions. When sea ice is not present, the winds develop waves which approach the Barrow beaches from either the west or north, the waves from the north being built by the east winds and refracted around Point Barrow. A southwest-flowing longshore transport system is developed by waves from the north, and a northeast transport by waves from the west. The net effective longshore transport, evidenced by growth on the southwest side of Point Barrow, is toward the northeast (Hume and Schalk 1967).

Near Barrow, erosion and slumping onto the beach of the deposits that form the bluffs southwest of the town appear to supply the materials transported northeast. Observations from the air and sea indicate little contribution of sediment to the beaches near Barrow from further southwest, because beaches there are narrow or missing (Hume et al. 1972).

The sediment resulting from erosion by waves and ice at the bluffs southwest of Barrow is transported both by longshore currents and by wave swash northeast to Point Barrow, where it is deposited on the west side of the Point. Wave swash is more important, because the beach sediments are generally too coarse to be carried in suspension. Some of the fine sediment bypasses the Point and continues as a muddy flow of water moving out to sea, where it finally settles — possibly in the Barrow Sea Valley (Rex 1964). Erosion of the low bluffs at Point Barrow, near the old Eskimo site of Nuwuk, provides the sediment which is building Eluitkak Spit to the southeast. Again, some of the fine material must bypass the end of that spit (Hume and Schalk 1967).

To the southeast of Point Barrow, waves approach the shore from the east and north and set up longshore transport flowing northwest or southeast. Apparently, currents and transport depend on local winds and may set in either direction along the coast.

A very important factor in the shallow-water environment is ice, which can stop the entire process of beach erosion, transportation, and deposition. It limits the size of waves and the patterns of currents. The position of the ice is extremely variable, especially west of Point Barrow. It may drift out of sight one day and be back on shore the next. As a result, one day there may be strong currents and waves, whereas the next day the waves are damped and the current entirely changed. Most years the ice probably stays near enough to land to affect wave and shore processes all the time. Occasionally, however, there is as much as 200 miles of open water off Barrow.

Because the ocean is free of ice for less than three months a year, shoreline processes in that region transport far less sediment than do the same processes in temperate regions. Even when the sea is relatively open, scattered ice often is present and acts as a wave damper, slowing wave-generated currents.

Along the beaches of the region, most significant gravel deposits are found on the Chukchi Sea shores and on the Beaufort Sea barrier island complex. The principal source of the gravels appears to be the surficial Gubik Formation — the Pleistocene deposit of mixed marine and alluvial silt, sand, and gravel that mantles the entire coastal plain. Coastal erosion and collapse contribute the sediments to the region's shores, where coastal currents and wave action winnow out the silt and fine sand, leaving behind the coarser sand and gravel fractions as lag deposits. These coarse deposits are then moved along the coast by wave swash longshore transport processes — forming beaches, spits, bars, and barrier islands.

Aside from normal wave and current deposition, several processes are at work which, though quantitatively less significant, do contribute to the sediment load on or near the beaches. The most significant of these is ice push. During times of onshore winds, the ice approaches the shore and often rides up onto it, bulldozing sediments from the foreshore high up onto the beaches (Hume and Schalk 1964a).

Deposition from ice push probably never amounts to more than 10 percent of the sediments above sea level, and is generally more in the range of one or two percent. Sediments carried by ice, and dropped when it melts, may account for up to three percent deposition nearshore (Hume and Schalk 1967).

Storms

Sedimentation on the beaches does not seem to be principally the result of a steady, slow addition of material over time, but rather of sudden large movements of material at times of major storms (Hume 1965). During the great storm of 1963 at Barrow, the amount of sedimentation near the tip of Point Barrow Spit amounted to 20 years' normal net deposition, while the tombolo was breached and cut off from the mainland. Sediment transport of similar magnitude occurred in other areas. Some barrier islands migrated appreciably, some totally disappeared — and, in some cases, separate islands were welded together to form a single island. The bluffs southwest of Barrow retreated 10 ft, and the beach fronting them retreated 60 ft. Though this storm was of particular severity, it appears that sediment movement along the coast is greatly accelerated during storms.

Inland processes

The scarcity of gravel resources within NPR4 is principally the result of the anomalous course of the Colville River. The river has its headwaters in the De Long Mountains west of 160° W longitude, almost at the western boundary of NPR4. Its course then runs east between the northern and southern foothills to nearly 151° W longitude before it turns north toward the sea. This unusual eastward trend results in the capture by the Colville of all drainage from that region of the Brooks Range. Between the Colville and Utukok rivers, all streams that cross the coastal plain have their headwaters in the northern foothills or the coastal plain, neither of which provides a source of gravel.

Within NPR4, processes such as stream erosion and deposition, development of thaw lakes, vegetational processes, and eolian deposition contribute very little coarse material to the region. Erosion, in the normal sense, takes place very slowly, because the streams and ground surface are frozen for up to nine months of the year. Even during the summer season, erosion takes place very slowly, as the frozen ground thaws slowly and only to shallow depths. Even where erosion is active, only fine sands, silts, and muds are transported and deposited.

FILL MATERIAL RESOURCES

Chukchi Sea shores

South of Barrow, narrow gravel beaches extend in a broken line for a distance of 40 miles, backed by low bluffs. The bluffs are highest and contain the highest percentage of gravelly sediments for the first few miles southwest of the city of Barrow. Most of the beach gravel in the vicinity of the city and northward is derived from these coastal bluffs. South of the higher bluffs, beach sediments are more sandy, and the gravel fraction smaller in size, than those found north of the bluffs.

Between Barrow and the base of Point Barrow Spit, gravel and sand beaches front the low tundra. From there, north to the tip of Point Barrow, low sand and gravel beaches front several higher, parallel beach ridges of coarse materials.

In protected areas, such as bays and sloughs, the texture of beach material is coarse sand and fine gravel; whereas in unprotected areas, where there are strong currents and wave action, the tendency is toward coarse gravel. Beach sediments are coarse grained in the surf zone, with coarse and medium gravel that grades into fine gravel and sand on the beach foreshore. Backshore areas usually consist of laminated sand and fine gravel. Beaches thaw to depths of about 5 ft by the end of summer.

The beaches along the entire length are composed chiefly of black and brown chert, with minor amounts of igneous and metamorphic rocks, notably quartzite. The percentage of quartz increases to the north and east, while the chert increases to the south and west.

The 5-mile-long Point Barrow Spit contains over 1,000,000 cu yds of sandy gravel, but borrow over the years has already amounted to nearly 300,000 cu yds. The last major removal, in 1972, resulted in the breaching of the spit during an autumn storm. There is some indication that coarse gravel resources may extend underwater north of Point Barrow as the extension of the longshore transport system of the Chukchi coast, but this has not been properly studied.

At Barrow and the nearby Naval Arctic Research Laboratory, gravel has been excavated for years from the nearby Chukchi beaches. The practice proved dangerous, as beach borrow has resulted in accelerated coastal retreat of up to five times the natural rate. Total retreat in the area from 1948 to 1969 was an average of nearly 150 ft. The major portion of the beach retreat has been attributed to fill material borrow (Hume and Schalk 1964b).

South of Barrow to Nunavak Bay, beaches abut against sea bluffs of the Gubik Formation. The beach foreshore is well developed, but the backshore is only developed where the beach crosses the mouths of estuaries. Beaches here are predominantly of dark, pebbly chert. The median diameter of particles is greater than those farther northeast because considerable quantities of gravel are present as lag deposits at the base of the bluffs. The bluffs contain sandy gravel and gravelly sand. About 35,000 to 40,000 cu yds of materials usually lie at the base of the bluffs, although they are occasionally stripped by severe storm waves. The materials act as an insulative cover for the ice-rich sediments in the bluffs. Their removal would result in thaw undercutting of the bluffs, and their collapse, accelerating coastal retreat.

At Nunavak Bay, 150,000 to 200,000 cu yds of fill materials have already been removed for the construction of the Barrow jet airport. Eighty thousand cubic yards or more of gravelly sand remain. The shore south from Nunavak Bay to Walakpa Bay is narrow and fronts on low banks. The material is mostly sand, in relatively small quantities. Three streams enter the sea in this area, and their mouths have small banks of gravel and sand. The northernmost stream mouth contains about 4000 cu yds of gravelly sand; the middle stream contains about 3800 cu yds of slightly gravelly sand; and the southern stream mouth contains about 1600 cu yds of sandy gravel. Walakpa Bay, the site of the Rogers-Post historical monument, contains about 150,000 cu yds of gravelly sand.

The shores south from Walakpa Bay to the Skull Cliff area are mostly sandy. Two small streams enter the sea here, followed by two larger bays. These stream mouths contain sandy gravel and gravelly sand resources. The northernmost stream contains about 7000 to 8000 cu yds of gravelly sand. The next stream to the south has about 15,000 to 20,000 cu yds of gravelly sand available. The bay near Nulavik contains 20,000 to 25,000 cu yds of sandy gravel. The bay near the old Skull Cliff Loran Station contains about 15,000 to 20,000 cu yds of gravelly sand.

Beaufort Sea shores

Shores east of Point Barrow are actively eroding by thaw action. Only sporadic, small accumulations of coarse materials are found on the mainland shore, on small, isolated beaches. East of Point Barrow, a spit and barrier island complex begins at Eluitkak Spit and runs east nearly to Cape Simpson. The entire complex has a maximum relief of 4 to 5 ft and contains nearly 4,000,000 cu yds of fill materials. At the western end, sediments are composed chiefly of coarse sand and fine gravel. Eastward the grade coarsens.

Eluitkak Spit, extending about 3 miles southeast from the tip of Point Barrow, consists of sediments lithologically similar to those of Point Barrow Spit. The spit contains about 10,000 cu yds of sandy gravel and gravelly sand.

The barrier islands east of Eluitkak Spit consist primarily of quartz sand and gravel, with minor amounts of chert. This change in lithologic makeup seems to indicate that the barrier islands are not a continuing part of the longshore transport system from the west, as are Point Barrow and Eluitkak Spit — but rather from the southeast.

The first barrier island east of Plover Point contains about 270,000 cu yds of gravelly sand. The Tapkaluk Islands, next in the group, contain about

800,000 cu yds of gravelly sand on the several islands. Cooper Island, the largest of the Plover Islands, contains over 2,000,000 cu yds of coarse materials. It is the prime prospective source for large quantities of gravel resources near Barrow.

The remainder of the island chain, southeast to near Cape Simpson, is composed of materials of approximately the same grade as Cooper Island. A total of about 700,000 cu yds of sandy gravel and gravelly sand are available there.

The beaches of Smith Bay consist only of sand and mud, including the extensive Ikpikpuk River delta. Farther east, narrow sand beaches border the coast of Avatanak Bight. From Avatanak Bight to Pitt Point, beaches contain about 1,100,000 cu yds of gravelly sand. From Pitt Point to the former Esook Trading Post, beaches contain about 550,000 cu yds of gravelly sand. From there to Cape Halkett, beaches are discontinuous and contain mostly sand with scattered short patches of gravel. Some gravelly sand does exist, however, in a short stretch of beach just north of Cape Halkett.

The shores of Harrison Bay, including the small offshore islands, have little beach development, and that which does exist is comprised totally of fine sand and mud. The extensive Colville Delta consists of only fine sand and mud.

Offshore

Offshore from the Chukchi beaches and the Beaufort barrier islands, coarse materials are found only to distances of about 30 ft from shore. Beyond that distance, only silts and muds exist. Several offshore bars, which migrate somewhat during storm periods, are composed only of silt and mud. Behind the barrier islands, the lagoons contain only fine materials except just offshore of the islands.

Inland

The Gubik Formation of the arctic coastal plain consists primarily of sandy silt, with only a scattering of gravel. There are some small concentrations of gravel within the Gubik, such as that found in the sea bluffs southwest of Barrow, but the sediments are in a permanently frozen condition. Their removal would be difficult, expensive, and would yield only small return in quantity.

Between Barrow and the Colville River, no inland streams contain any significant quantities of gravel. These sediments all originate either in the northern foothills or the coastal plain, and they all consist predominantly of medium to fine sand and mud.

The coastal plain land surface is interrupted here and there by topographic features of somewhat higher relief. Many of these were described by early investigators as ancient beach ridges, which were expected to contain sand and gravel resources. Most of them, however, proved to be initial surface residuals — remnants of the old initial coastal plain surface before it was dissected by thaw lake basins. The residuals are composed of the same sandy and silty materials of the Gubik Formation that make up the rest of the coastal plain.

One important exception to this is the arcuate relief feature just south of Point Barrow Spit, which has the same general outline as the present Point Barrow. Known as Central Marsh Ridge, it does appear to be an ancient beach ridge, and is composed of gravelly sand and sandy gravel overlain by about 18 inches of silt. This formation thaws only to a depth of about a foot by the end of summer, and excavation of its resources would be difficult and expensive.

Investigation was made of numerous other topographic highs, many having linear or arcuate shapes suggestive of beach ridges. Inspection was possible in most cases, because streams or thaw lakes had dissected part of the feature allowing exposure of the underlying sediments. None of these were underlain by any significant quantities of gravel; most were composed of sand and silt with an occasional scattering of gravel pebbles through the finer sediments.

Several cubic miles of medium to fine sand mantle much of the coastal plain south of the latitude of Teshekpuk Lake, from the Meade River to the Colville River. Active sand dunes lie along the western shores of many lakes and streams, and old, vegetated, stabilized sand dunes mantle most of the area. The predominant particle size within the dunes is medium to fine sand.

Beaches and bottoms of almost all thaw lakes are composed of sand and silt, but a large gravel beach containing about 900,000 cu yds of sandy gravel exists on the northwest shore of Teshekpuk Lake. Westward longshore transport along the north shore of the lake has built the beach from winnowed Gubik sediments bordering the lake. Most of the northern shore of the lake consists of a very narrow gravelly beach, but quantities there are too small to be of significance as a resource.

By far the greatest resources of gravel in the region are contained in the bars and beaches of the Colville River. At least 35,000,000 cu yds of sandy gravel and gravelly sand lie within the active river bed north of latitude 70° N, and more lies within the old floodplain. Gravel bars lie in the river as far north as the apex of the Colville Delta. North of there, only small scattered concentrations of a few dozen cubic yards each exist in the delta proper, the result of the winnowing of local Gubik sediments in the river banks. The predominant materials in the delta are sand and silt. South from the delta apex, the material coarsens.

REFERENCES

BLACK, R. F.

    1951   Eolian deposits of Alaska. *Arctic* 4: 89-111.

    1964   Exploration of Naval Petroleum Reserve No. 4 and adjacent areas, northern Alaska, 1944-53. Part 2, Regional studies: Gubik Formation of Quaternary age in northern Alaska. U. S. Geological Survey Professional Paper 302-C, pp. 59-91.

HOPKINS, D. M.

1967 Quaternary marine transgressions in Alaska. In *The Bering land bridge*, edited by D. M. Hopkins. Stanford Univ. Press, Stanford, California, pp. 47-90.

HUME, J. D.

1965 Shoreline changes near Barrow, Alaska caused by the storm of October 3, 1963. Subcontract ONR-343, Final report to Arctic Institute of North America, Washington, D. C.

HUME, J. D., and M. SCHALK

1964a The effects of ice-push on arctic beaches. *American J. of Science* 262: 267-273.

1964b The effects of beach borrow in the Arctic. *Shore and Beach* 32: 37-41.

1967 Shoreline processes near Barrow, Alaska: a comparison of the normal and the catastrophic. *Arctic* 20: 86-103.

HUME, J. D., M. SCHALK, and P. W. HUME

1972 Short-term climate, changes and coastal erosion, Barrow, Alaska. *Arctic* 24(4): 272-278.

HUSSEY, K. M, and R. W. MICHELSON

1966 Tundra relief features near Barrow, Alaska. *Arctic* 19: 162-184.

LaBELLE, J. C.

1973 Fill materials and aggregate near Barrow, Naval Petroleum Reserve No. 4, Alaska. ONPR Contract NOd-9915, Arctic Institute of North America, Washington, D. C., 147 pp.

1974 Fill materials and aggregate in the Cape Halkett region, Naval Petroleum Reserve No. 4, Alaska. ONPR Contract NOd-9915, Arctic Institute of North America, Washington, D. C., 101 pp.

LEWELLEN, R. I.

1970 Permafrost erosion along the Beaufort Sea coast. Littleton, Colorado. Published by the author.

O'SULLIVAN, J. B.

1961 Quaternary geology of the arctic coastal plain. Ph.D. Thesis, Iowa State University, Iowa City, Iowa.

REX, R. W.

1953 Uplifted beach ridges and first generation lakes in the Barrow, Alaska area. Final report, section 2, Contract NONR 225 (09).

1964 Arctic beaches, Barrow, Alaska. In *Papers in marine geology,* edited by R. L. Miller, MacMillan Press, New York, pp. 384-400.

CHAPTER **14**

Sediment transported by ice-rafting in southcentral Alaska

C. M. HOSKIN[1] *and* S. M. VALENCIA[2]

Abstract

Transport of sediment by ice-rafting includes such mechanisms as icebergs calved from glaciers, floes in polar seas, anchor-ice and overflow ice in rivers, and the shore ice of bays and beaches. Mobilization of sediment from ice by melting, settling of sediment through the water column, and concomitant lateral transport of suspended sediment by fjord circulation results in physical separation of grainsize classes. In this chapter, the authors describe the grainsize characteristics of accumulated sediment, which is supplied in large part by icebergs calved from modern valley glaciers, and compare the grainsize distribution of sediment on and in icebergs with that of bay-floor sediment in an Alaskan coastal environment.

INTRODUCTION

The fjord environment of Yakutat-Disenchantment bays, Alaska, (Fig. 14.1), contains all elements essential for sediment ice-rafting and, because of the density of sampling, is presented as a convenient example. This coastal region has been reviewed by Rosenberg (1972) and by Wright and Sharma (1969). Thickness (about 240 m) and distribution of sediment are known from continuous seismic reflection profiles obtained with a 5000-joule arcer (Wright 1972).

As shown in Figure 14.2 (a-d), the Turner and Hubbard glaciers actively calve icebergs into Disenchantment Bay. Abundance of icebergs in Disenchantment Bay is largely a function of wind; onshore wind action concentrates the bergs, and offshore winds disperse them over a period of a

---

[1] *Institute of Marine Science, University of Alaska, Fairbanks, Alaska 99701.*
[2] *Department of Geology, Butte College, Oroville, California 95965.*

174

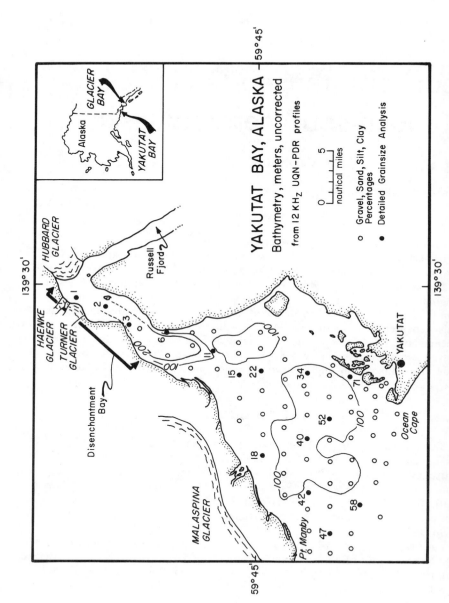

Fig. 14.1 Location of sediment samples and generalized bathymetry of Yakutat Bay. Adapted from Wright (1972); base from Coast & Geodetic Survey, U.S. Dept. of Commerce (1968).

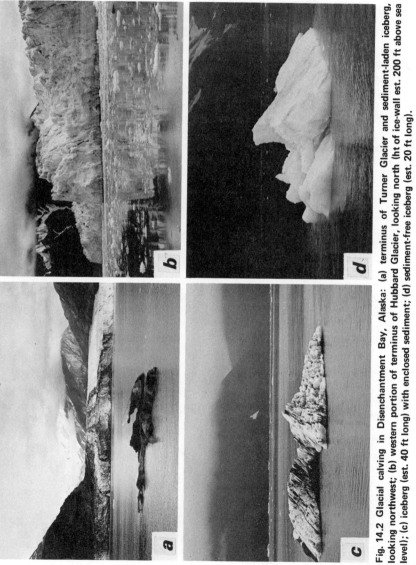

Fig. 14.2 Glacial calving in Disenchantment Bay, Alaska: (a) terminus of Turner Glacier and sediment-laden iceberg, looking northwest; (b) western portion of terminus of Hubbard Glacier, looking north (ht of ice-wall est. 200 ft above sea level); (c) iceberg (est. 40 ft long) with enclosed sediment; (d) sediment-free iceberg (est. 20 ft long).

few hours (R. D. Muench, personal communication, 1974). Hubbard Glacier is advancing at an unusually rapid rate (Miller 1964) which, if continued, could cause closure of the connection between Russell Fjord and Disenchantment Bay (W. S. Reeburgh, personal communication, 1974). Calved bergs move down-bay, away from the glaciers, due to the combined effects of surface freshwater flow, wind, and tides. Most of the bergs disintegrate in Yakutat Bay, and few escape to the Gulf of Alaska. Sediment accumulated in Yakutat Bay can be confidently assigned a result of ice-rafting because of the observed supply, movement, and melting of sediment-laden icebergs. Other sediment sources are not known to be important; however, Malaspina Glacier may contribute sediment by east-flowing streams from Malaspina Lake (Gustavson 1974). Burbank (1974) showed, from density-slicing of ERTS-A imagery, a relative increase in suspended sediment in surface water of western Yakutat Bay. This suggests that Malaspina Glacier supplies sediment to Yakutat Bay by stream flow from Malaspina Lake. Although biogenic particles are rare in the bulk of bay-floor sediment, living brachiopods and stony red corals were observed attached to pebbles recovered in grab and dredge hauls near islands in southeastern Yakutat Bay.

BAY-FLOOR SEDIMENT

Sediment was collected by Shipek grab or pipe dredge on a 2 x 2–mile grid (Fig. 14.1) and stored wet, without preservatives, in polyethylene bags. Grainsize analyses were made on aliquots of salt-free and organic-free sediments using sieves for gravel and sand, and pipets for mud (Folk 1968). Data for abundance of gravel, sand, silt, and clay are available for 69 samples (Wright 1972).

Size classes. The presence of gravel in these 0.5-kg samples limits the usefulness of grainsize data for some samples. Three of fifteen samples contained one piece of gravel, constituting 59, 33 and 29 percent by weight of each sample, respectively. Folk (1968) suggested a minimum of 1 kg of gravel to insure meaningful grainsize data for gravels, and Slatt (1971) found it necessary to sieve 5 to 10-kg samples in the field to determine grainsize characteristics for gravelly ice-contact sediments of modern valley glaciers. Therefore, emphasis in this chapter is placed on sand and mud analysis.

Average sand (2.00 to 0.0625 mm) content is 13.2 percent, with 44 of 69 samples containing an average of 2.5 percent sand. Two samples contain no sand.

Average silt (0.0625 to 0.0039 mm) content is 31.5 percent. A frequency distribution plot shows the most common silt content to be between 30 and 40 percent; four samples contain no silt. Clay (<0.0039 mm) content averages 37.7 percent, with about one-fifth of the samples containing 50 to 60 percent clay. Six samples have no clay. Average mud (silt plus clay) content is 69.2 percent. Geographic variations in grainsize characteristics of surface sediment are not striking, except for greater gravel and sand content at the bay mouth. Wright (1972) has suggested this is due to the presence of a bay-mouth moraine.

Grainsize names. The most common sediment type (26 of 69) is slightly gravelly mud, using the nomenclature of Folk (1954); 52 of 69 are muds, 15 are gravels, and 2 are sands.

Grainsize modes. Complete grainsize analyses are available for 15 samples; 30 grainsize modes were identified in frequency curves generated from arithmetic-ordinate cumulative curves for these samples (Fig. 14.3). The most common particle size in the spectrum between 64.00 and 0.002 mm occurs at 0.0055 mm (silt) in 7 of 30 modes. No recurrent modal sizes were seen for sand (six modes). For sand, five samples have well-developed grainsize modes, eight have no modes, and two samples do not contain particles through the sand range. Of 15 samples, 12 are bimodal or polymodal.

Statistical parameters. Probability-ordinate cumulative curves were used to derive descriptive statistical parameters (Table 14.1) using the formulas of Folk and Ward (1957). As necessary, cumulative curves were linearly extrapolated from 10.0 ϕ to 100.0 percent at 14.0 ϕ (Folk 1966).

The most common graphic mean size is about 8.3 ϕ (0.0032 mm); the mean size for 10 of 15 samples ranges between 7.65 to 8.82 ϕ (0.005 to 0.0022 mm). Total range of graphic mean size is -2.37 to 8.82 ϕ (5.2 to 0.0022 mm). These sediments are very poorly sorted with the average graphic standard deviation for 12 of 15 samples 2.12 ϕ, ranging from 1.73 to 2.38 ϕ. Total range of sorting is 1.73 to 5.79 ϕ. About half the samples are negatively skewed, six are near-symmetrical, and two are positively skewed. Total range of skewness is extreme: -0.41 to +0.97. All but one sample is platykurtic, with the average centering near a normalized graphic kurtosis of 0.45; total range of normalized kurtosis is 0.39 to 0.59.

ICEBERG SEDIMENT

Seven 1 to 2-kg samples and two 8 to 10-kg samples of ice were chipped from nine different icebergs, and the enclosed sediment was recovered by melting. Two samples (8 and 10) were lag deposits on flat-surfaced icebergs. Storage and analysis of iceberg sediment were identical to that for bay-floor sediment.

Size classes. The average amount of gravel found in these iceberg sediments was 29.0 percent. Sample size limits the usefulness for gravel data, as only 3 of 11 samples exceeded 1.5 kg dry weight sediment (Table 14.1). Sand content ranged between 18.5 and 55.8, averaging about 40 percent. Silt and clay contents averaged 22.6 and 8.3 percent, respectively, with the average mud content being 30.9 percent.

Grainsize names. The most common type of sediment (5 of 11 samples) was muddy or muddy sandy gravel. Four samples were gravelly muddy sand, and two samples were gravelly muds.

TABLE 14.1 Sediment descriptors for Yakutat Bay

Sample no.	Sample wt(g)	Weight percent				Grainsize descriptors[5]				Grainsize name[6]	Grainsize modes (mm)
		Gravel[1]	Sand[2]	Silt[3]	Clay[4]	$M_z\phi$	$\sigma_I\phi$	Sk_I	K'_G		
Sediment on icebergs (lag deposit)											
213-8	1687.6	61.0	18.5	13.0	7.5	-0.98	4.73	-0.34	0.44	muddy gravel	0.053
213-10	1628.1	31.4	44.1	16.7	7.8	1.61	4.35	+0.15	0.49	muddy sandy gravel	0.25, 0.044
Sediment in icebergs											
213-1	49.1	15.9	48.3	24.5	11.3	2.71	3.84	+0.26	0.48	gravelly muddy sand	0.053
213-2	48.8	27.5	39.8	20.9	11.8	2.03	4.46	+0.21	0.47	gravelly muddy sand	none
213-4	1535.5	35.0	35.8	19.2	9.9	1.33	4.63	+0.16	0.46	muddy sandy gravel	8.0, 0.125, 0.053
213-5	62.1	15.6	40.4	26.6	17.4	3.51	4.58	+0.20	0.50	gravelly mud	0.053
213-6	153.1	32.1	40.8	22.4	4.8	1.15	4.03	-0.16	0.43	muddy sandy gravel	0.15
213-7	269.2	24.1	55.8	17.4	2.7	1.15	3.54	-0.33	0.53	gravelly muddy sand	0.18, 0.053
213-9	98.5	54.0	26.5	15.9	3.6	-0.64	4.08	-0.23	0.40	muddy sandy gravel	0.053
213-11	40.0	29.6	37.4	31.0	2.0	1.41	3.46	+0.23	0.42	gravelly muddy sand	0.053, 0.026
213-12	265.2	27.6	35.9	25.7	10.8	2.03	4.75	+0.03	0.46	gravelly mud	0.35, 0.053
			$\bar{x}$ 40	$\bar{x}$ 23	$\bar{x}$ 8	s:m = 40:31; about 1:1.					
Bay-floor sediment[7]											
1		0	14.3	53.8	31.9	7.65	2.28	-0.08	0.39	sandy mud	0.15, 0.037, 0.016, 0.003
2		.02	.9	50.7	48.4	8.04	2.13	+0.10	0.42	slightly gravelly mud	0.026, 0.008
3		.9	3.6	48.2	47.3	8.06	2.11	-0.29	0.52	slightly gravelly mud	0.053, 0.0055

TABLE 14.1 (continued)

Sample no.	Sample wt(g)	Weight percent				Grainsize descriptors[5]				Grainsize name[6]	Grainsize modes (mm)
		Gravel[1]	Sand[2]	Silt[3]	Clay[4]	$M_z\phi$	$\sigma_I\phi$	Sk_I	K'_G		
Bay-floor sediment[7] (Continued)											
6		.8	1.8	48.5	48.5	8.08	2.04	+0.04	0.39	slightly gravelly mud	0.053, 0.019, 0.008
11		0	25.2	31.3	43.6	6.04	4.43	-0.38	0.46	sandy mud	0.037, 0.0055
15		1.1	2.7	41.1	55.1	8.36	2.21	-0.06	0.44	slightly gravelly mud	0.008
18		1.7	1.9	33.2	63.3	8.80	1.95	-0.33	0.45	slightly gravelly mud	0.0055
22		0	1.4	39.2	59.5	8.82	1.98	-0.25	0.43	mud	0.0055
34		0	2.0	40.3	57.7	8.76	2.38	-0.26	0.44	mud	0.0055
40		.02	1.7	43.7	55.5	8.27	2.37	-0.13	0.43	slightly gravelly mud	0.053, 0.0055
42		65.3	2.8	17.0	14.9	-1.19	5.79	+0.97	0.38	muddy gravel	0.044
47		0	38.0	47.8	14.3	6.43	3.50	-0.01	0.39	sandy silt	0.15, 0.004
52		0	4.5	39.9	55.6	8.48	2.11	-0.04	0.42	mud	0.074, 0.0055
58		82.3	17.8	0	0	-2.24	1.73	-0.41	0.39	gravel	2.38, 1.00
71		80.5	16.0	2.0	1.5	-2.37	2.20	+0.58	0.59	gravel	9.4, 4.8, 0.71, 0.15

$\bar{x}$ 9 $\bar{x}$ 36 $\bar{x}$ 40 s:m = 9:16, about 1:8.

[1] Particles: > 2.00 mm
[2] 2.00 to 0.0625 < mm
[3] 0.0625 to 0.0039 mm
[4] Particles < 0.0039 mm
[5] After Folk and Ward (1957)
[6] After Folk (1954)
[7] See Wright (1972, Table 1) for size-class data on 69 samples from Yakutat Bay.

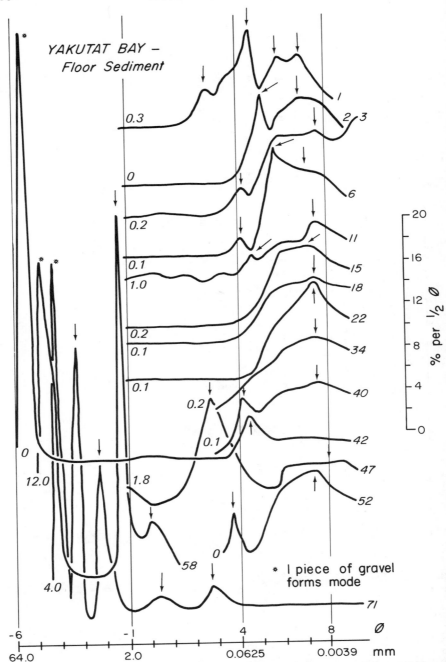

Fig. 14.3 Size-frequency distribution for Yakutat bay-floor sediments. Number at left of each curve is the y-axis intercept in percent per one-half φ, arrows along each curve mark position of grainsize modes, and number at right of curve is the sample number.

Grainsize modes. Frequency curves were generated from arithmetic-ordinate cumulative curves (Fig. 14.4), with due regard given to gravel data being limited by sample size. Inspection of these curves gives an overall impression of relatively flat distributions. Sixteen modes occur in these eleven samples; eight modes are believed to be real, and eight (at 0.053 mm) are suspected to be artifacts due to the change in analytical method from sieves to pipets at 0.0625 mm. Pipet analysis was done with the hydrometer cylinders immersed in a 27° C thermostatically-controlled water bath. Reproducibility of pipetting was excellent; three separate aliquots of sample 213-4 gave data points which were indistinguishable on the cumulative curve. This grainsize mode (0.053 mm) appeared in 3 of 15 bay-floor sediments (Fig. 14.3).

Statistical parameters. Probability-ordinate cumulative curves were used for these computations, extrapolated to 100 percent at 14.0 ϕ. Average graphic mean size was 0.32 mm, with the total range occurring between 1.98 and 0.088 mm. Average graphic standard deviation was 4.15 ϕ (extremely poorly sorted), with all data clustering between 3.46 and 4.73 ϕ. Data for skewness were scattered; seven samples were positively skewed, and four were negatively skewed. Most samples were platykurtic, four were mesokurtic, and one was leptokurtic.

DISCUSSION

Comparisons between sediment in-transit via icebergs in Yakutat Bay with sediment on the floor of Yakutat Bay yielded two observations: bay-floor sediment was enriched in mud, and most bay-floor muds had well-developed grainsize modes.

Mud enrichment of bay-floor sediment was documented in terms of the relative abundance of sand to mud on a weight basis. The sand:mud ratio was approximately 1:1 (ranging from 0.9 to 2.8:1) for sediment recovered from icebergs, but there was a shift to much larger proportions of mud in bay-floor sediment. Using all data for bay-floor sediment (65 samples, as 4 samples contained either no sand or no mud), the sand:mud ratio fell between 0.03 and 249. The distribution of this ratio was skewed, and the largest number of samples (26, or 40 percent of the total) were found in the class 0 to 9.99. For these 26 samples, the sand:mud ratio was the same as that for sediment recovered from Yakutat Bay icebergs. However, the sand-to-mud ratio exceeded 10 ($\overline{X}$ = 82) for 39 bay-floor sediments, and these are the basis for mud enrichment. Simply stated, if sediment were delivered dumptruck style to the bay-floor, there should be no difference in the sand:mud ratio for iceberg and bay-floor sediment. Observing that more than half of the bay-floor sediments have larger proportions of mud than are found in their source, we are led to the proposition that some process of mud enrichment has been active.

It may be inaccurate to compare the upper few centimeters of bay-floor sediment with the bulk of sediment from icebergs. A possible mechanism for mud enrichment of the bay-floor sediment is that gravel and sand settling

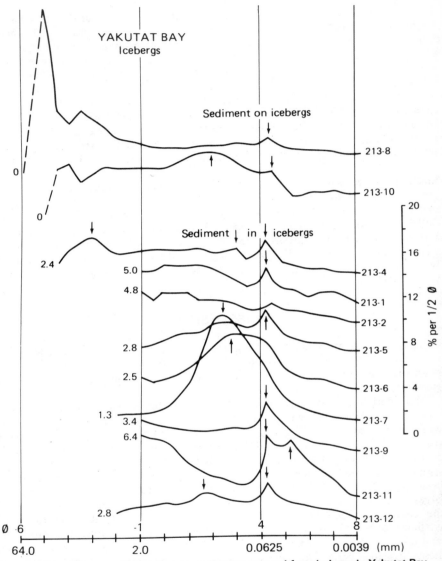

Fig. 14.4 Size-frequency distributions for sediment recovered from icebergs in Yakutat Bay.

from a melting iceberg may fall into the fluid-like layer of mud at the water-sediment interface. Measurements of *in situ* mud strength, experiments with pebbles and sand dropped from known heights through a water column, and information on sedimentary structures as obtained from box cores will prove useful in determining effects of the fluid mud layer on sediment distribution. Such data are lacking at present. Also, there may be net transport of mud suspended in the water column towards the bay head due to fjord circulation, and there is possibly a tendency for mud to be readily mobilized from icebergs with coarser particles tending to accumulate on iceberg surfaces. The general tendency of icebergs to move towards the bay mouth may result in mud enrichment of the bay head, with coarser sediment being more abundantly supplied toward the bay mouth.

Sediment from modern Alaskan valley glaciers follows one of two main pathways from glacier to fjord. For those glaciers having a terrestrial terminus, sediment is mostly transported to the marine environment by melt-water streams (Hoskin and Burrell 1972). In studying unreworked ice-contact sediment at the snout of Alaskan terrestrially-terminated glaciers, Slatt (1971, Fig. 2, p. 832) found this sediment to be nearly free of grainsize modes. For glaciers with a tidewater terminus, sediment enclosed in ice is calved directly into the sea at the glacier snout. Our data for sediment recovered from icebergs in Yakutat Bay show strong grainsize modes in 3 of 11 samples (excluding the gravel range). The origin of grainsize modes in iceberg sediments is presently not known. Previous transport by wind or englacial streams seems to be the likely reason, although there are insufficient data to evaluate the idea. The presence of well-developed grainsize modes in Yakutat bay-floor sediments, especially the muds, suggests that these modes are the result of some process, as yet unidentified, operating in the period between release of sediment from the ice and deposition on the bay- floor.

Ovenshine (1970, p. 893) suggested that "clusters of rafted stones" would indicate ice-rafting in ancient sedimentary rocks. Pebbles in mud have been found in Yakutat bay-floor sediment. Almost certainly, these pebbles were ice-rafted, as, with the exception of sea lions, other pebble-transporting mechanisms (reviewed by Emery 1963) appear to be unimportant in Yakutat Bay.

Ovenshine (1970) also suggested that 2- to 3-mm pellets of sediment, formed as intra-crystal fillings in glacial ice, would be diagnostic of ice-rafted sediment. Pellets were not observed in the Yakutat Bay sediments studied here, although wet-sieving was done carefully in a specific search for the pellets. Pellets similar to those described by Ovenshine (1970) were seen in two of nine sediment samples from a stranded iceberg in Queen Inlet, Glacier Bay, and in one of seven samples from icebergs in Derickson Bay, western Prince William Sound. Different bedrock lithologies or different glaciers may yield more abundant pellets. Some of the mud pellets may be of biological origin, formed by sediment-eating invertebrates on the bay-floor and transported to icebergs in the alimentary tracts of seals.

CONCLUSIONS

Ice-rafted sediment accumulating on the floor of Yakutat Bay is dominantly mud, with an admixture of gravel and sand. Sediment recovered from icebergs in Yakutat Bay contains about equal proportions of sand and mud.

Assuming these data are comparable gives rise to the suggestion that some mechanism(s) causing separation of grainsize classes is operating between the time that sediment is released from icebergs and finally settles on the bay-floor.

Conference Discussion

Unidentified — Since icebergs move primarily along the northwest shore of Yakutat Bay, did you notice a higher proportion of larger grainsizes along that shore?

Hoskin — No, our data show no increased gravel or sand abundance in western Yakutat Bay.

Acknowledgments

This study was supported, in part, through the Heavy Metal Program of the U. S. Geological Survey, Contract no. 14-08-001-10885 to the Institute of Marine Science, University of Alaska, and by the National Science Foundation, Grant no. 40159X. We wish to thank F. F. Wright, R. Von Huene, and the officers and crew of the R/V *Acona* for skillful assistance in the field. S. Short and R. Nelson helped with laboratory analyses. W. S. Reeburgh reviewed the manuscript, and we thank him for his helpful suggestions. Contribution no. 282, Institute of Marine Science, University of Alaska.

REFERENCES

BURBANK, D. C.

    1974   Suspended sediment transport and deposition in Alaskan coastal waters. M.S. Thesis, Univ. Alaska, Fairbanks, p. 42 (Fig. 21).

EMERY, K. O.

    1963   Organic transportation of marine sediments, In *The sea*, edited by M. N. Hill. Interscience, New York, Vol. 3, pp. 776-793.

FOLK, R. L.

    1954   The distinction between grainsize and mineral composition in sedimentary-rock nomenclature. *J. Geology* 62: 344-359.

    1966   A review of grainsize parameters. *Sedimentology* 6: 73-93.

    1968   *Petrology of sedimentary rocks*. Hemphill's Book Store, Austin, Texas, 170 pp.

FOLK, R. L., and W. C. WARD

1957   Brazos River Bar: a study in the significance of grainsize para-
       meters. *J. Sed. Pet.* 27: 3-26.

GUSTAVSON, T. E.

1974   Sedimentation of gravel outwash fans, Malaspina Glacier
       Foreland, Alaska. *J. Sed. Pet.* 44: 374-389.

HOSKIN, C. M., and D. C. BURRELL

1972   Sediment transport and accumulation in a fjord basin, Glacier
       Bay, Alaska. *J. Geology* 80: 538-551.

MILLER, M. N.

1964   Inventory of terminal position changes in Alaskan coastal
       glaciers since the 1750s. *Proc. American Philos. Soc.* 108:
       257-273.

OVENSHINE, A. T.

1970   Observations of iceberg rafting in Glacier Bay, Alaska, and the
       identification of ancient ice-rafted deposits. *Geol. Soc. America
       Bull.* 81: 891-894.

ROSENBERG, D. H. (Ed.)

1972   A review of the oceanography and renewable resources of the
       northern Gulf of Alaska. Report No. R72-23, Inst. Mar. Sci.,
       Univ. Alaska, Fairbanks, 690 pp.

SLATT, R. M.

1971   Texture of ice-cored deposits from ten Alaskan valley glaciers.
       J. Sed. Pet. 41: 828-834.

UNITED STATE DEPARTMENT OF COMMERCE

1968   Yakutat Bay, Coast and Geodetic Survey, Chart 8455, 8th ed.,
       1: 800,000 Washington, D. C.

WRIGHT, F. F.

1972   Marine geology of Yakutat Bay, Alaska. U. S. Geol. Surv.
       Prof. Paper 800-B, pp. B9-B15.

WRIGHT, F. F., and G. D. SHARMA

1969   Periglacial marine sedimentation in southern Alaska. Union
       Internationale pour l'Étude du Quaternaire, VIII Congrès
       Inqua, Paris, pp. 170-185.

dynamic physical processes

Assessment of the Arctic Marine Environment: Selected Topics
Copyright © 1976 by Institute of Marine Science, University of Alaska, Fairbanks

CHAPTER **15**

Dynamic sedimentological processes along the Beaufort Sea coast of Alaska

J. A. DYGAS *and* D. C. BURRELL [1]

Abstract

Processes of coastal erosion, sediment transport, and relevant environmental impact considerations are reported for Simpson Lagoon, a representative part of the Beaufort Sea (Arctic Ocean) coastline of Alaska. The study area, which lies just east of Harrison Bay and the Colville River delta, is bordered seaward by the Jones Islands, and further eastward by the Kuparuk River and Gwyder Bay. Principal coastal erosional processes include formation and collapse of thermo-erosional niches along beach cliffs by the effects of storm waves currents, and high tides. These processes have resulted in a mean rate of erosion of 1.4 m/yr along the Simpson Lagoon coast. The longshore movement of sandy gravel sediment along this coast is predominantly from east to west. In addition, a significant linear correlation coefficient of +0.5 was obtained between the east-west components of longshore sediment transport rates and wind speeds. Aerial photographic and ground truth observations indicate a net westward migration of some of the offshore barrier islands, particularly Pingok and Thetis islands, at rates of 6.0 to 40.0 m/yr and 26.0 m/yr, respectively. Frequent late summer storms have the potential to cause flooding on both barrier islands and the low-lying coast, thereby potentially endangering coastal facilities such as roads and gravel foundation pads by erosion. The movement of sediment in these shallow coastal waters, as well as the effects of wind tides on local water depths, are conducive to shifting shoals and consequently pose a potential danger to coastal navigation and shipping routes.

---

[1] *Institute of Marine Science, University of Alaska, Fairbanks, Alaska 99701.*

INTRODUCTION

The Institute of Marine Science of the University of Alaska has conducted a multi-disciplinary study (Alexander et al. 1974; Dygas 1974) of the near-shore aquatic environment along a part of the Beaufort Sea coast of Alaska (Fig. 15.1). As part of this program, studies of coastal erosional processes and sediment transport in an arctic environment were undertaken. The objectives of this study were to describe the general arctic coastal erosional processes; rates and directions of movement of the nearshore sediments in relation to the local wind regime in this specific locality; and the potential relationship between these dynamic sedimentological processes and man's environmental activities. Sedimentological processes are most active during the interim between the thaw and break-up of lake, river, and shorefast ice in late May and June and their freezeup in early October. The long (7 to 8 month), subfreezing arctic winter essentially preserves nearshore geomorphological features until the following spring breakup.

Arnborg et al. (1966, 1967) and Reimnitz and Bruder (1972) have described local morphological and sediment dispersal processes of river and shorefast ice in the vicinity of the Colville River delta at the time of the spring thaw, and Barnes and Reimnitz (1972) have described the breakup process of the Kuparuk River, which lies just to the east of the Simpson Lagoon study area (Fig. 15.1). In addition, Wiseman et al. (1973) have detailed the events occurring at this time of year at Point Lay and Pingok Island, in their comparative study of geomorphological processes along the northwest and northern arctic coasts of Alaska.

Schrader (1904) and Leffingwell (1919) initiated geological mapping along Alaska's arctic coast, but it was not until the work of MacCarthy (1953) that specific studies concerning rates of shore erosion were first made in the Point Barrow area. R. W. Rex and E. J. Taylor (unpublished manuscript, 1953) also studied the annual beach cycle at this locality. The effects of beach borrow, ice push and storm effects on beaches at Point Barrow have been subsequently reported by Hume and Schalk (1964, 1967). More recently, Lewellen (1970, 1972) has made studies of permafrost erosion along parts of the Beaufort Sea coast of Alaska, and Walker (1969) and Walker and Arnborg (1963) have also studied riverbank erosion in the Colville River delta.

STUDY LOCALITY

The beaches studied in this report are those of the Simpson Lagoon coast and some of the Jones Islands (Fig. 15.2). Coastal waters here are generally quite shallow, with depths of less than 10 m up to 7 km from the coast-line. Water depths in Simpson Lagoon are generally less than 3 m, increasing to 8 m in some places within the channels between the islands. Elevations along the Simpson Lagoon coast and on the Jones Islands average less than 4 m. A few of the Jones Islands (such as Pingok Island) are partially covered by tundra, with permafrost exposed in low beach cliffs. Other islands, such as Spy and Thetis, are surfaced by gravelly sands, and tundra is absent. Most of the major headlands along the coast, including Oliktok,

191

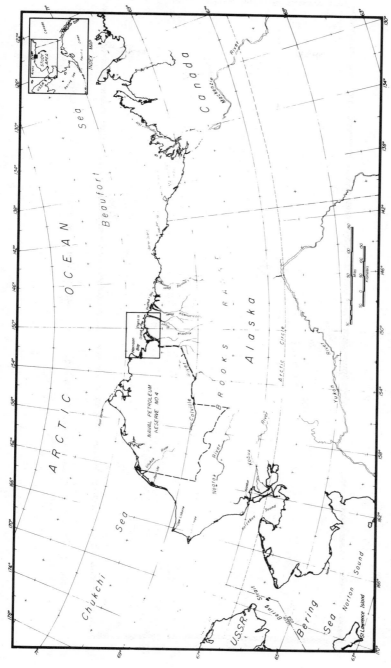

Fig. 15.1 Locality of study area on Beaufort Sea (Arctic Ocean) coast of Alaska.

192

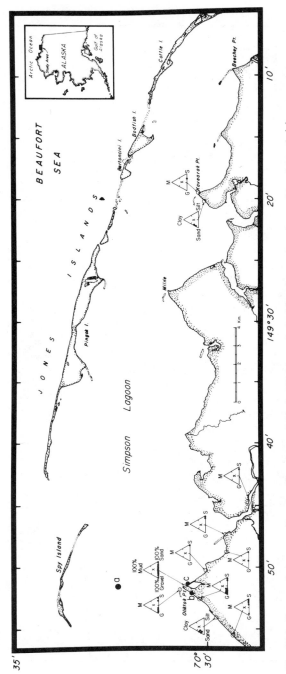

Fig. 15.2 Simpson Lagoon on Beaufort Sea coast of Alaska showing principal beach sampling sites and localities of: **(a)** current meter; **(b)** anemometer and wind vane, **(c)** wave recorder.

Milne, Kavearak, and Beechey Points (Fig. 15.2), have elongated spits and submerged bars which trend westward or northwestward. The two major promontories on the lagoon side of Pingok Island also have bars and spits extending in a southwesterly direction.

The beach sediment sampling sites are marked on Figure 15.2 with triangular sediment classification diagrams. Point *a* in Simpson Lagoon indicates the locality of an Aanderaa current meter installation (in 2.5 m of water); point *b* shows the locations of an R. M. Young anemometer and wind vane mounted 10 m above sea level on the northwest shore of Oliktok Point. A single Interstate Electronics continuous-resistance wire wave staff was mounted at point *c* in shallow water on the northeast shore of this same promontory.

METHODS

Mean rates of sediment erosion over the previous several decades along the Simpson Lagoon coast have been determined from a comparative study of aerial photos taken in 1949 and 1955 by the U. S. Navy and in 1971 and 1972 by personnel of the Naval Arctic Research Laboratory at Point Barrow. Short-term and seasonal changes occurring in 1971 and 1972 were determined in beach topography from a series of profile measurements using the modified horizon technique of Emery (1961) at four sites along both shores of Oliktok Point, and also at the western end of the spit extending from this point.

At least twice daily over the periods 24 July to 22 August 1971 and 15 August to 12 September 1972, measurements were made of breaking wave heights, wave period, angle of approach of the breaking waves, longshore current velocities, and wind speed and direction. From these data, together with beach profiles and sediment size analyses, estimates of the longshore energy flux and Inman and Bagnold's (1963) immersed weight energy flux were calculated as described in Dygas et al. (1972). Approximations of the volume of longshore sediment transported were determined empirically using a design curve which correlated volume of sediment transport with the longshore component of wave energy flux (U. S. Coastal Engineering Research Center 1966).

RESULTS AND DISCUSSION

Coastal erosion

Permafrost horizons and ice wedges are exposed in beach cliffs and banks along the entire Beaufort Sea coast, and the thawing of this material, in response to extended periods of above-freezing ambient temperatures during the summer months, gives rise to the major, non-localized, erosional processes. The shallow surface layer of tundra and the immediately underlying sediment, which freezes and thaws annually, is termed the active layer. The latter was observed to be 1 m thick landward of Simpson Lagoon.

Thermo-erosion gives rise to such geomorphological features (Lewellen 1972) as frost heave and subsidence, thermokarst topography, niche formation, and saturated soil flow. Frost heave is a direct consequence of the volume increase accompanying growth of ice crystals from interstitial water in the active layer.

The amount of frost heave depends on factors such as heat loss, sediment particle size, moisture content, and soil matrix conditions. Thermokarst topography is due to the irregular settling and collapse of the soil and underlying rock layers following thawing of subterranean ice.

Rates of coastal erosion along the arctic coast are accelerated by the formation of thermo-erosional niches and slumping of beach cliff material. Walker and Arnborg (1963) have described the formation of such niches as the result of the thawing of ice in an ice-cemented sediment. This material is subsequently eroded and transported away by high storm surges and associated wave and current action on the beach. Erosion of beach cliffs in this fashion is increased further by the effect of saturated soil flow. The latter is a response to the thawing of ice-cemented surficial sediment and subsequent saturation of the soil because of the increased moisture content. On an exposed surface, such as a beach cliff, the soil looses its cohesiveness and flows downslope onto the back shore of the beach.

It is calculated that the Simpson Lagoon coast is eroding at an approximate mean rate of 1.4 m/yr (see data given in Fig. 15.3), a value comparable with those reported by other investigators along the Beaufort Sea coast. Lewellen (1972) has given 1.3 m/yr as the mean rate of coastal erosion along the Elson Lagoon coast, which lies eastward of Point Barrow. This author also cites maximum erosion rates of 10.4 m/yr at the latter locality, and 10.0 m/yr on the north side of Flaxman Island, 300 km to the east of Point Barrow.

Wiseman et al. (1973) have suggested that the eastern end of Pingok Island (Fig. 15.2) is eroding at some 40.0 m/yr, based on observations during the summer of 1972. Dygas et al. (1972), however, have given an equivalent value of 6.0 m/yr, as shown in Figure 15.4, which represents mean erosion over a 22-year period. Sediment eroded from the east end of this island is generally being transported to the western end, thus resulting in a net westward migration of the entire island.

Further clear evidence for the net westward transportation of sediment in this area is provided by physiographic changes on Thetis Island off the Colville River delta (Fig. 15.1). This island lacks a tundra cover and is a depositional sink for sediment transported from the east along the Jones Islands and for sediment discharged from the Colville River. Thetis Island has grown at a mean rate of approximately 26.0 m/yr over the 22-year period to 1971 (Fig. 15.5). Orientation in response to the prevailing summer northeasterly winds should be noted.

Longshore sediment transport

During the observation period, 24 July through 22 August 1971, mean breaking wave heights were measured in the range of 5.8 to 32.2 cm, with

195

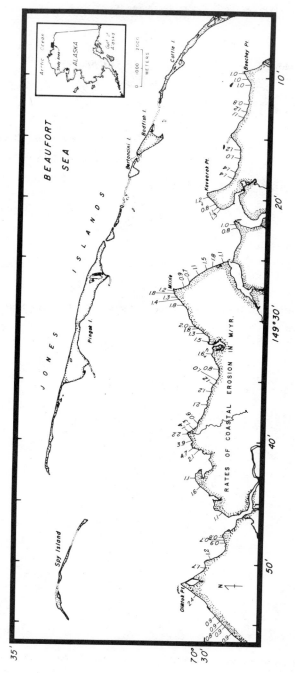

Fig. 15.3 Erosion rates (m/yr) along the Simpson Lagoon coast.

PINGOK ISLAND

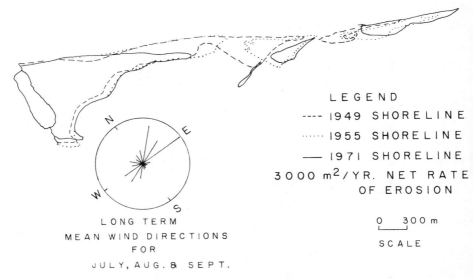

LEGEND

---- 1949 SHORELINE

······ 1955 SHORELINE

——— 1971 SHORELINE

3000 m^2/YR. NET RATE OF EROSION

LONG TERM
MEAN WIND DIRECTIONS FOR
JULY, AUG. & SEPT.

0   300 m

SCALE

Fig. 15.4 Migration of Pingok Island over the period 1949-1971. Wind data from Searby and Hunter (1971).

THETIS ISLAND

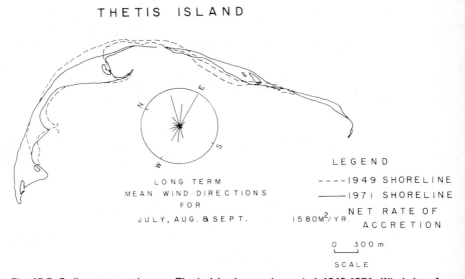

LONG TERM
MEAN WIND DIRECTIONS FOR
JULY, AUG. & SEPT.

LEGEND

- - - - 1949 SHORELINE

——— 1971 SHORELINE

1580M^2/YR. NET RATE OF ACCRETION

0   300 m

SCALE

Fig. 15.5 Sediment accretion on Thetis Island over the period 1949-1971. Wind data from Searby and Hunter (1971).

periods of 1.3 to 3.6 secs and 0 to 30° approach angles. Mean longshore current velocities ranged up to 75.5 cm/sec towards Oliktok Point along the northwest shore, and up to 58.0 cm/sec westward along the northeast shore of this promontory. The longshore component of breaking wave energy was determined to be 0.1 to 11.0 and 0.02 to 2.12 x 10^6 ergs/cm-sec on the northwest and northeast shores, respectively. Figure 15.6 demonstrates both graphically and statistically the empirical relationship between observed longshore sediment transport rates and east-west wind speeds. Although there is a fair degree of scatter, particularly at the higher wind speeds and sediment transport rates, the +0.5 linear correlation coefficient is significant at the 0.95 probability level. The prevailing easterly winds (see the wind roses information of Figs. 15.4 and 15.5) are considered to be the major driving force in the generation of waves and longshore currents, and these in turn are primarily responsible for the net westward transport of sediment along the Simpson Lagoon coast and the westward migration of the barrier islands.

Industrial activity

There have been scattered examples of military and industrial development along the Alaskan Beaufort Sea coast for several decades. In recent years, such impingement has measured dramatically with the local oil exploitation activity, and this is likely to continue long into the foreseeable future. It would seem timely, therefore, to consider the relationship between such

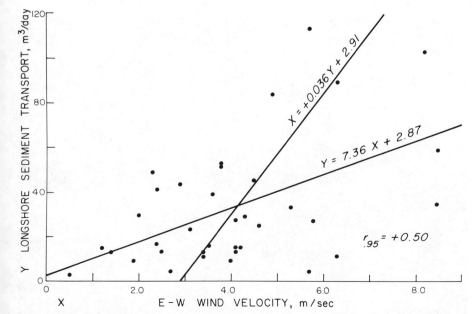

Fig. 15.6 Relationship between longshore sediment transport rates and east-west component wind speeds at Oliktok Point.

present and projected development and some aspects of the arctic coastal environment.

Current activities in Simpson Lagoon include local transportation and various shore-based facilities.

The movement of sediment, both along shore and as bedload, in Simpson Lagoon (and other similar lagoons) can potentially affect the size of various shoals or small navigable channels. For example, according to local residents at Beechey Point (Dygas et al. 1972), Gwyder Bay, adjacent to this lagoon, shoaled to such an extent over the past few decades that shallow draft vessels can no longer navigate readily through it. The barge and tugboat shown in Figure 15.7 are approaching the northeast shore of Oliktok Point in order to resupply the DEW-line station at Oliktok Point. During the approach to Oliktok Point through Simpson Lagoon, these resupply barges are often grounded on various submerged bars and shoals, and considerable local disturbance to the bottom sediment results before the vessels are freed. The possibilities of seiche effects and changing water levels because of shifting winds and wind tides (as described by Dygas and Burrell in Chapter 19 of this volume) can readily affect water depths, and therefore, the extent to which particular areas in the lagoon are navigable.

Structures built on shore adjacent to the beaches are obviously subject to the effects of both thermal and hydraulic coastal erosion, and especially to the intense wave and current action associated with late summer storms (Dygas and Burrell: Chapter 19, this volume). The sea-walls shown in Figure 15.8 have been constructed from used oil drums to give protection to the

Fig. 15.7 Resupply barge at Oliktok Point.

Fig. 15.8 Sea-walls constructed of oil-drums on the northwest shore at Oliktok Point.

Fig. 15.9 The beach at the end of Oliktok Point.

fuel storage tanks at Oliktok Point. The beach debris also appears to function as a sediment trap, hindering material erosion process. The area beyond the termination of the sea-wall shown in Figure 15.8 is also subject to thermo-erosion.

Particularly severe storms may do considerable damage, often transporting more sediment than would be moved over several decades of steady-state longshore activity. During the September 1970 storm described by Hurst (1971) and Dygas et al. (1972), sediment transportation was likely to be towards the east, opposite to the prevailing longshore movement (Dygas and Burrell: Chapter 19, this volume), and the tundra surface above and beyond the highest oil-drum sea-wall was flooded to such an extent that only those facilities built on gravel pads remained high above water. The erosive impact of this storm-generated flood water was sufficient to breach and wash out about 100 m + of a gravel road leading to the fuel storage tank at the tip of Oliktok Point in Figure 15.8. Figure 15.9 is another view of the fuel tanks at the end of Oliktok Point, where strand lines of driftwood and debris can be seen quite close to the base of the storage pad. These strand lines have been left by storms of less severity than the September 1970 storm cited above.

SUMMARY

The general patterns of sediment movement observed at Oliktok Point and around Simpson Lagoon are probably reasonably representative of conditions along much of this coastline. Observations specific to this locality indicate that the Simpson Lagoon coast is eroding at a mean rate of 1.4 m/yr. Measurements conducted on some of the barrier islands show net transport of sediment westward at rates of 6.0 to 40.0 m/yr and 26.0 m/yr for Pingok and Thetis Islands, respectively. Similarly, the net longshore movement of (sandy gravel) sediment along the Simpson Lagoon coast is from east to west, and there is a significant correlation (+0.5 coefficient) between transport rates and east-west component wind speeds.

Navigation within the shallow lagoon is rendered more hazardous because of migrating shoals and variations in local water depths due to wind tides. Shore-based structures are particularly subject to (usually) late-summer wave and current storm activity which, over a very short time frame, may move sediment.

Acknowledgments

Dr. A. S. Naidu is sincerely thanked for critical review of the manuscript; gratitude is expressed to Dr. R. I. Lewellen for his field and laboratory assistance. This work is a result of research sponsored in part by the Alaska Sea Grant Program (NOAA Office of Sea Grant, Department of Commerce) under Grant no. 04-3-158-41; the U. S. Environmental Protection Agency, Office of Research and Monitoring, under Grant no. R-801124; and by the State of Alaska. Contribution no. 288, Institute of Marine Science, University of Alaska.

REFERENCES

ALEXANDER, V., ET AL.

1974 Environmental studies of an Arctic estuarine system. Report R74-1, Inst. Mar. Sci., Univ. Alaska, Fairbanks, 359 pp.

ARNBORG, L., H. J. WALKER, and J. PEIPPO

1966 Water discharge in the Colville River. *Geografiska Annaler* 48A: 195-210.

1967 Suspended load in the Colville River, Alaska. *Geografiska Annaler* 49A: 131-144.

BARNES, P. W., and E. REIMNITZ

1972 River overflow onto the sea ice off the northern coast of Alaska. *Trans. Amer. Geophys. Union*, 53 pp.

DYGAS, J. A.

1974 A study of wind, waves and currents in Simpson Lagoon. *In* Environmental studies of an arctic estuarine system, edited by V. Alexander et al. Report 74-1. Inst. Mar. Sci., Univ. Alaska, Fairbanks, pp. 15-44.

DYGAS, J. A., R. TUCKER, and D. C. BURRELL

1972 Heavy minerals, sediment transport and shoreline changes of the barrier islands and coast between Oliktok Point and Beechey Point. *In* Baseline data study of the Alaskan arctic aquatic environment, edited by P. J. Kinney et al. Report 72-3, Inst. Mar. Sci., Univ. Alaska, Fairbanks, pp. 61-121.

EMERY, K. O.

1961 A simple method of measuring beach profiles. *Limnol. Oceanogr.* 6: 90-93.

HUME, J. D., and M. SCHALK

1964 The effects of beach borrow in the Arctic. *Shore and Beach* 32: 37-41.

1976 Shoreline processes near Barrow, Alaska: A comparison of the normal and catastrophic. *Arctic* 20: 86-103.

HURST, C. K.

1971 Investigation of storm effects in the Mackenzie delta region. Eng. Programs Branch, Canadian Dept. Public Work, Ottawa, Canada.

INMAN, D. L., and R. A. BAGNOLD

    1963   Littoral processes. In *The sea*, Vol. 3, edited by M. N. Hill. Interscience, New York, pp. 529-533.

LEFFINGWELL, E. K.

    1919   The Canning River region, northern Alaska. U. S. Geol. Survey, Prof. Paper 109.

LEWELLEN, R. I.

    1970   Permafrost erosion along the Beaufort Sea coast. Dept. Geog. and Geol., Univ. Denver.

    1972   *Studies on the fluvial environment — Arctic Coastal Plain Province, northern Alaska*. R. I. Lewellen Publ., Littleton, Colorado, 282 pp.

MacCARTHY, G. R.

    1953   Recent changes in the shoreline near Point Barrow, Alaska. *Arctic* 6: 44-51.

REIMNITZ, E., and K. F. BRUDER

    1972   River discharge into an ice-covered ocean and related sediment dispersal, Beaufort Sea, coast of Alaska. *Geol. Soc. Amer. Bull.* 83: 861-866.

SEARBY, H. W., and M. HUNTER

    1971   Climate of the north slope Alaska. NOAA Technical Memorandum AR-4. U. S. Dept. of Commerce, 53 pp.

SCHRADER, F. C.

    1904   A reconnaissance in northern Alaska. U. S. Geol. Survey, Professional Paper 20.

U. S. COASTAL ENGINEERING RESEARCH CENTER

    1966   Shore Protection Planning and Design. Technical Report No. 4.

WALKER, H. J.

    1969   Some aspects of erosion and sedimentation in an Arctic delta during break-up. Assoc. Internatl. d'Hydrologie Scientifique. Actes du Colloque de Bucarest, Hydrologie des deltas. Geoscience Report 70: 202-219.

WALKER, H. J., and L. ARNBORG

    1963   Permafrost and ice wedge effect on river bank erosion. *In*

Proc. Int'l Permafrost Conf. National Acad. Sci., National Res. Council 1287: 164-171.

WISEMAN, W. J., ET AL.

1973   Alaskan arctic coastal processes and morphology. Tech. Report No. 149, Coastal Studies Institute, Louisiana State Univ., 171 pp.

Assessment of the Arctic Marine Environment: Selected Topics
Copyright © 1976 by Institute of Marine Science, University of Alaska, Fairbanks

CHAPTER **16**

Circulation and sediment distribution in Cook Inlet, Alaska

L. W. GATTO [1]

Abstract

The purpose of this investigation was to analyze surface circulation, suspended sediment distribution, water-type migration, and tidal flushing mechanisms, utilizing medium and high altitude aircraft and repetitive synoptic satellite imagery with corroborative ground truth data. This approach provides a means of acquiring synoptic information for analyzing the dynamic processes of Cook Inlet in a fashion not previously possible. LANDSAT-1 and -2 and NOAA-2 and -3 imagery provided observations of surface currents, water type migrations and sediment and sea ice distributions during different seasons and tides. NASA NP-3A and U-2 aircraft multispectral imagery was used to analyze coastal processes, i.e., currents and sediment dispersion in selected areas. Ground truth data were utilized in the interpretation of the aircraft and satellite imagery and verified many of the regional circulation patterns inferred from the suspended sediment patterns apparent on the imagery. Several local circulation patterns not previously reported were identified.

INTRODUCTION

The Cook Inlet area in southcentral Alaska (Fig. 16.1) is currently undergoing the most rapid development in the state. Anchorage, the state's most

---

[1] *U.S. Army Cold Regions Research and Engineering Laboratory, Box 282, Hanover, New Hampshire 03755.*

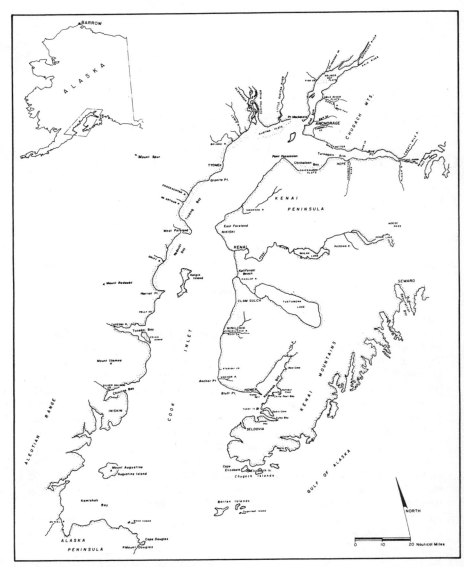

Fig. 16.1 Regional map of Cook Inlet.

populated city, located at the head of Cook Inlet, is presently the center of transportation, commerce, recreation, and industry. The use of Cook Inlet as a water road to this growing region will increase as the areal development continues. To prepare for this future development, an improved understanding of inlet circulation, especially tidal flushing processes as natural mechanisms for dissipating pollutants, is required to more intelligently and efficiently plan and perform construction, maintenance, design and related engineering activities in this marine environment.

The inlet is oriented in a northeast-southwest direction and is approximately 330 km long, increasing in width from 37 km in the north to 83 km in the south. It is considered a positive, tidal estuary characterized by greater runoff and precipitation than evaporation, resulting in dilution of seawater by fresh water. Mean diurnal tide range varies from 4.2 m at the mouth to 9.0 m at Anchorage. It varies within the lower portion of the inlet from 5.8 m on the east side to 5.1 m on the west (Wagner et al. 1969; Carlson 1970). The extreme tidal range produces currents typically 4 knots and occasionally 6 to 8 knots (Horrer 1967). The strong tidal currents, high Coriolis force at this latitude, and the inlet geometry produce considerable cross currents and turbulence within the water column (especially along the eastern shore) during both ebb and flood tide (Burrell and Hood 1967).

Cook Inlet is geographically divided into a northern and southern region by the East and West Forelands and is bordered by extensive tidal marshes, lowlands with many lakes, and glacier-carved mountains (Fig. 16.2). The coast from the head of Kachemak Bay to Turnagain Arm is characterized by sea cliffs and pocket beaches. An extensive coastal plain borders a low-lying, marshy coastline with scattered sea cliffs along the northwest and west shore from Point MacKenzie to Harriet Point. Steep mountains slope directly into the inlet in most locations from Harriet Point to Cape Douglas. Bayhead beaches have formed in many of the small embayments, and sea cliffs are found on the promontories along this coast. Tidal flats border much of the inlet coast but are prevalent in the north.

Climate of the area is transitional between the interior, with its cold winters, hot summers, low precipitation and moderate winds, and the maritime with cool summers, mild winters, high precipitation and frequent storms with high winds (Evans et al. 1972). The Matanuska, Knik, and Susitna rivers contribute approximately 70 percent of the fresh water annually discharged into the inlet. These rivers originate in glaciers, consequently showing large seasonal fluctuations in discharge, and they are the main contributors of suspended sediment to the inlet (Rosenberg et al. 1967).

METHODS AND PROCEDURES

The Multispectral Scanner (MSS) on LANDSAT was the primary source of satellite imagery for the investigation of circulation patterns, suspended sediment distribution, water type migration, and tidal flushing mechanisms. The NOAA-2 and -3 Very High Resolution Radiometers (VHRR) provided imagery in two bands, visible red (0.6-07 μm) and thermal infrared (10.5-12.5 μm). The thermal IR imagery was useful in sea ice studies. NASA NP-3A and U-2 aircraft multispectral imagery, acquired on 22 July 1972 and 23 June 1974, respectively, was used to verify patterns observed on LANDSAT imagery and to aid in analyzing surface currents, mixing patterns,

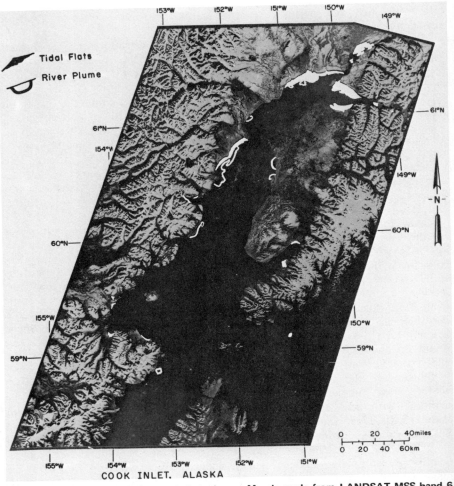

Fig. 16.2 Tidal flat distribution and river plumes. Mosaic made from LANDSAT MSS band 6 (0.7-0.8 μm) images acquired 3 and 4 November 1972.

and diffusion processes at river mouths and along the coast, tidal flat morphology, and coastline configuration.

Ground truth data used as corroborative information in interpreting satellite and aircraft imagery were obtained from two principal sources: the National Ocean Survey, NOAA, and the Institute of Marine Science, University of Alaska. Tidal, current, and hydrographic surveys were performed by NOAA during the summer of 1973, 1974, and 1975. Additional temperature, salinity, pH, dissolved oxygen, nutrient and suspended sediment concentration measurements were provided from cruise data collected by the Institute of Marine Science, University of Alaska, from July 1966 to June 1973.

RESULTS AND DISCUSSION

Tidal flats

The high sediment load in glacial rivers discharging into the inlet is the primary source of material for the tidal flats. LANDSAT MSS band 5 (0.6-0.7 μm) and 7 (0.8-1.1 μm) images were found to be ideal for a regional analysis of these tidal flats. The most extensive flats are found north of the Forelands (Fig. 16.2). In the lower inlet, tidal flats usually occur as bayhead bars in embayments along the western shore and northeast of Homer in Kachemak Bay. Strong, variable tidal currents cause significant changes in tidal flat configuration. Migration of some of the major tidal channels in Knik and Turnagain arms was substantiated by comparing Figure 16.2 with the National Ocean Survey Navigational Chart 8553 (Cook Inlet, Northern Part).

Suspended sediment distribution and circulation

Turbid fresh water discharging into the inlet, especially from the north and west, produces sediment plumes which appear lighter in tone than less turbid water in LANDSAT MSS band 4 (0.5-0.6 μm) and 5 images. Suspended sediment concentrations in Cook Inlet vary from greater than 1700 mg/liter near Anchorage (Wright et al. 1973) to as low as 0.4 mg/liter near the inlet mouth (Kinney et al. 1970). The suspended sediment acts as a natural tracer by which circulation patterns and water types with different suspended sediment loads, temperatures, and salinities are visible from satellite and aircraft altitudes. Descriptions of surface flow near river mouths were made based on the shape of sediment plumes and on mixing patterns observed on aircraft and satellite imagery. Figure 16.3, a LANDSAT-1 image acquired on 7 August 1972, shows the sediment plumes from the Drift (3) and Big (4) rivers during flood tide at Seldovia. The shape and location of the plumes are convenient markers to determine current directions along the west shore between McArthur River (1) and Tuxedni Bay (2). The plumes are clearly moving in a northerly direction. Relict sediment plumes from earlier tidal stages, visible far offshore, indicate water movement through several tidal cycles. Relative difference in sediment concentration of the inlet water can also be distinguished. The darker tones apparent farther from shore indicate that the water is less turbid. Tidal flats (5) appear as a gray border along the coastline. Mt. Spurr (6), one of the many volcanoes in the Chigmit Mountains; Lake Chakachamna (7); and numerous glaciers (8) are also visible.

Figure 16.4 shows the movement of near-surface and surface water in the area of the Drift River tanker terminal during late flood tide at Seldovia, approximately the same tidal stage as shown in Figure 16.3. Tidal range at this location generally varies from 3-6 m, and surface current velocities are 1-3 knots. Currents at depths of 8 and 15 m are generally lower but periodically reach velocities equal to those at the surface (Marine Advisers 1966). Turbid fresh water from the Drift (1) and Big (2) rivers forms a distinct surface layer riding over and mixing with the saline inlet water (8). Turbulence caused by local tidal and wind currents produces the mixing as

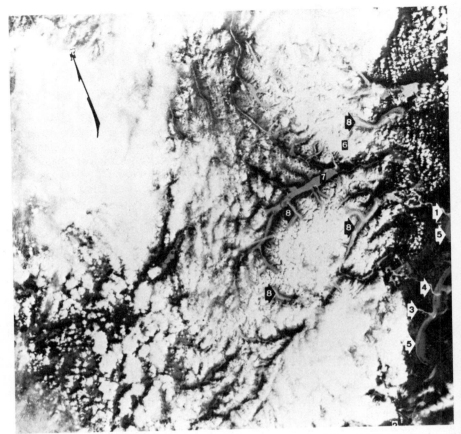

Fig. 16.3 West shore of Cook Inlet between McArthur River and Tuxedni Bay. MSS band 5 image acquired 7 August 1972. Numeric designators mark locations of (1) McArthur River, (2) Tuxedni Bay, (3) Drift River; (4) Big River, (5) tidal flats, (6) Mt. Spurr, (7) Lake Chakachamna, and (8) glaciers.

the fresh water is diverted to the north by the flood tide. Several mixing boundaries are evident between sediment-laden and clearer, oceanic water. Turbid water (5) from the rivers rapidly mixes with the less turbid oceanic water (8) to form different water types of variable temperatures, salinities and sediment concentrations (6, 7) within the mixing zone. More complete mixing has occurred in the southern portion of the area, and the Drift River plume has migrated from the south during flood tide. The tanker at the terminal (4) roiled the more buoyant fresh water and brought deep, clear water up to the surface (dark area of clear water off the ship's stern). The surface foam lines (9) are common where different water types converge and mix. Internal waves (linear patterns) near density boundaries are also common in these locations and are obvious southeast of the tanker terminal.

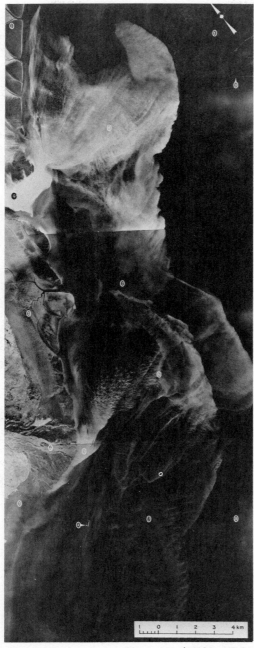

Fig. 16.4 Drift and Big rivers area during flood tide (NASA NP-3A photograph). Note similarity to plume shape in Figure 16.3.

Compare the detail of tidal flats (3) to that observed on the LANDSAT image in Figure 16.3.

Repetitive LANDSAT imagery was used to analyze the generalized surface circulation patterns throughout the inlet. Imagery from October and November 1972 (Fig. 16.5) showed changes in patterns previously determined from data acquired by conventional ship-board techniques. The clear oceanic water from the Alaska Current entered the inlet at flood tide along the east side around the Barren Islands. This clear water became less distinct toward the north as it dispersed and mixed with the turbid water in the middle inlet around Ninilchik. The tide front progressed up the inlet primarily along the east shore, being diverted in that direction by the Coriolis force. A back eddy not previously reported (dotted arrows, Fig. 16.5) was apparent just

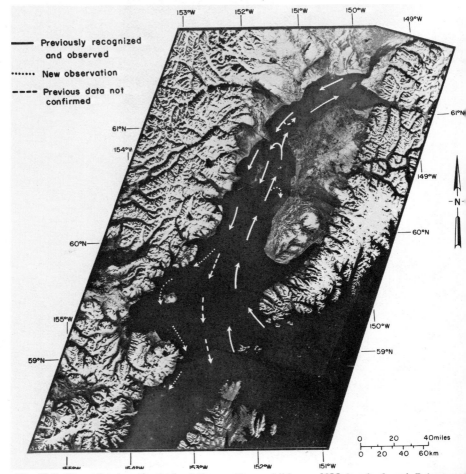

Fig. 16.5 Generalized surface circulation patterns visible on MSS bands 4 and 5 imagery acquired in October and November 1972.

offshore from Clam Gulch. The eddy formed in the slack water northeast of Cape Ninilchik during flood. The tide front continued past the East Foreland and was partially diverted across the inlet. Part of the front was diverted south of the West Foreland; the remainder moved north. This resulted in a counterclockwise circulation pattern around Kalgin Island. This pattern was verified by direct observation from 1800 m altitude during an aircraft underflight at the time the satellite passed. However, circulation patterns near Kalgin Island and the Forelands are variable with local tidal stage.

Surface circulation north of the Forelands at this time appeared to be similar to that previously reported by Evans et al. (1972). The ebbing water in the inlet moved predominantly along the northeastern shore past the Forelands. However, the previously reported counterclockwise pattern north of the Forelands (formed as ebbing waters strike the West Foreland, are diverted across the inlet and become incorporated into the flood current along the east shore) was not observed (Anderson et al. 1973). South of the Forelands, as the sediment-laden ebbing water moved past Chinitna Bay, a portion appeared to flow along the shoreline and circulate around the west side of Augustine Island in Kamishak Bay; the remainder of the water moved past the bay and continued parallel to the coast past Cape Douglas and progressed through Shelikof Strait. This circulation pattern in Kamishak Bay was not previously reported but was evident on the MSS imagery and verified with ground truth data.

Changes in this generalized circulation and sediment distribution pattern were also detected and mapped (Fig. 16.6). Figure 16.6a shows differences in the position of the boundary between oceanic and inlet water in the southern inlet on two successive days; this boundary separated these two water types during late ebb tide in Anchorage and late flood in Seldovia. The irregularities and changes in the location of this boundary may be due to changes in the mixing zone between the water types. Near the time of low water in Seldovia, the boundary is located more toward the southeast; near high water it is more to the northwest. Changes in the boundary over an 18-day period are shown in Figure 16.6b. The 17-18 October imagery showed the position of the boundary during mid-flood tide in Anchorage and early ebb in Seldovia. The boundary location was generally comparable to that shown in Figure 16.6a, and the general relationships of the two water types appear from these observations to be consistent during the period October through November.

LANDSAT-1 imagery from 24 September 1973 (Fig. 16.7) clearly showed many of the aforementioned sediment and circulation patterns. The distribution of and areas of mixing between the less turbid, more saline, oceanic water in the southeast and the turbid, fresher inlet water in the north and southwest were apparent during late flood tide in Seldovia and late ebb tide in Anchorage. The oceanic water (1) on the southeast has migrated as far north as Kenai. The boundary (2) between oceanic and inlet waters (3) previously observed was less obvious on this imagery. Surface runoff was probably reduced, and less suspended sediment was being discharged into the inlet. Mixing between the two major water types appeared to be more extensive in the middle and southern inlet.

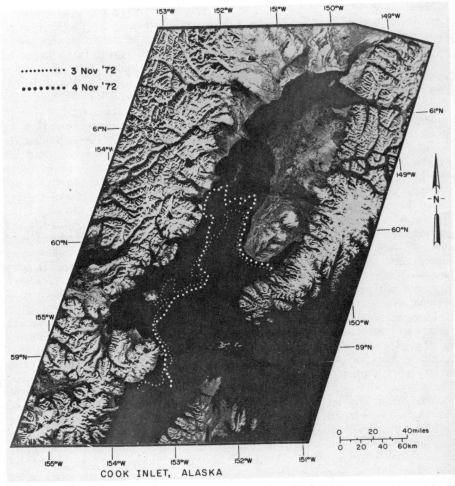

········· 3 Nov '72

········· 4 Nov '72

COOK INLET, ALASKA

Fig. 16.6a Daily changes in boundaries separating oceanic and inlet water.

The entire inlet south of the Forelands was dominated by oceanic water, and complex surface circulation patterns existed near Kalgin Island. The patterns differed from those observed on the November 1972 imagery. Flooding oceanic water appeared to bifurcate near the latitude of Tuxedni Bay and move northward east and west of Kalgin Island. A nearshore counter current appeared along the east coast of Kalgin Island. The high sediment concentration along the east coast between Cape Ninilchik (4) and Cape Kasilof (5) was likely caused by scouring of the bottom or a southerly counter current along the coast. North of the Forelands, the oceanic water mixed extensively with the turbid inlet water but appeared to continue northerly along the west shore and in the middle inlet. The McArthur River

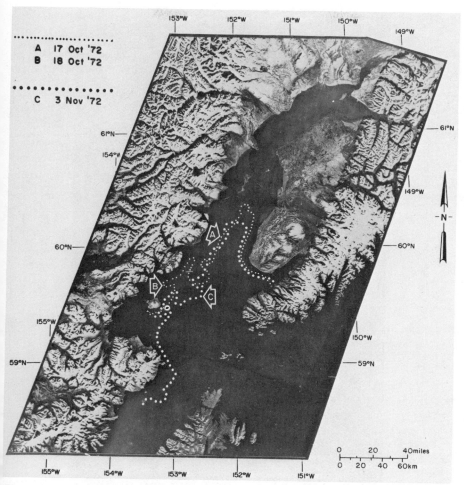

Fig. 16.6b Changes over an 18-day period in boundaries separating oceanic and inlet water.

plume (8) indicated that coastal water here was moving north. Surface salinity data from August 1972 (Fig. 16.8) showed a similar pattern north of the Forelands. The Beluga River (9) is a meltwater stream flowing from Beluga Lake (10), a periglacial lake at the terminus of the Triumvirate Glacier (11). The river plume (6) indicated southerly nearshore currents. This coastal water may continue along the coast to the North Foreland (7), then move across the inlet and become mixed with flooding water.

The generalized suspended sediment distributions (Fig. 16.9) are controlled primarily by the circulation described above. Suspended sediment, mostly of glacial origin, is concentrated in the well-mixed northern inlet; it is nearly absent in the water of the east-central and eastern portions of the inlet

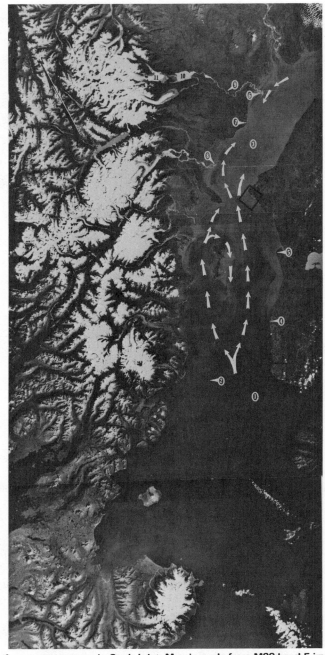

Fig. 16.7 Surface water patterns in Cook Inlet. Mosaic made from MSS band 5 images acquired 24 September 1973. (See text for marker designations).

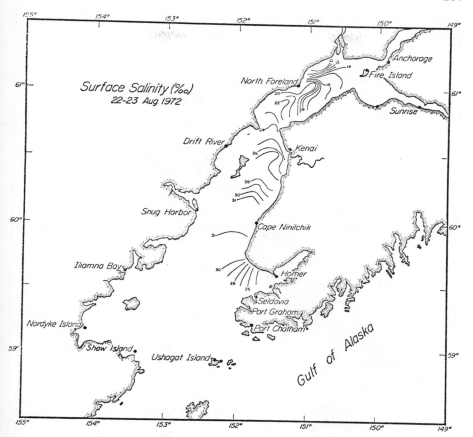

Fig. 16.8 Surface salinity (°/oo) distribution in August 1972 (adapted from unpublished data provided by F. F. Wright).

mouth. This regional distribution is maintained year-round, but the total suspended sediment load varies with season and with depth (Murphy et al. 1972, Kinney et al. 1970, Rosenberg et al. 1967). Subsurface measurements show that suspended load normally increases with depth (Sharma et al. 1974). Maximum values generally occur at approximately 10 m near the head of the inlet; concentrations increase with depth south of the Forelands.

Repetitive ground truth observations of water characteristics correlated with and verified many of the interpretations and observations made from the imagery. Ground truth surface temperature and salinity patterns (Figs. 16.10 and 16.11, respectively) show that cold water (≈ 5.0° C) enters the inlet on the southeast side while warmer water moves seaward along the west shore. The warmest water is found in Kamishak Bay and appears to circulate out of the bay around Augustine Island. This circulation was observed on LANDSAT imagery acquired in October and November 1972

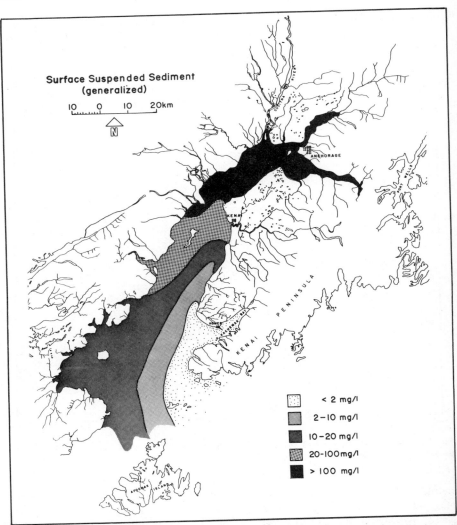

Fig. 16.9 Generalized surface suspended sediment distribution (after Sharma et al. 1974).

(Fig. 16.5) but was not previously recognized. The more saline water is concentrated in the southeast, and dilution by fresh water occurs primarily from the north and southwest. A very distinct shear zone is indicated by the 1973 salinity data.

Inlet ice

Sea ice is a navigational hazard particularly in the upper inlet north of the Forelands for 4 to 5 months of the year (Marine Advisers 1964; Alaska District, Corps of Engineers 1948; Rosenberg et al. 1967). The movement

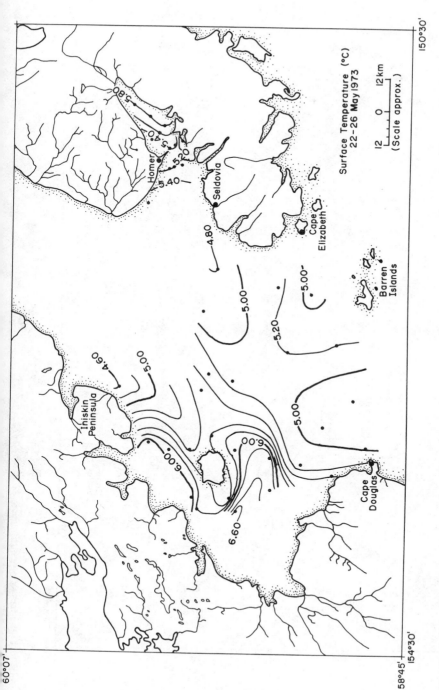

Fig. 16.10 Surface temperature distribution, 22-26 May 1973.

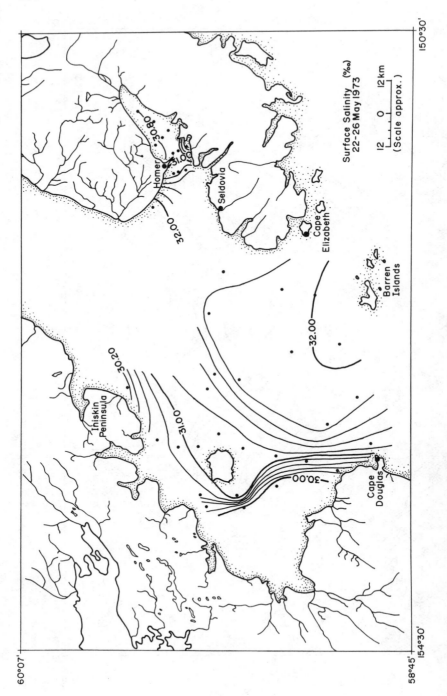

Fig. 16.11 Surface salinity distribution, 22-26 May 1973.

and strength of the ice are important aspects of the inlet environment to be considered, especially in planning offshore construction. The ice exists as large floes which are commonly greater than 320 m across with individual blocks generally less than 1 m thick. Pressure ridges up to 6 m in depth occasionally form on the floe peripheries due to frequent collisions with other floes (Blenkarn 1970). Large ice floes become scattered and move primarily up and down the upper inlet with the 6-8 knot tidal currents. Some move as far south as Anchor Point on the east side (Wagner et al. 1969). Large floes are commonly carried by winds and tides along the west side as far south as Kamishak Bay and beyond to Cape Douglas. Brash and frazil ice are common between the rounded floes. Previous reports (U. S. Department of Commerce 1964) suggest that ice occasionally closes Iliamna Bay for brief periods; however, repetitive LANDSAT and NOAA-2 and -3 imagery indicates that large floes persist in Kamishak Bay for long periods

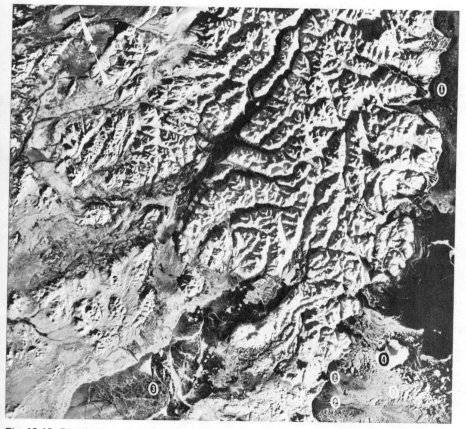

Fig. 16.12 Distribution of inlet ice along the southwest shore from MSS band 5 image acquired 29 January 1974: (1) western coast from Kamishak Bay north to Redoubt Bay; (2) out of Kamishak Bay past Augustine Island; (3) Lake Iliamna; and (4) Augustine Island.

and that many of the small embayments bordering the bay appear ice-covered for a major portion of the winter.

Most of the ice forms by the freezing of river water as it flows over the tidal flats. Smaller amounts of ice are formed from seawater left on the flats during low tide (Sharma and Burrell 1970). Much of the ice on the flats is picked up during flood tide, moved out into the inlet and incorporated into the floe. The remainder is left on the flats and repeatedly refreezes to form sheets or stacks of ice (stanukhi), some as thick as 12 m. Some of these thick sheets are eventually transported to the floe while others remain on the flats throughout the winter.

Cloud-free LANDSAT imagery acquired on 29 January 1974 (Fig. 16.12) shows the distribution (1) of ice floes and frazil ice along the western coast from Chenik Head in Kamishak Bay north to Redoubt Bay. Westerly winds

Fig. 16.13 Distribution of sea ice north of the Forelands from MSS band 7 image acquired 22 March 1974; (1) ice floes, (2) frazil ice, (3) large ice blocks (4) city of Anchorage, (5) linear ice, (6) tidal flats, and (7) tidal channels.

are moving the ice southeast (2) out of Kamishak Bay past Augustine Island (4), while winds and tidal currents produce complex patterns along the inlet coast farther north. Note ice-covered Lake Iliamna (3).

The ice floe distribution north of the Forelands (1) is shown in Figure 16.13. Frazil ice (2) and large blocks (3) formed the floe which was moving south during ebb tide at Anchorage (4). The ice appeared linear (5) as it conformed to the rectilinear flow patterns of the surface water. Note the ice- and snow-covered upper tital flats (6), well defined tidal channels (7) and the snow-enhanced road net and airports in Anchorage.

NOAA-2 imagery acquired on 18 February 1973 (Fig. 16.14) shows the distribution of sea ice throughout the inlet. The sea ice patterns were similar

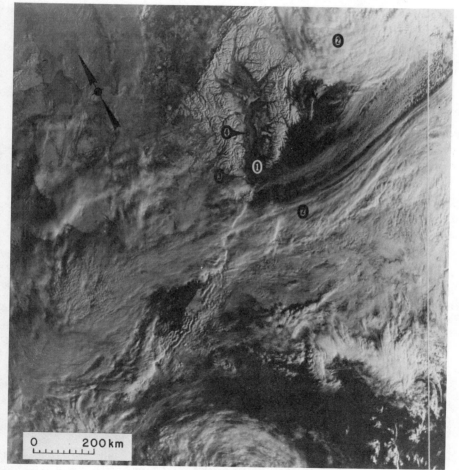

Fig. 16.14 Sea ice in Cook Inlet on 18 February 1973, NOAA-2 VHRR visible image: (1) near mouth of Cook Inlet, (2) clouds obscuring Kamishak Bay, (3) Augustine Island, and (4) Kalgin Island.

to the patterns of suspended sediment as observed on LANDSAT imagery. The highest concentrations of sea ice occurred in the northern inlet and in the western portion of the southern inlet. The eastern portion south of the Forelands was generally ice-free because this area is characterized by the intrusion of seawater, which is warmer than inlet water during winter months. The area near the inlet mouth (1) was ice-free, but ice was concentrated around Kalgin Island (4) and in Kamishak Bay around Augustine Island (3). Clouds (2) obscured portions of Kamishak Bay.

The presence of this mobile ice cover was useful in making comparisons between summer and winter circulation patterns. The highest concentrations of sea ice generally occur in the western portion of the lower inlet, while the eastern portion remains ice-free. Predominantly north winds in the winter also move the ice to the west and southwest side of the inlet. Although large ice floes are a navigation hazard as they move up and down the inlet between the Forelands during flood and ebb tide, and numerous reports of shipping difficulties have been made (Alaska District, Corps of Engineers 1973), extensive damage to shipping has generally been avoided.

SUMMARY AND CONCLUSIONS

Synoptic interpretations made from repetitive aircraft, LANDSAT Multispectral Scanner (MSS), and NOAA-2 and -3 Very High Resolution Radiometer (VHRR) imagery with corroborative ground truth data provide for the first time a means of analyzing, on a regional basis, the estuarine surface circulation, sediment distribution, water type movement, coastal processes, tidal flat distribution, coastal landforms, and sea ice distribution in Cook Inlet. The distribution and configuration of tidal flats were mapped with MSS band 5 and 7 images. Coastal landforms and configuration are most apparent in band 6 and 7 images. The distribution of suspended sediment and surface circulation patterns were mapped on MSS bands 4 and 5. Comparisons of these features and processes were made with color IR and thermal IR aircraft imagery. Sediment distribution, current directions, mixing patterns along river plume and water type boundaries, tidal flats and coastal landforms show best on the color IR photography. The thermal IR scanner imagery was most useful in interpreting mixing zone patterns.

Based on this investigation, the following observations can be made: Semidiurnal (mixed) tides, Coriolis effect, inlet configuration, and the Alaska current govern the estuary circulation. Clear oceanic water enters the inlet on the southeast during flood tide at Seldovia, progresses northward along the eastern side with minor lateral mixing, remains as a distinct water mass to the latitude of Kasilof-Ninilchik, then mixes extensively with the turbid inlet water in the area near the Forelands. Ebbing water moves south primarily along the north and west shore, and a distinct zone between the two water masses forms in mid-inlet south of Kalgin Island. Turbulence and vertical mixing appear most active along the east shore, while stratification is pronounced in Kamishak and Kachemak bays, especially during periods of high freshwater runoff. Bottom scouring is evident along the east shore south of Pt. Possession. Most of the sediment discharged into the inlet is deposited along the extensive tidal flats, and some suspended sediment is transported along the west side and out of the inlet past Cape Douglas. Complex circulation patterns were observed near Kalgin Island; at this location, ebbing and flooding waters converge, current velocities are high, and coastline configuration causes strong cross-inlet currents. Several local circulation patterns not previously reported were identified: a clockwise back eddy observed during flood tide in the slack water area west of Clam Gulch; a counterclockwise current north of the Forelands during ebb tide at Anchorage, and sediment-laden water ebbing west of Augustine Island, out the inlet around Cape Douglas and through Shelikof Strait.

The water in the upper inlet is well mixed, due to the very large tidal fluctuations and high current velocities in this shallow, narrow basin. During summer, when surface runoff is high, there is a net outward movement of water from the upper inlet; with reduced runoff in the winter, there is virtually no net outflow (Murphy et al. 1972). The middle inlet has a net inward circulation of cold, saline oceanic water up the eastern shore and a net outward flow of warmer and fresher inlet water along the western shore (Evans et al. 1972). These water types are well mixed vertically along the eastern shore and are separated laterally by a well defined shear zone. In the lower inlet a lateral temperature and salinity separation is maintained; but in the western portion, vertical stratification occurs with colder, saline oceanic water underlying warmer, less saline inlet water. During tidal inflow the deeper oceanic water rises to the surface at the latitude of Tuxedni Bay and mixes with the inlet water (Kinney et al. 1970).

Wise utilization of our natural resources with minimal environmental degradation is now acknowledged to be a primary concern of our society. Cook Inlet is the fastest growing area in industrial development in the state of Alaska. Increased pressures of population and industrialization make it imperative that investigations of the water resources of Cook Inlet be continued in order to manage these resources in the most efficient manner possible.

Acknowledgments

The author expresses appreciation to Dr. Duwayne Anderson for assistance, critical suggestions and support throughout the project; to Dr. Frederick Wright, previously of the Institute of Marine Science, University of Alaska, and to Messrs. Isaiah Fitzgerald, Coastal Mapping Division, and Robert Muirhead, Oceanographic Division, National Ocean Survey, National Oceanic and Atmospheric Administration, for providing ground truth data. The work was funded by the National Aeronautics and Space Administration under NASA-Defense Purchase Request W-13, 452, SR/T Project 160-75-89-02-10, "Tidal Flushing in Cook Inlet, Alaska", and by the Office of Chief of Engineers, Civil Works, under the Coastal Engineering, Inlets and Estuaries work unit, "Remote Sensing of Tidal Flushing in Cook Inlet" (0204/31064).

226   GATTO

REFERENCES

ALASKA DISTRICT, CORPS OF ENGINEERS

1948 Engineering data, survey of Cook Inlet and Anchorage Harbor, Alaska. Preliminary report. Anchorage, Alaska, 21 pp.

1973 Operation and maintenance of the Anchorage Harbor, Anchorage, Alaska. Draft of environmental impact statement, 56 pp.

ANDERSON, D. M., L. W. GATTO, H. L. McKIM, and A. PETRONE

1973 Sediment distribution and coastal processes in Cook Inlet, Alaska. Proc. Symposium on Significant Results obtained from the Earth Resources Technology Satellite-1, NASA SP-327, March, pp. 1323-1339.

BLENKARN, K. A.

1970 Measurement and analysis of ice forces on Cook Inlet structures. Offshore Technology Conference, 22-24 April, Houston, Texas, pp. 365-378.

BURRELL, D. C., and D. W. HOOD

1967 Clay-inorganic and organic-inorganic association in aquatic environments, Part 2. Report 68-5, Inst. Mar. Sci., Univ. Alaska, Fairbanks, 114 pp.

CARLSON, R. F.

1970 The nature of tidal hydraulics in Cook Inlet. *The Northern Engineer* 2(4): 4-7.

EVANS, C. D., ET AL.

1972 The Cook Inlet environment, a background study of available knowledge. Resource and Science Service Center, Univ. Alaska, Anchorage.

HORRER, P. L.

1967 Methods and devices for measuring currents. In *Estuaries*, edited by G. Lauff, Publication No. 83, Amer. Ass. Adv. Sci., Washington, D. C., pp. 80-89.

KINNEY, P. J., J. GROVES, and D. K. BUTTON

1970 Cook Inlet Environmental data, R/V *Acona* Cruise 065-May 21-28, 1968. Report R-70-2, Inst. Mar. Sci., Univ. Alaska, Fairbanks, 120 pp.

MARINE ADVISERS, INC.

1964 Oceanographic conditions at Beshta Bay, Cook Inlet, Alaska. Report prepared for Humble Oil Company of California. La Jolla, California, May, 37 pp.

1966 Currents near the mouth of Drift River, Cook Inlet, Alaska. Report prepared for Cook Inlet Pile Line Company. La Jolla, California, August, 7 pp.

MURPHY, R. S., R. F. CARLSON, D. NYQUIST, and R. BRITCH

1972 Effect of waste discharges into a silt laden estuary, a case study of Cook Inlet, Alaska. Report IWR-26, Inst. Water Resources, Univ. Alaska, Fairbanks, 42 pp.

ROSENBERG, D. H., ET AL.

1967 Oceanography of Cook Inlet with special reference to the effluent from the Collier Carbon and Chemical Plant. Tech. Report No. R67-5, Inst. Mar. Sci., Univ. Alaska, Fairbanks, 80 pp.

SHARMA, G. D., and D. C. BURRELL

1970 Sedimentary environment and sediments of Cook Inlet, Alaska. *Amer. Assoc. Pet. Geologists Bull.* 54(4): 647-654.

SHARMA, G. D., F. F. WRIGHT, J. J. BURNS, and D. C. BURBANK

1974 Sea surface circulation, sediment transport, and marine mammal distribution, Alaska continental shelf. Final report of ERTS Project 110-H, Univ. Alaska, Fairbanks, 77 pp.

U. S. DEPARTMENT OF COMMERCE

1964 U. S. Coast Pilot 9, Pacific and Arctic Coasts, Alaska, Cape Spencer to Beaufort Sea. U. S. Government Printing Office, Washington, D. C., 347 pp.

WAGNER, D. G., R. S. MURPHY, and C. E. BEHLKE

1969 A program for Cook Inlet, Alaska for the collection, storage and analysis of baseline environmental data. Report No. IWR-7, Inst. Water Resources, Univ. Alaska, Fairbanks, 284 pp.

WRIGHT, F. F., G. D. SHARMA, and D. C. BURBANK

1973 ERTS-1 observations of sea surface circulation and sediment transport, Cook Inlet, Alaska. Proc. Symposium on Significant Results Obtained from the Earth Resources Technology Satellite-1, NASA SP-327, March, pp. 1315-1322.

Assessment of the Arctic Marine Environment: Selected Topics
Copyright © 1976 by Institute of Marine Science, University of Alaska, Fairbanks

CHAPTER **17**

Mesoscale thermal variations along the arctic North Slope

W. J. WISEMAN, JR., *and* A. D. SHORT [1]

Abstract

Initial statistical and spectral analyses have been performed on atmospheric records from seven DEW-line stations taken over a period of 10 years. The characteristic difference between the winter continental climate and the summer maritime climate is readily apparent in the statistics. Spectrum analysis, although not conclusive, suggests the importance of a periodic variation at a 45-day period. Anomalous warming trends associated with low-pressure intrusions from the Bering Sea are found to occur at least once a year.

INTRODUCTION

As the population and development of the North Slope of Alaska grow, a significant increase in our knowledge of the physical environment will be needed. Past studies by the Coastal Studies Institute of the coastal morphology and oceanographic processes (e.g., Wiseman et al. 1973) have reinforced our opinion that the meteorological and climatological understanding of this region is woefully inadequate. In order to fill this void, we have begun to analyze the daily weather data from the DEW-line stations, directing our attention initially to the mesoscale thermal variations along the coast. By mesoscale, we signify length scales between 160 and 750 km and time scales between one day and one year. These variations are important not only for their effect on the planning of any large-scale operation involving both men and machinery, but also on numerous physical processes — most notably thermal erosion of tundra and the timing of freezeup and breakup.

---

[1] *Coastal Studies Institute, Louisiana State University, Baton Rouge, Louisiana 70803.*

THE DATA BASE

The data were collected at the seven presently operating DEW-line sites (Fig. 17.1). These are spaced at approximately 160–km intervals along the arctic coast from Cape Lisburne to Barter Island. Maximum and minimum temperatures were recorded daily. When these data were not available, we approximated the temperature extremes from the hourly or six-hourly temperature readings which were recorded. Depending on the station, the data set began between January 1956 and August 1957 and, for this study, ended in 1965.

Data analysis

The entire data set was used to compute the mean degree day for each month that data were available and for each station. For our purposes, the degree day was defined as the mean of the maximum and minimum temperatures for each day. The maximum and minimum temperatures for each month and the standard deviation of degree day for each month and each station were also computed (Figs. 17.2 and 17.3). For each station, all degree day estimates for each date of the year were then averaged to give a mean annual degree-day curve (Fig. 17.4). These curves were then subtracted from the original data sets to obtain deviation time series, which were subjected to further variance and spectrum analyses. For these additional analyses, only the synoptic data from the years 1958 through 1963 were used. Autocovariances and cross covariances were computed for the deviation time series with a linear trend removed for lags up to 200 days. These covariances were then used to compute autospectra and phase and coherence estimates between series using the Blackman-Tukey (1950) method.

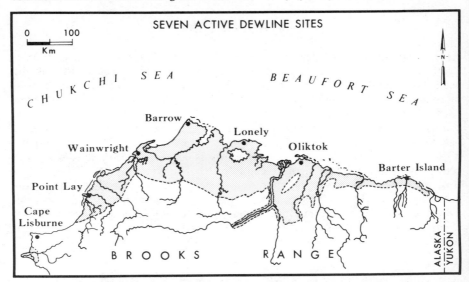

Fig. 17.1 Location map for the seven active DEW-line sites. Stippled area is coastal plain.

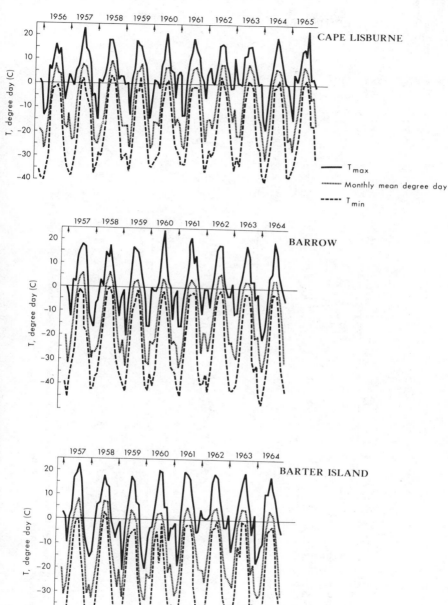

Fig. 17.2 Examples of the curves of mean degree day, maximum temperature, and minimum temperature for each month are shown for the stations at Cape Lisburne, Barrow, and Barter Island.

232

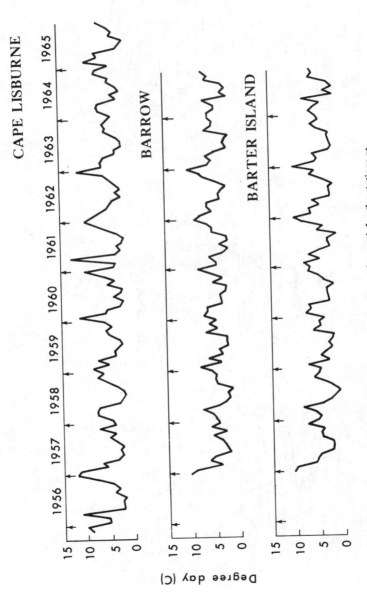

Fig. 17.3 Examples of the standard deviation of degree day for each month for the stations at Cape Lisburne, Barrow, and Barter Island.

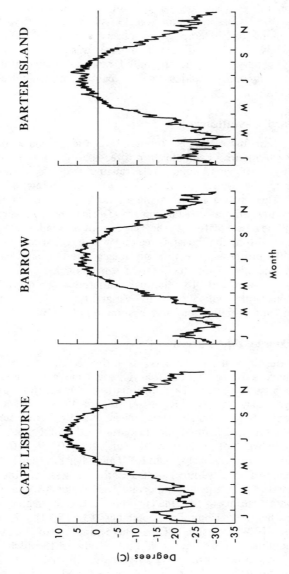

Fig. 17.4. Examples of the mean annual degree day curve for the stations at Cape Lisburne, Barrow and Barter Island.

RESULTS

Monthly means

The maximum monthly mean degree day in our data was 10°C at Cape Lisburne in July 1958; the minimum was -36° C at Wainwright in February 1964. The maximum daily temperature observed was 24°C at Wainwright and Barrow in July 1960, and the minimum was -51° C at Oliktok in February 1961. Of course, this rather limited range of temperatures does not reflect the human discomfort which can be caused by winds accompanying the colder temperatures.

Monthly standard deviations

More interesting than these means, though, are the standard deviations of degree day for each month. Examples for three stations are shown in Figure 17.3. These clearly show an annual cycle ranging from high values between 4 and 13°C during the winter months to low values below 4°C during the summer months. This change in the standard deviations reflects the buffering effects of open water on the coastal climate. During the summer, open-water season, the large heat capacity of the shelf waters tends to restrict large thermal variations within the coastal region, whereas during the ice-covered winter season, this buffering effect is no longer present and large temperature changes are possible. Thus at different times of the year, the Alaskan arctic coastline is subjected to thermal variations appropriate to both maritime and continental climates. This variability is important for both equipment design and specification and accommodations for human comfort.

Mean annual degree day curves

Of particular interest for us, because of its influence on coastal processes, is the timing of above-zero weather — that is, the time when melt and break-up of the ice and snow begin. This information may be obtained from the mean annual degree day curves (Fig. 17.4). Figure 17.5 shows that the first day that the degree day estimate rises above and thereafter remains above freezing is May 23 and 24 for Cape Lisburne and Point Lay, respectively. The same event does not occur until June 6, 10, 7, 10, and 7 for Wainwright, Point Barrow, Lonely, Oliktok, and Barter Island, respectively. At the other end of the summer season, the arrival of the continuous below-zero temperatures is equally important, as it heralds the refreezing of the land and water surfaces. Starting in the east, the degree day average falls and remains below freezing on September 13, 15, 14, 7, 14, 18, and 29 at the seven DEW-line sites. Thus there appear to be two thermal regimes — one containing the stations east of Wainwright, which are also the northernmost stations and have between 89 and 99 consecutive above-freezing degree days per year on the average and another comprised of the two most southwesterly stations, Point Lay and Cape Lisburne, which have 117 and 122 consecutive above-freezing degree days during an average year.

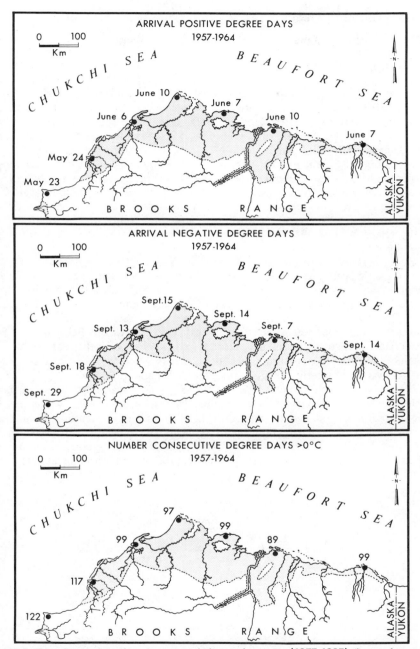

Fig. 17.5 Top: Date when the mean annual degree day curve (1957-1965) rises and remains above freezing. Middle: Date when the mean annual degree day curve (1957-1964) falls and remains below freezing. Bottom: Maximum number of continuous above-freezing degree days as determined from the mean annual degree day curve (1957-1964). Stippled area is coastal plain.

Winter warming trends

Quite often between late fall and mid-spring, there will be an anomalous rise in the mean degree day and in the daily maximum temperature. These trends may attain maximum excursions of 10° C or more in the average curve (Fig. 17.4) and significantly greater values in any particular year. For example, in the 1960 degree day record from Cape Lisburne (Fig. 17.6), an abrupt 30°C rise in temperature occurs near day 45, and two less intense rises are noted near days 20 and 340. One such typical warming trend, which occurred in January and February 1963, is illustrated in Figure 17.7. On January 10, cold, high-pressure winter conditions dominate the entire North Slope. By January 12, the front of a low-pressure system has moved as far north as Cape Lisburne, raising the temperature relative to that of the rest of the slope. By January 16, the low pressure is still stalled at Cape Lisburne, where the warming effect of the southerly air flow has raised the temperature to 0° C and is obviously affecting Point Lay as well. By the 18th, relatively low pressure dominates the entire slope, and southerly winds raise temperatures across the slope.

Four days later, returning high pressure and cold easterly winds push the low pressure cell back to the Cape Lisburne area and lower the coastal temperatures. However, by the 28th, the southerly flow of air has again forced its way across the entire slope, raising temperatures to and above zero. Finally, on the 30th, rising pressure and easterly air flow cool the slope, pushing the warm air west. Two days later, on February 1, cold temperatures, high pressure, and easterly air flow return the entire slope to the more typical winter conditions, three weeks after the low-pressure system and its warm, southerly air flow first reached Cape Lisburne.

Over the 10-year period 1956 to 1965, similar warming trends occurred each winter. An average of two warming trends occurred each winter season. The duration of the trends ranged from 3 to 18 days, tending to decrease to the east, with a mean duration of 8 days. The longer trends, such as the one just examined, resulted from two pulses of low-pressure air — the second tending to have a longer and more pronounced warming effect.

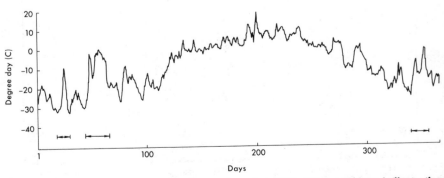

Fig. 17.6 Degree day record from Cape Lisburne for 1960. Horizontal bars indicate three periods of anomalous winter heating.

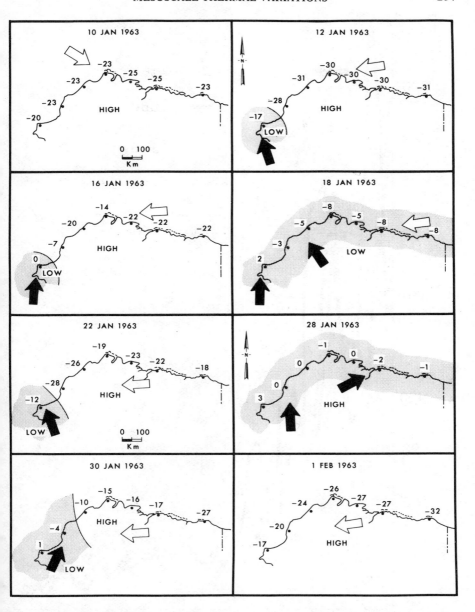

Fig. 17.7 Schematic illustration of the progression of an anomalous winter warming trend in 1960. Degree days are shown next to station locations in degree Celsius. Stippled area is region of low pressure. Arrows indicate wind direction.

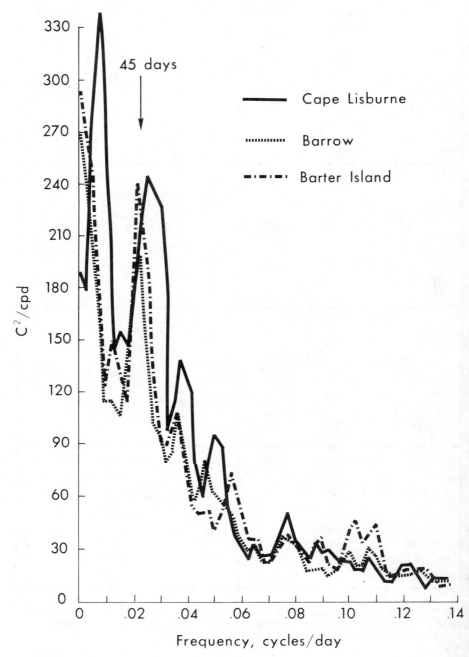

Fig. 17.8 Examples of power spectrum density estimates at the stations at Cape Lisburne, Barrow, and Barter Island.

Autocovariances and autospectra

The autocovariances of the deviation time series suggest that the thermal patterns along the North Slope are extremely variable. The e-folding times for the series from the seven stations, moving westward from Barter Island, are 4, 4.5, 5, 5, 4, 4.5, and 6 days, respectively. The autospectra were also computed from these covariance functions. The method we used to create the annual mean curve was very effective in removing any spectral peak at a period of one year. Although this cannot be readily verified on the plots from the Blackman-Tukey (1950) routine, discrete Fourier transforms were also computed — and no annual peak was observed in the deviation time series. The high-energy densities at low frequencies are thus probably due to poor resolution of the long-period trend. Recurrent peaks are found near periods of 45, 26, 19, and 13 days (Fig. 17.8). The two longer periods may be related to the anomalous warming trends mentioned above and may thus be, in part, artifacts of the analysis techniques. The warming trends are present in all the mean annual degree day curves in an attenuated and widened form. Therefore, when these mean curves are subtracted from the instantaneous records to produce the deviation series, a periodic fluctuation will be introduced into the series. The shorter periods are probably associated with local storm systems.

Cross-spectrum analysis was also performed on pairs of deviation time series, but the results were inconclusive.

SUMMARY

From this preliminary analysis of the DEW-line temperature records, a number of conclusions may be drawn concerning the mesoscale thermal variations along the coast of arctic Alaska.

During winter, a cold, continental-type climate exists along the coast, characterized by large variations in temperature on time scales less than one month — whereas during the summer, open-water season the same area experiences a cool maritime climate and small variations in temperature.

Large warming trends occur between late fall and early spring. These trends are related to the passage of low-pressure systems arriving from the North Pacific and bringing a strong flow of warm, southerly air which may raise temperatures significantly up to and above 0° C. They originate in the southwest, propagate eastward across the slope, and occur on an average of twice each winter.

The summer thermal conditions may be divided into two regimes. Cape Lisburne and Point Lay, the most southerly and westerly of the stations, experience longer, warmer periods of consecutive degree days above zero, on the order of 120 days. The remaining stations to the north and east have a shorter, cooler season, ranging from 90 to 100 days.

Preliminary spectrum analysis of the degree day deviations from the averaged daily mean indicates that a dominant 45-day period exists at all stations. We suspect, though, that this may be an artifact of the technique used to generate the time series.

Analysis presently underway of the spectra and phase relationship of both the deviations and of the actual degree day records will shed more light on the space-time variations of thermal conditions across the slope. This analysis will also include the daily atmospheric pressure, wind velocity, and wind direction.

Acknowledgments

This work was supported under Contract NOOO14-69-A-0211, Project NR388–002, with the Geography Programs, Office of Naval Research. Mrs. G. Dunn and Mr. K. Lyle are responsible for the illustrations. Ms. M. Erickson assisted with the computer programming. Ms. B. Julien and P. Pettigrew performed the tedious task of transferring the data from microfilm to IBM coding sheets.

REFERENCES

BLACKMAN, R. B., and J. W. TUKEY
    1950   *The measurement of power spectra.* Dover Publishing Co., New York, 190 pp.

WISEMAN, WM. J., ET AL.
    1973   Alaskan Arctic coastal processes and morphology. Coastal Studies Inst., Louisiana State Univ., Tech. Rept. 149.

Assessment of the Arctic Marine Environment: Selected Topics
Copyright © 1976 by Institute of Marine Science, University of Alaska, Fairbanks

CHAPTER **18**

Duration of sea states in northern waters

O.G. HOUMB[1] *and* I. VIK[2]

Abstract

The duration of sea states, defined in terms of the significant wave height, is investigated, and a Poisson model for estimation of the duration of sea states is developed. Estimates of duration of sea states in the North and Norwegian seas are made using the Poisson model.

INTRODUCTION

In wave statistics applied to engineering problems, efforts are concentrated on extreme value and long-term distributions of wave heights.

It has been verified (e.g., Nordenstrøm 1969, Battjes 1970 and Nolte 1973) that the Weibull distribution gives a good description of the distribution of wave heights. The Frechet and Gumbel distributions seem to give the best estimates of the extreme value distributions of wave heights (see, e.g., Thom 1973).

Information derivable from the statistics mentioned above is widely used for design in coastal and ocean engineering.

For the operation of offshore structures, information on the duration of storms is also needed. The most important work on this problem was done by Jensen et al. (1967) and by Rijkoort and Hemelrijk (1957). Jensen investigated the risk of damage to Danish dikes from storm surges.

Rijkoort and Hemelrijk considered whether twin storms occur more often than expected under the hypothesis of randomness. This problem can also be stated in another way: Is the probability of occurrence of a storm increased if another storm has occurred a few days before? Their conclusion was that twin storms appear more frequently than expected, assuming storms to be generated by a stochastic process.

---

[1] *Division of Port and Ocean Engineering, Norwegian Institute of Technology, University of Trondheim, N-7034 Trondheim, Norway.*

[2] *The Ship Research Institute of Norway, Paul Fjermstads Vei 26, 7000 Trondheim, Norway.*

Houmb studied the occurrence and duration of storms in the North Sea (1971). His definition of a storm was in terms of a certain wave intensity, in that a storm is said to occur when the visually observed wave height exceeds 2.5 m.

This paper contains a derivation of a Poisson model for the duration of sea states where a few parameters have to be determined to arrive at practical results. Sea states are defined in terms of H_S (significant wave height). Estimates based on this model are compared to results obtained from instrumental wave data.

DERIVATION OF THE MODEL

A statistical model by which estimates are made of the duration of sea states above or below certain levels is derived. The sea state is described by the significant wave height H_S, and measurements of H_S will be used as input in the model. It is assumed that the number of storms per unit time can be described by a Poisson process, which means that the duration of the storms follows an exponential distribution.

In Figure 18.1 the variation of H_S is shown for a random period of time. Measured values of H_S are plotted and connected by a curve, the ordinates being Weibull distributed.

The duration of a storm above a particular level, H'_S, is defined as the time when consecutive values of H_S equals or exceeds H'_S. Such an event is called a storm of level H'_S. The intervals in Figure 18.1, which are defined as storms of level H'_S are from a to b, from c to d and from e to f.

According to this definition, the frequency of occurrence of a storm will be equal to the frequency with which the curve of H_S crosses the level H'_S with a positive slope (i.e., up-crossing). It is assumed that the probability of the curve crossing the level H'_S during an interval of time is described by the Poisson process. The assumptions defining the Poisson process are the following:

1. The probability of the curve crossing the level H'_S in any non-overlapping time interval is independent and time-invariant.

2. The probability of crossing the level H'_S in a small time interval is proportional to the size of the interval.

3. The probability of multiple crossings in a small time interval is negligible relative to the probability of a single crossing.

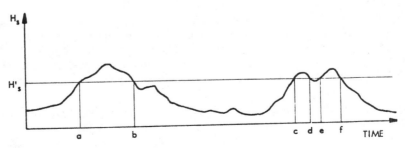

Fig. 18.1 H_S as function of time showing three storms.

The Poisson distribution function is given by:

$$p(N(t) = n) = \frac{(\lambda t)^n}{n!} \exp(-\lambda t)$$

(1)

with mean and variance

$$E(N(t)) = Var(N(t)) = \lambda t$$

(2)

The average number of events per unit time is λ. One of the elementary properties of the Poisson process is that the probability distribution of distances τ between consecutive events is exponential, having the density distribution

$$p(t) = \frac{1}{\lambda} \exp\left(-\frac{1}{\lambda} t\right)$$

(3)

The time duration of a sea state of level H'_S is the time between an upcrossing and a downcrossing of this level.

Setting λ equal to $\bar{\tau}(H'_S)$, which is the estimated average time duration of sea states of level H'_S, one obtains the density distribution of the duration of storms:

$$p(t) = \frac{1}{\bar{\tau}(H'_S)} \exp\left(-\frac{t}{\bar{\tau}(H'_S)}\right)$$

(4)

For a given level H'_S, $p(t)$ is known when $\bar{\tau}(H'_S)$ is determined.

The average frequency with which H_S crossed the level H'_S with a positive slope which equals the average frequency of sea states of level H'_S is given by

$$f(H'_S) = \int_0^\infty \dot{H}_S \cdot p(H'_S, \dot{H}_S)\, d\dot{H}_S$$

(5)

where $p(H'_S, \dot{H}_S)$ denotes the joint probability density function of H_S and its time derivative $\dot{H}_S$.

If L represents the total measurement period, the time that H_S exceeds H'_S during the same period is $(1-P(H'_S))$ L where $P(H'_S)$ is the probability that H_S is less than or equal to H'_S. The average number of upcrossings, i.e., the average number of sea states, during the measurement period is $f(H'_S)L$. The average duration of the sea states, is the amount of time the curve of H_S exceeds the level H'_S, divided by the expected number of sea states of this level.

$$\bar{\tau}(H'_S) = \frac{(1-P(H'_S))\, L}{f(H'_S)L} = \frac{1-P(H'_S)}{f(H'_S)}$$

(6)

The cumulative distribution of H_S is given by the Weibull distribution

$$P(H_S) = 1 - \exp\left(-\frac{H_S - H_O}{H_C - H_O}\right)^\gamma$$

(7)

The expression for the function $f(H'_S)$ in Eq. 5, contains the probability density function $p(H'_S, \dot{H}_S)$, which is unknown. From general statistics one knows that for a differentiable stationary real process, the ordinates of a random function and its derivatives taken at the same instant of time are uncorrelated random variables. For simplicity, H_S and $\dot{H}_S$ will be assumed to be independent. Physically, this means that we assume the time rate of change of H_S to be independent of H_S.

We then have

$$p(H_S, \dot{H}_S) = p(H_S)p(\dot{H}_S) \tag{8}$$

where $p(H_S)$ and $p(\dot{H}_S)$ are the probability density functions of H_S and $\dot{H}_S$, respectively.

The probability density function of H_S is obtained from Eq. 7:

$$p(H_S) = \frac{dP(H_S)}{dH_S} = \frac{(H_S-H_O)^{\gamma-1}}{(H_C-H_O)^{\gamma}} \gamma \exp(-(\frac{H_S-H_O}{H_C-H_O})^{\gamma}) \tag{9}$$

The function $p(\dot{H}_S)$ is unknown. When a random function is at a given level, however, an increase must on the average be compensated by a decrease, and the mean of its time derivative has to be zero; that is, the mean of $p(\dot{H}_S)$ is zero. The velocity with which H_S increases is not necessarily equal to the velocity with which it decreases, because generation and attenuation of waves are different physical mechanisms. The integration (Eq. 5) is taken from 0 to ∞; that is, for the positive values of $\dot{H}_S$, which are for waves in the generation phase. It is assumed that if the function $p(\dot{H}_S)$ fits a normal distribution with zero mean and standard deviation determined from the positive values of H_S measurements $_S$, the $p(\dot{H}_S)$ can be applied to positive values of $\dot{H}_S$ only. These are also the values of interest. We have

$$p(\dot{H}_S) = \frac{1}{\sigma\sqrt{2\pi}} \exp(-\frac{(\dot{H}_S)^2}{2\sigma^2}) \tag{10}$$

where σ is the standard deviation of positive values of $\dot{H}_S$, which must be determined from the data.

By use of Eqs. 5-10, the average time duration for sea states of any level H'_S is obtained:

$$\bar{\tau}(H'_S) = \frac{1 - (1-\exp(-(\frac{H'_S-H_O}{H_C-H_O})^{\gamma}))}{\int_0^\infty \gamma \dot{H}_S \frac{(H'_S-H_O)^{\gamma-1}}{(H_C-H_O)^{\gamma}} \exp(-(\frac{H'_S-H_O}{H_C-H_O})^{\gamma}) \frac{1}{\sigma\sqrt{2\pi}} \exp(-\frac{(\dot{H}_S)^2}{2\sigma^2}) d\dot{H}_S}$$

$$= \frac{\sqrt{2\pi}(H_C-H_O)^{\gamma}}{\gamma\sigma(H'_S-H_O)^{\gamma-1}} \tag{11}$$

From Eq. 11 it is seen that, in order to obtain the average time duration of any level, only one parameter, σ, must be determined in addition to the parameters of the long-term distribution of H_S.

When $\bar{\tau}(H'_S)$ is determined from Eq. 11, the average frequency of level H'_S, $f(H'_S)$, can be derived from Eq. 6:

$$f(H'_S) = \frac{1-P(H'_S)}{\bar{\tau}(H'_S)}$$

The probability for the number m of sea states of level H'_s that would be encountered in a length of time t can be predicted in terms of the Poisson model as

$$p(M(t) = m) = \frac{(f(H'_s) \cdot t)^m}{m!} \exp(-f(H'_s) \cdot t) \tag{12}$$

The probability that the duration is longer than t can be obtained by integrating Eq. 4, resulting in

$$Q(H'_s, t) = \exp\left(\frac{t}{\bar{\tau}(H'_s)}\right) \tag{13}$$

The frequency and duration of sea states below a certain level H'_s are also of interest and can easily be obtained; these will of course be equal to the frequency above the same level, which is $f(H'_s)$. Again let L represent the total measurement period. The time that H_s exceeds H'_s during the same period is $(1-P(H'_s))L$. The time that H_s is lower than H'_s is $L-(1-P(H'_s))L = P(H'_s)L$. The average time duration of sea states below the level H'_s is the amount of time that H_s is lower than H'_s, divided by the expected number of sea states below this level. Thus, the average time duration of sea states below the level H'_s is

$$\bar{\tau}_b(H'_s) = \frac{P(H'_s) \cdot L}{f(H'_s) \cdot L} = \frac{1}{f(H'_s)} - \bar{\tau}(H'_s) = \bar{\tau}(H'_s)\left(\frac{1}{1-P(H'_s)} - 1\right) \tag{14}$$

The probability that the time duration of a sea state below the level H'_s is greater than t is obtained from Eq. 13, using $\bar{\tau}_b(H'_s)$ instead of $\bar{\tau}(H'_s)$:

$$Q_b(H'_s, t) = \exp\left(-\frac{t}{\bar{\tau}_b(H'_s)}\right) \tag{15}$$

Experience showed poor agreement between average time duration of storms calculated from data and those predicted by Eq. 11.

The relation between H_s and σ was therefore investigated. $\sigma(H_s)$ was calculated for the two actual data bases (Figs. 18.2-18.4). For H_s greater than 5 m, values of σ are uncertain because of few observations. For H_s smaller than 5 m, the function $\sigma(H_s)$ for all three sites are approximately fitted by a straight line. There is a question whether these lines can be extrapolated beyond 5 m or if the functions $\tau(H_s)$ will tend to flatten out for higher values of H_s. In Figures 18.2-18.4, straight lines are fitted by eye to the plotted points. In Figures 18.2 and 18.3, the plotting positions for H_s higher than 6 m were ignored, because these points are based on very few observations.

It may be questioned whether this linear relationship is valid beyond the range covered by the data. There has to be a limit for the velocity with which the sea can receive energy from the wind and also for the velocity with which wave energy can be dissipated. According to the relationship in Figures 18.2-18.4, these velocities increase with increasing sea state. However, at a certain sea state the dissipation of wave energy has to equal the energy input from the wind. It is therefore reasonable to assume that the relashionship between σ and H_s flattens out for higher values of H_s.

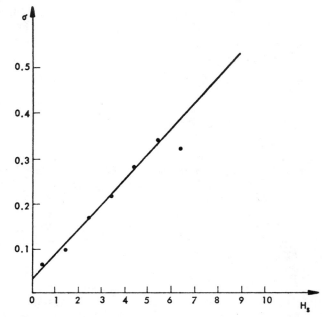

Fig. 18.2 Standard deviation (σ) of H_s as function of H_s at Utsira.

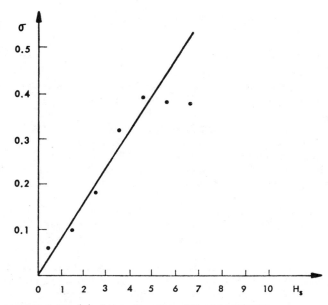

Fig. 18.3 Standard deviation (σ) of H_s as function of H_s at Lopphavet.

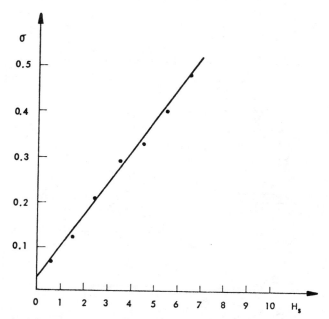

Fig. 18.4 Standard deviation (σ) of H_s as function of H_s at Halten.

The equation corresponding to Eq. 11 is then

$$\bar{\tau}(H'_s) = \frac{\sqrt{2\pi}}{\sigma(H'_s) \cdot \gamma} \frac{(H_c - H_o)^{\gamma}}{(H'_s - H_o)^{\gamma - 1}} \tag{16}$$

It was found that the duration of sea states was well fit by a Weibull distribution.

$$P(t) = 1 - \exp(-(\frac{t}{t_c})^{\beta}) \tag{17}$$

where $P(t)$ is the probability that the duration is shorter than or equal to t. The parameters are β and t_c. Figure 18.5 shows a typical fit of durations to the Weibull distribution.

We then have

$$\bar{\tau}(H'_s) = m_1 = \int_o^\infty t \frac{dP(t)}{dt} dt \tag{18}$$

where

$$m_n = \int_o^\infty t^n \frac{dP(t)}{dt} dt$$

248

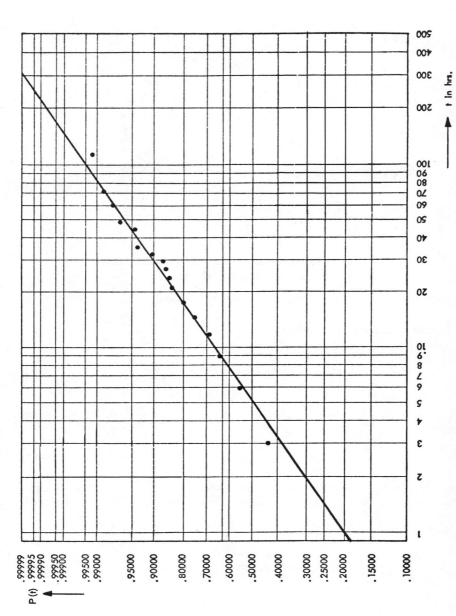

Fig. 18.5 Weibull distribution of duration of sea states above 3 m at Utsira. H′$_s$ = 3 m; N = 129 storms; β = 0.68; t$_c$ = 8.5 hrs.

is the n'th order moment. This gives

$$\overline{\tau}(H'_s) = t_c \cdot \Gamma\left(\frac{1}{\beta} + 1\right)$$

(19)

The variance of the duration is given by

$$\sigma^2(t) = m_2 - m_1^2$$

From Eq. 18 one obtains

$$\sigma^2(t) = t_c \left\{\Gamma\left(\frac{2}{\beta} + 1\right) - \left(\Gamma\left(\frac{1}{\beta} + 1\right)\right)^2\right\}$$

(20)

DISCUSSION

Data base

The data used to test the model were recorded on strip charts using the Dutch Datawell Waverider. The recording sites were Utsira, Halten and Lopphavet at depths of 100, 140 and 80 m, respectively (Figure 18.6). At all sites waves were recorded for 20 min every 3 hrs, and H_s was calculated by the Tucker method (Draper and Tucker 1964). The Utsira data cover approximately 20 months from the years 1969 to 1972, where each month of the year had approximately the same representation. From Halten, only 3 months of data were available, for the period 10 October 1972 to 5 April 1973, whereas Lopphavet covers the period from 3 September 1971 to 3 July 1972. Among the two last mentioned data bases, the Halten data are dominated by winter conditions. The $\dot{H}_s$ values were calculated taking the difference between two consecutive values of H_s and dividing this by 3 hrs, which is the time interval between two records.

Test of the duration model

Table 18.1 shows values of t_c and β from the three sites for different levels of H'_s. The parameter β is always less than 1, which should be the value of β if the time duration of storms according to the first definition were exponentially distributed. Further, the values of β are relatively constant for all locations and sea states except for $H'_s = 1$ m. If for each location the β-values are weighted with the size (measurement period) of the respective data base, mean values of β for the different sea states are obtained as shown in Table 18.2. For $H'_s = 1$ m, we use $\beta = 0.5$, and for higher values of H'_s, $\beta = 0.7$.

Table 18.3 contains values of t_c calculated by Eq. 19 compared to values obtained from plotting durations on Weibull probability paper. Also, average durations given by Eq. 11 are compared to those calculated from data.

We find the model estimates of the duration parameters to be in satisfactory accord with those calculated directly from data.

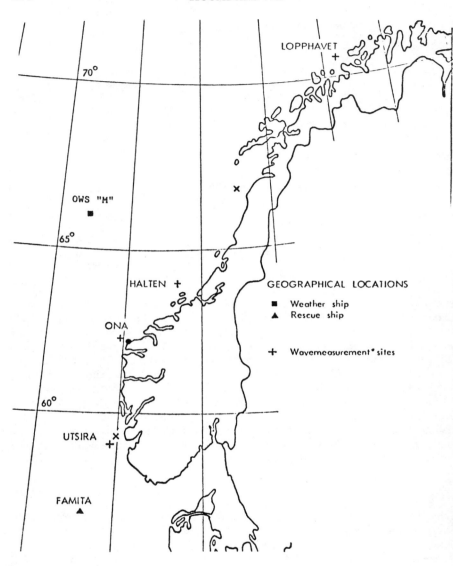

Fig. 18.6 Location map.

Return period and encounter probability

The average number of sea states per year exceeding a certain duration t is given by

$$F(H'_s, t') = F(H'_s)(1-P(t'))$$ (21)

TABLE 18.1   Parameters of the Weibull distribution of storm durations, calculated from data

Location	H'_s (m)	Number of storms	β	t_c
Utsira	1	231	0.45	18.0
	2	253	0.68	13.0
	3	129	0.68	8.5
	4	64	0.68	6.0
	5	26	0.61	4.7
Halten	1	7	0.41	240.0
	2	38	0.71	40.0
	3	58	0.79	14.7
	4	49	0.69	7.6
	5	26	0.95	7.0
Lopphavet	1	69	0.57	46.0
	2	90	0.72	14.0
	3	43	0.67	10.0
	4	25	0.67	9.7
	5	16	0.77	10.0

TABLE 18.2   Weighted values of β for different sea states

H'_s	1 m	2 m	3 m	4 m	5 m
β_{mean}	0.47	0.69	0.69	0.68	0.68

TABLE 18.3   Parameters estimated by the model compared to those calculated from data

Location	H'_s (m)	$F(H'_s)$ per year		$\bar{\tau}(H'_s)$		t_c		$\sigma(t)$
		Model	Data	Model	Data	Model	Data	
Utsira	1	153	143	40.5	44.3	20.3	18.0	99.0
$\gamma = 1.22$	2	143	156	19.1	17.5	15.0	13.0	34.0
$H_c = 1.4$	3	81	79	12.5	12.8	9.8	8.5	22.0
$H_0 = 0.4$	4	38	39	9.2	8.8	7.2	6.0	16.0
	5	16	16	7.1	7.9	5.6	4.7	12.5
Halten	1	54	34	149.0	320.0	75.0	240.0	367.0
$\gamma = 2.02$	2	144	182	44.0	45.8	35.0	40.0	78.0
$H_c = 3.5$	3	198	277	21.2	17.0	17.0	14.7	38.0
$H_0 = 0$	4	194	234	12.2	10.4	9.6	7.6	22.0
	5	137	124	8.0	8.1	6.3	7.0	14.1
Lopphavet	1	99	113	37.0	56.1	18.5	46.0	91.1
$\gamma = 0.90$	2	83	147	19.8	17.7	15.6	14.0	35.0
$H_c = 1.14$	3	55	70	14.3	13.0	11.3	10.0	25.0
$H_0 = 0$	4	35	41	10.9	12.0	8.6	9.7	19.3
	5	22	26	8.9	10.7	7.0	10.0	15.7

The return period $Rp(H'_s, t')$ can now be obtained as

$$R_p (H'_s, t') = \frac{1}{F(H'_s)(1-P(t'))} \qquad (22)$$

The fact that durations are Weibull distributed indicates that sea states may be considered as not generated by a Poisson process. Consecutive events in a process of this type were the beginning and the end of a certain sea states. If instead, the beginning of one sea state and the beginning of the next are regarded as consecutive events, we would probably be more in common with the Poisson process. Considering extreme sea states, their duration will be small compared to the respective return periods. Then it is possible to have an approximate Weibull distribution of the duration of extreme sea states and an approximate exponential distribution of the periods between them, because a significant change in a certain duration only gives relatively little change in the time period between them. Therefore, the good fit of a Weibull distribution to the durations cannot be used as proof against the assumption that the time durations between sea states are governed by a Poisson process. Considering extreme sea states as events more than durations, it still will be reasonable to assume that these events are governed by a Poisson process. Then one can calculate the probability $P(n)$ that a number of n occurrences of the sea state considered can be expected during a given time period. This probability is assumed to be given by Poisson's law:

$$p(N(t)=n) = \frac{(\lambda t)^n}{n!} \exp(-\lambda t) \qquad (23)$$

where λ is the average number of sea states per unit of time. If λ represents the average number of occurrences per year and t the length of a considered time period in years, the average number of exceedances m during the period t is then given by the relation

$$m = \lambda t$$

If $m = 1$, we have $t = Rp$ which combined with $n = 0$ gives

$$Q = 1 - P(0) = 1 - e^{-1} = 0.63 \qquad (24)$$

This means that during Rp years there is a 63% probability of at least one exceedance of a duration of return period Rp.

The relation between duration and return period for this risk level (63%) is given by Eq. 22. Because the data from Halten cover only winter months, these data cannot without modifications be used in calculation of return periods of several years. Therefore, only the data from Utsira will be considered in this case. Figures 18.7 and 18.8 show plots of Eq. 22 for H_s levels of 4 and 6 m at Utsira. Lines for different values of m are drawn in the same figures. Each value of m corresponds to the risk level given by the probability obtained using Poisson's law.

The information given by the different values of m can also be obtained in another way. For convenience, the encounter probability will be introduced. The encounter probability, Ep, is defined as the probability that a given duration will be

253

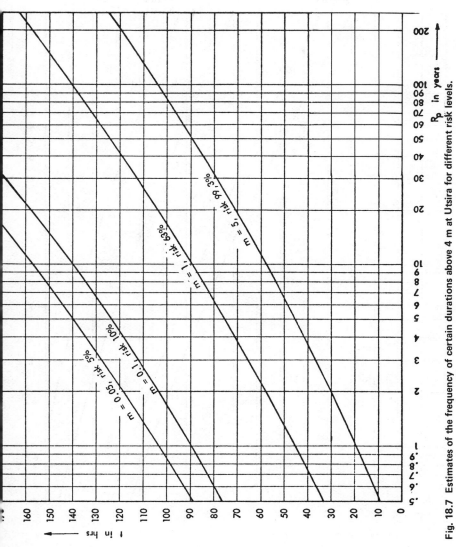

Fig. 18.7 Estimates of the frequency of certain durations above 4 m at Utsira for different risk levels.

254

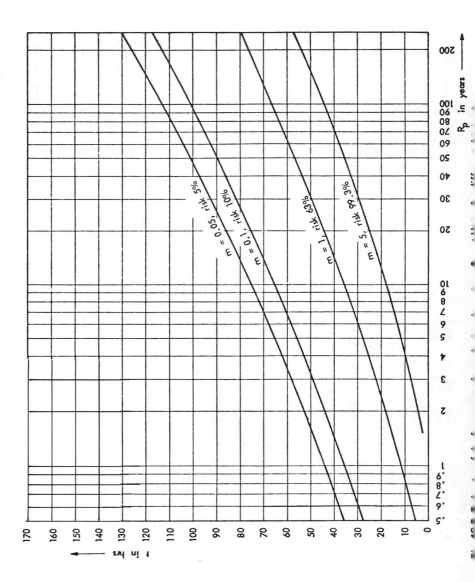

exceeded in a given period of time. If n = 0 in Poisson's law, we have the probability of no exceedance:

$$P(0) = e^{-(\lambda t)} = e^{-(\frac{t}{Rp})}$$

The encounter probability Ep is now given by:

$$Ep = 1 - P(0) = 1 - e^{-(\frac{t}{Rp})} \tag{25}$$

Figure 18.9 shows a plot of Eq. 25. The use of this figure can be illustrated by an example: We want the probability that in a time period of 5 yrs a storm will occur in which the significant wave height is greater than 6 m for a time duration longer than 40 hrs. From Figure 18.8, the return period corresponding to a duration of 40 hrs (m = 1) is about 14 yrs. That is, t/Rp in Figure 18.9 is 5/14 = 0.36, and the desired probability, the encounter probability, is 0.30.

Durations based on visual data

Due to the scarcity of instrumental wave data, visual observations of waves are used extensively in estimating long-term distributions of wave height, for example.

The relation between H_V and H_S was investigated by Pedersen (1971), who found this given by Figure 18.10 for values of H_S and H_V at the same level of probability. For some applications the assumption

$$H_V = H_{1/3}$$

gives a sufficient accuracy.

Visual wave data can be used as input data for the model. There is one problem, however, concerning the relation of the standard deviation between $\dot{H}_S$ and H_S.

In the determination of this relation one needs exact information of the change in sea state with time. H_V are reported in intervals of 0.5 m, and for high values of H_V the reports show that the observers prefer to give the wave length in intervals of 1.0 m. This procedure leads to an overestimation of the standard deviation of the time derivative of H_V. This overestimation will probably be significant, and the function $\sigma = \sigma(H_V)$ obtained from visual observations will be unuseful without modifications.

Two data bases will now be considered: weather ship station M in the North Atlantic at 66°00'N, 02°00'E (Fig. 18.6) and the weather and rescue vessel *Famita* in the North Sea at 57°30'N, 03°00'E (Fig. 18.6). Visual observations of wave height for the period 1949-1972 are available from the M station.

The Weibull distribution of H_V for these data is given in Figure 18.11. Data from *Famita* are available for the winter months October to March from 1959. The Weibull distribution of visual heights reported from *Famita* is given in Figure 18.12.

Due to the difficulty in obtaining a reliable relation between σ and H_V, as mentioned above, a mean value of this relation in the instrumental data from Utsira, Halten and Lopphavet was adopted for use in connection with the visual data (Fig. 18.13).

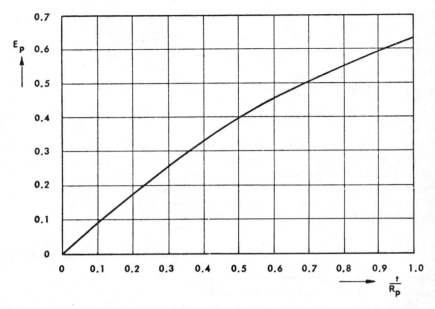

Fig. 18.9 The encounter probability Ep as function of $\dfrac{t}{R_p}$.

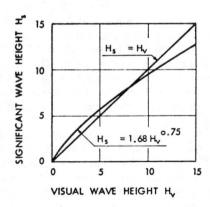

Fig. 18.10 Connection between H_s and H_y at the same probability level (Pedersen 1971).

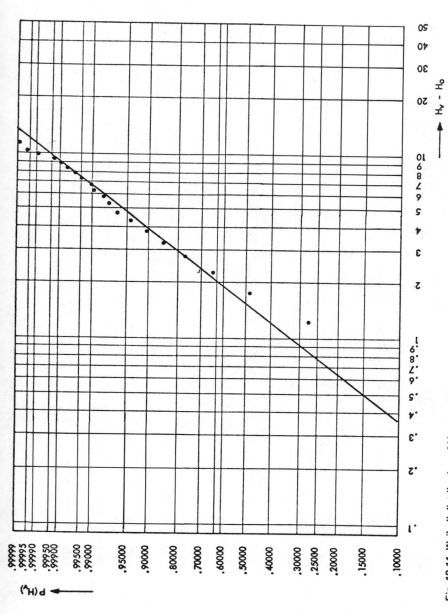

Fig. 18.11 Weibull distribution of H_v at weather station M, 1949-1972. $H_o = 0$; $H_c = 2.0$ m; $\gamma = 1.29$.

258

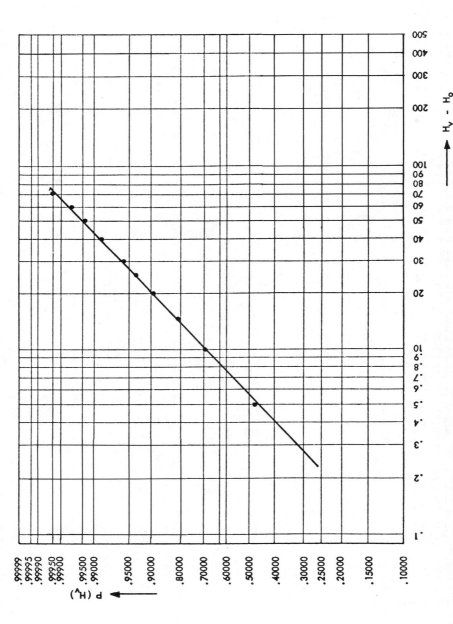

Fig. 18.12 Weibull distribution of H_v at *Famita*, 1959-1969. $H_o = 1.0$ m; $H_c - H_o = 0.86$ m; $\gamma = 0.91$.

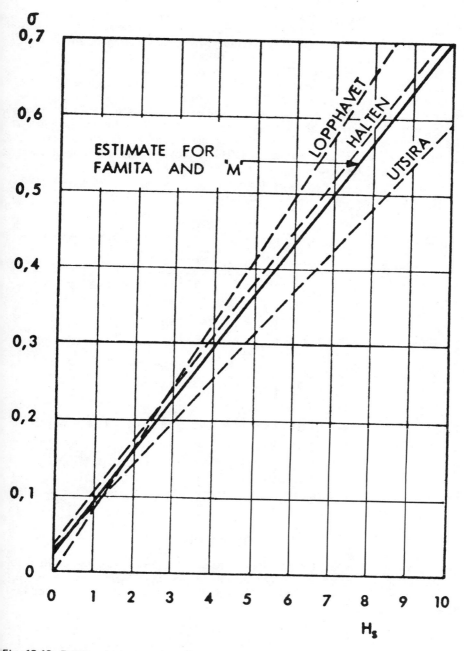

Fig. 18.13 Relation between σ and H_s used in connection with visual data from *Famita* and weather station M.

This adoption will of course introduce an uncertainty in the calculations, and the results should only be regarded as approximations. The relation $H_V = H_S$ is therefore assumed to give sufficient accuracy. The parameter β in the Weibull distribution of storm durations is assumed to be 0.7.

The parameters $\bar{\tau}(H'_V)$, t_c, $F(H'_V)$ and σ_T are calculated and listed in Table 18.4.

The weather ship station M in the North Atlantic is probably slightly more exposed to severe storms than *Famita* in the North Sea. One should therefore expect both longer mean durations and higher frequencies of storms in this area. According to Table 18.4, this is usually the case. But it is not true for mean durations above levels of H_V' equal to 7 m and higher, and the difference between the data bases is not as great as expected. This is probably due to the observation periods at *Famita*. Here, waves are reported only in the months October to March; therefore, these data are not representative for the whole year and will overestimate durations and frequencies of storms in this area.

TABLE 18.4   Input and output parameters in model for station M and *Famita*.

	H'_V (m)		2	3	4	5	6	7	8	9	10
Station M $H_0=0$, $H_C=2.0$ $\gamma=1.29$	$\sigma(H'_V)$		0.16	0.23	0.30	0.36	0.43	0.50	0.57	0.64	0.70
	$\bar{\tau}(H'_V)$	(hrs)	24	15	11	8.2	6.6	5.4	4.6	4.0	3.5
	t_c	(hrs)	19	12	8.5	6.5	5.2	4.3	3.6	3.1	2.7
	$F(H'_V)$	(per year)	136	110	73	43	23	11	5.0	2.0	0.88
	σ_T	(hrs)	43	27	19	14.5	11.6	9.5	8.1	7.0	6.1
Famita $H_0=1.0$, $H_C-H_0=0.86$ $\gamma=0.91$	$\sigma(H'_V)$		0.16	0.23	0.30	0.36	0.43	0.50	0.57	0.64	0.70
	$\bar{\tau}(H'_V)$	(hrs)	15	11.2	8.9	7.5	6.4	5.6	5.0	4.6	4.2
	t_c	(hrs)	11.7	8.8	7.1	5.9	5.1	4.4	4.0	3.6	3.3
	$F(H'_V)$	(per year)	189	90	42	20	8.7	4.0	1.9	0.81	0.38
	σ_T	(hrs)	26	20	16	13	11	9.9	8.9	8.0	7.3

CONCLUSION

The model for estimating the duration of sea states of levels H'_S gives results that are in satisfactory accord with results obtained from data. To improve the model, the relation of the standard deviation between $\dot{H}_S$ and H_S should be investigated for extreme sea states.

REFERENCES

BATTJES, A.

1970 Long-term wave height distribution at seven stations around the British Isles. NIO Int'l Report A-44.

DRAPER, L., and M. J. TUCKER

1964 Derivation of a "design wave" from instrumental records of sea waves. Institution of Civil Engineers, vol. 29, session 1963-1964.

HOUMB, O. G.

1971 On the duration of storms in the North Sea. First Int'l Conf. on Port and Ocean Engineering under Arctic Conditions (POAC), Trondheim, Norway, August 1971.

JENSEN, A., K. SCHMIDT, N. E. JENSEN, and S. WEYWADT

1967 Analyse af Stormflodsrisiko. Report on storm surges. Technical University of Denmark, Copenhagen.

NOLTE, K. G.

1973 Statistical methods for determining extreme sea states. Second Int'l Conf. on Port and Ocean Engineering under Arctic Conditions (POAC), 1973, Iceland.

NORDENSTRØM, N.

1969 Methods for predicting long-term distributions of wave loads and probability of failure for ships. Appendix 1. Long-term distributions of wave height and period, Det norske Veritas, Report no. 69-21-S, Oslo.

PEDERSEN, B.

1971 Predictions of long-term wave conditions with special emphasis on the North Sea. First Int'l Conf. on Port and Ocean Engineering under Arctic Conditions (POAC), Trondheim, Norway, August 1971.

RIJKOORT, P. J., and J. HEMELRIJK

1957 The occurrence of twin storms from the northwest on the Dutch coast. Statistica Neerlandica 1957, vol. 11-3.

THOM, C. S.

1973 Extreme wave height distributions over oceans. ASLE vol. 99, no. WW3 (August).

CHAPTER **19**

Response of waves and currents to wind patterns in an Alaskan lagoon

JOSEPH A. DYGAS *and* DAVID C. BURRELL [1]

Abstract

The response of waves and currents to the prevailing summer wind regime is reported for a portion of the Beaufort Sea (Arctic Ocean) coastline of Alaska. The study site, immediately east of the Colville River delta, is bounded seaward by a barrier island complex. Observation periods in 1972 for wind, currents, and waves were 17 July to 20 October, 11 August to 18 September, and 22 August to 1 September, respectively. Northeasterly winds predominate, with prevailing currents towards the west or northwest. Linear correlation coefficients of +0.73 and -0.52 were determined for wind and current velocity and direction, and current direction reversals within the lagoon occur (with a relatively short time lag) when the prevailing wind shifts from an easterly to a westerly pattern. Mean wind velocities and the frequency of westerly storms increase from July to October. Power spectral analysis of low-pass filtered components of the lagoon current record gives low-frequency, wind drift currents with periodicities of 4 to 5 days. Wind drift currents have a greater effect than tidal currents on the general circulation in this restricted area. Spectral analysis of wave records shows that wind-waves with significant periods of 1.9 to 2.1 sec predominate; 7.5 to 15.0-sec period swell is also present, but with less energy.

---

[1] *Institute of Marine Science, University of Alaska, Fairbanks, Alaska 99701.*

INTRODUCTION

Coastal physical processes under polar climatic conditions have received less
attention to date than comparable processes in more temperate environments.
However, with the present thrust to exploit the resources of these arctic
regions — and concurrently to understand the environmental impact of this
activity — such work has assumed considerable urgency. This chapter describes the nearshore wind, wave, and current patterns observed over a
relatively brief period in the summer of 1972 within one lagoon on the
arctic (Beaufort Sea) coast of Alaska. The work originally constituted part
of an interdisciplinary program (Kinney et al. 1972; Alexander et al. 1974)
designed to establish baseline oceanographic conditions in this specific
locality (and hopefully applicable in part to other areas on the Beaufort Sea
coast) prior to any large-scale industrial development. Since the initiation of
this project in 1970, a number of complementary efforts have been reported, e.g., by Reed (1974) and most especially — a comparative study by
Wiseman et al. (1973) of physical processes and geomorphology along
portions of both the Beaufort and Chukchi sea coasts of Alaska.

LOCALITY AND ENVIRONMENT

The study area is centered upon Oliktok Point at the west end of Simpson
Lagoon and about 30 km northeast of the Colville River, midway along the
Beaufort Sea (Arctic Ocean) coast of Alaska (Fig. 19.1). Simpson Lagoon is
approximately 7 km wide (N-S) and 25 km long (E-W), bordered seaward
by a chain of low relief barrier islands. Water depths within the lagoon are
generally less than 2 m.

The climate in this region is characterized by: a relatively cold and stable
high pressure center, the fluctuating periphery of which affects the passage
of local cyclones and anti-cyclones; a temperature inversion; and long (eight
to nine months) dark winters and short summers. Sater et al. (1971) have
discussed the development along the northeast coast of Siberia of easterly
moving surface cyclones which bring frontal storms into the study area. A
pronounced seasonal temperature gradient develops between warm air over
the coastal plain and cold air over the oceanic ice-pack after the summer
thaw and break-up of snow and ice along the arctic coast. Subsequently,
ice-free coastal water supplies moisture to the local atmosphere, contributing
to the development of low clouds, fog, and precipitation.

METHODS

The procedure adopted for this study has been to describe the prevailing
currents within Simpson Lagoon during the ice-free months of July to
October 1972 and to statistically compare the wind regime with wave and
current patterns. Only a brief description of the methods of data observation, collection, and analysis are included here. Additional details concerning

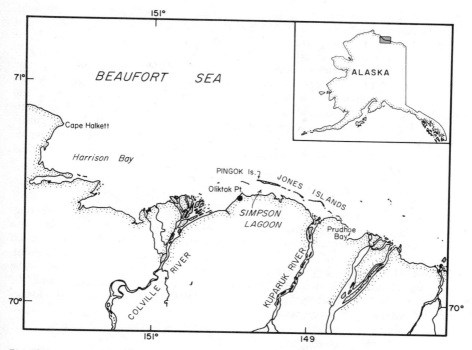

Fig. 19.1 Locality of Simpson Lagoon on Beaufort Sea (Arctic Ocean) coast of Alaska.

field methods and instrumentation for measuring wind, currents and breaking waves have been discussed by Dygas et al. (1972) and Alexander et al. (1974).

Wind

Wind direction and velocity data were recorded in analog form over the period 17 July through 20 October using a three-cup anemometer and wind-vane instrument package (R. M. Young Co.) mounted 10 m above sea level on the western shore of Oliktok Point. These records were subsequently digitized at 1-hr intervals, and wind roses, histograms of wind direction, cumulative probability distributions of wind speeds, and linear correlation and regression analysis with the Simpson Lagoon current records were computed.

Currents

Current directions and speeds were sampled at 10-min intervals using an Aanderaa digital-recording meter moored 3.6 km north of Oliktok Point and 1 m off the bottom of the lagoon. Current direction and speed were recorded from 11 August to 18 September, as well as for the 25 August to 1 September period of the previous year (1971). The 1972 digital current record was computer-analysed to yield linear correlation and regression with wind data, and a description of the frequency composition of the current record was made using time series analysis. The procedure for the latter analysis consisted of: separation of the velocity vector into N-S and E-W components, time-averaging these components over periods ranging from one sample per minute to one per hour, filtering the samples using a low-pass (Doodson and Warburg 1941) filter with a 24.0-hr cutoff periodicity, and performing a power spectral analysis of the filtered component output using the time series spectral analysis program described by Fee (1969). A cutoff frequency was selected in order to separate the tidal components, which have periods of 24 hrs and less, from longer-period components of wind drift currents.

Waves

Wave heights and periods were obtained from 22 August through 1 September using a portable wave recorder with a continuous-resistance wire wave-staff. The wave sensor was located just seaward of the breaker zone on the northeast shore of Oliktok Point. A spit, westerly from Oliktok Point, shielded the sensor from waves originating from SW-NW directions. The resulting analog wave records were digitized at 0.3-sec intervals, and the power spectral analysis program was employed to compute and display spectral energy against frequency.

RESULTS AND DISCUSSION

Wind

The wind direction histograms obtained during August and September (Fig. 19.2) indicate that prevailing winds at Oliktok Point are primarily from the northeast and secondarily from the northwest. Searby and Hunter (1971) note that the prevailing winds at Barter Island (about 250 km east of Oliktok Point; no more suitable record exists) are easterly or westerly. These latter data, collected over the period 1953 to 1970, are included in Figure 19.2. It must be re-emphasized that the Oliktok values reported here are for a relatively short sampling period during one year only; nevertheless, the winds in the study area appear to correspond closely with the long-term Barter Island pattern. Some minor deviations are probably attributable to the wide spatial separation between these stations. Both localities appear to demonstrate similar basic meteorological characteristics, and on this basis it would appear that the short record obtained at Oliktok Point in 1972 is typical of summer conditions in this area.

The histograms (Fig. 19.1) and the wind direction frequencies listed in Table 19.1 illustrate that westerly to northwesterly winds increase from July to October. A larger variability of the wind velocities (expressed in Table 19.1 as a percentage of the total variance of wind velocities for the indicated groups of wind directions during each month) is also apparent over the same period. These trends are consistent with observations that cyclonic

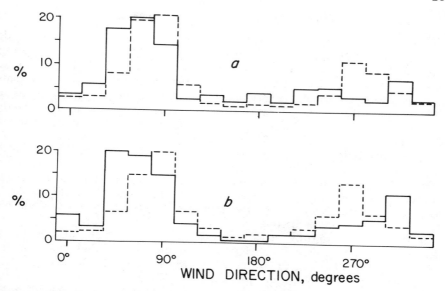

Fig. 19.2 Oliktok Point and Barter Island wind direction histogrammes: a. August; b. September. Solid line is 1972 Oliktok Point data; dashed line gives mean directions recorded 1953 to 1970 at Barter Island (Searby and Hunter, 1970).

storms travel eastward along the northern coast of Alaska with increasing frequency during late summer through early winter. Hurst (1971), for example, has documented the effects of one such storm (September 1970) in the Mackenzie delta region. At this time, wind velocities approaching 130 km/hr were recorded at the DEW-line station located at Oliktok Point, and the storm surge on the western shore of the Point was estimated to be approximately 3 m (J. DiMaio, personal communication). Dygas et al. (1972) have discussed the residual sedimentological and physiographic effects of this storm in the vicinity of Simpson Lagoon.

TABLE 19.1   Distribution of wind directions and variance of wind velocities at Oliktok Point, Alaska, 1972

	July		August		September		Total frequency of occurrence %
Wind direction	Frequency of occurrence %	Variance of wind speeds %	Frequency of occurrence %	Variance of wind speeds %	Frequency of occurrence %	Variance of wind speeds %	
N - E	27.4	45.0	19.3	21.4	19.2	27.6	65.9
E - S	1.0	19.9	4.0	29.5	2.2	4.4	7.8
S - W	2.0	18.6	5.0	24.0	3.3	22.6	10.4
W - N	2.4	16.4	5.0	25.0	8.5	45.3	15.9

Currents

Initial observation of the current patterns in Simpson Lagoon were made during 1–11 September 1971 (Dygas et al. 1972). These data, together with studies of currents along the shores of Pingok Island (Fig. 19.1) by Wiseman et al. (1973), suggested a significant correlation between currents and the prevailing wind system. In addition, it was noted by Dygas et al. (1972) that the height of the meteorological tide within the lagoon commonly exceeded the lunar tidal range. Matthews (1970) notes the lunar tidal range to be about 0.3 m at Point Barrow. Dygas et al. (1972) observed that wind stress acting across the shallow waters of Harrison Bay and Simpson Lagoon (Fig. 19.1) tended to pile water up on the downwind side. Dye studies in the deeper (5 m) inlets between the barrier islands showed some outflow of bottom water from the lagoon to the Beaufort Sea, while surface water under the influence of easterly winds was moving shoreward into Simpson Lagoon.

This report is primarily concerned with an analysis of the distribution of periodicities, and statistical correlations with concurrent wind data, for the initial 29 days of the current record recovered from the lagoon in 1972. Results of these analyses are presented as filtered, low-frequency current components and corresponding power spectra (Figs. 19.3–19.6), and high-frequency records and corresponding power spectra in Figures 19.7–19.10.

The low-period (greater than 24 hrs) current records (Figs. 19.3–19.6) indicate the presence of wind-drift or other long-period currents. Comparison of Figures 19.3 and 19.5 shows that the range of E-W wind-drift current speeds is greater than that for the N-S component. The E-W component record (Fig. 19.3) has some 80 percent of the spectral energy density confined within the 4 to 5 day period range, and this feature is more distinctive than in the N-S speed record of Figure 19.5. This marked 4 to 5 day periodicity within the low-frequency current record suggests a correlation with 3 to 7 day period atmospheric Rossby waves (Rossby 1937, 1939). It thus appears likely that periodic current reversals in Simpson Lagoon are a function of periodic fluctuations in the atmospheric weather cycles.

The short-period current components (less than 24 hrs) have spectral energy peaks at 33.0 and 11.0 hrs in the E-W component records (Figs. 19.7 and 19.8) and at 20.0 to 24.0 hrs and 8.3 hrs in the N-S records (Figs. 19.9 and 19.10). The 33.0 hr E-W energy peak suggests that the digital filter did not completely separate the low- and high-frequency components at the assigned cutoff period of 24 hrs. The 20.0 to 24.0-hr periodicity present in the N-S data indicates the possible occurrence of diurnal tidal currents. Spectral energy peaks at semi-diurnal tidal periodicities (11.97 to 12.91 hrs) and inertial current periods of about 12 hrs are not very distinctive (Figs. 19.8 and 19.10), possibly because of side-lobe effects of the strong, low-frequency peak at 20 to 24 hrs. There is, in fact, only a slight inflection of the spectral curves at the 12.0-hr period (Figs. 19.8 and 19.10). The distinct energy peak at 8.3 hrs in the N-S spectrum (Fig. 19.10) is attributable to neither tidal nor inertial current periods, but it is due possibly to a natural period of oscillation within this specific lagoon.

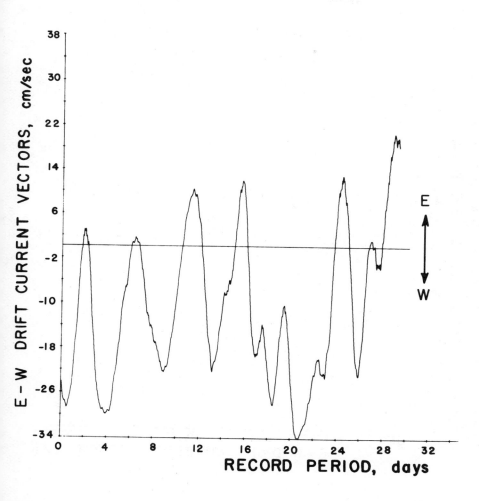

Fig. 19.3 Low frequency (periods $>$ 1 day) east-west drift currents. Oliktok Point, August 11 to September 8, 1972.

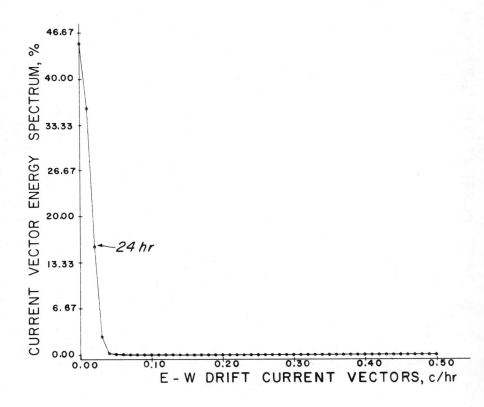

Fig. 19.4 Energy spectrum of east-west components of the filtered current record.

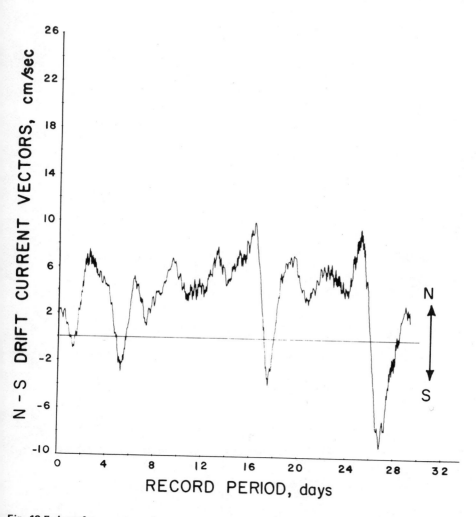

Fig. 19.5 Low frequency north-south drift currents. Oliktok Point, August 11 to September 8, 1972.

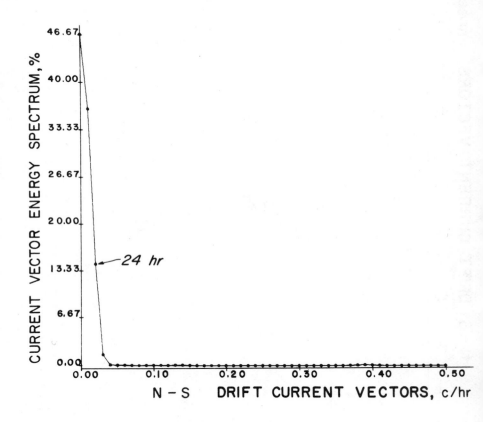

Fig. 19.6 Energy spectrum of north-south components of the filtered current record.

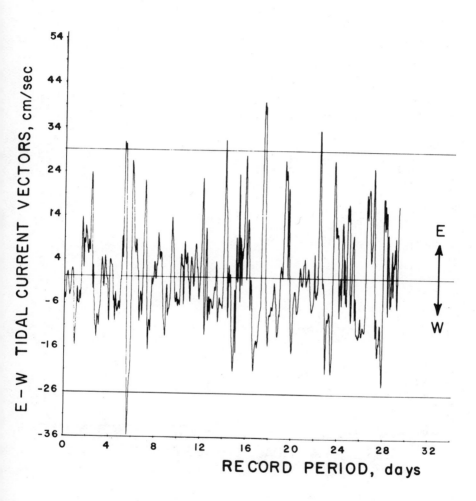

Fig. 19.7 High frequency (periods < 1 day) east-west tidal currents. Oliktok Point, August 11 to September 8, 1972.

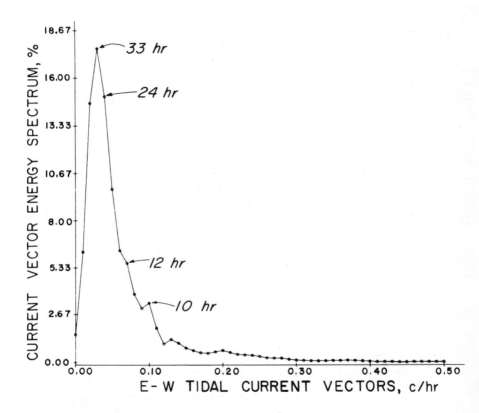

Fig. 19.8 Energy spectrum of the east-west components of the filtered current record.

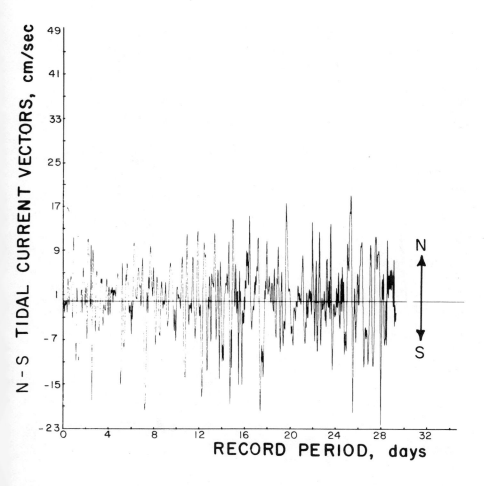

Fig. 19.9 High frequency north-south tidal currents. Oliktok Point, August 11 to September 8, 1972.

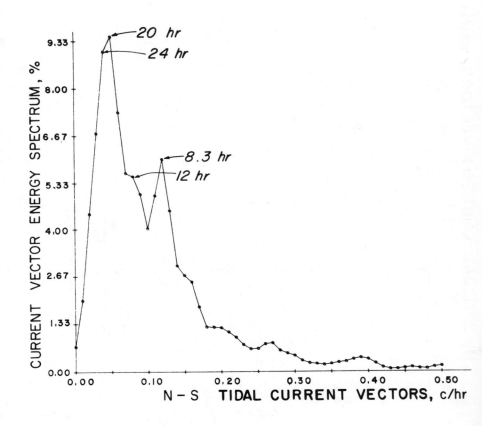

Fig. 19.10 Energy spectrum on the north-south components of the filtered current record.

A plot of progressive current vectors for Simpson Lagoon (Fig. 19.11) yields a net water transport distance of 358 km over the period 11 August to 18 September at a mean vector velocity of 10.6 cm/sec in a WNW direction. The absence of any distinctive oscillatory pattern suggests that the amplitudes of tidal current velocities are small in relation to mean flow.

Linear correlations between wind and current speed and direction (Figs. 19.12 and 19.13) yield correlation coefficients of +0.73 and -0.52 (at the 0.95 confidence level) between the speed and direction, respectively. These relationships are of greater significance than is indicated by the numerical coefficients, since the effects of time lags in the response of currents to changes in the wind regime are not taken into account. Future research on this topic might profitably be directed towards a study of this response time via a cross-power spectral correlation between concurrent wind and current records.

This analysis of the current record taken in the summer of 1972 shows that the water in Simpson Lagoon, from the surface to 1 m from the bottom, flows predominantly in a WNW direction under the influence of prevailing easterly winds and that this movement is reversed with little observable time lag when the wind is from the west or northwest. It has been demonstrated previously (Dygas et al. 1972) that winds from the latter quadrant increase in frequency as summer progresses.

Waves

Waves inside Simpson Lagoon consist of local, wind-generated waves and small-amplitude swell which filters through passages between the barrier islands from the Beaufort Sea. Except during severe westerly frontal storms (such as that of September 1970, noted above), breaking wave heights along the coast of the lagoon are typically less than one meter, and of the plunging variety. Observations of breaking waves during August 1971 at Oliktok Point (Dygas et al. 1972) gave mean and significant breaker heights of 17.7 and 27.3 cm, respectively, with a mean wave period of 2.2 sec.

A spectral analysis of wave records collected in August 1972 on the eastern shore of Oliktok Point is presented in Figure 19.14. Significant wave period and concurrent wind data are listed in Table 19.2. These wave periods (corresponding to the peak spectral wave energy) range from 1.9 to 2.1 sec, but a number of lower spectral energy peaks — from low energy swell — with equivalent periods of 7.5 to 15.0 sec, are also present in the spectra. The occurrence of swell with a visually distinctive wave height at Oliktok Point is unusual (J. DiMaio, personal communication of non-systematic observation over the decade 1962 to 1972). However, in early September 1972, the authors and Wiseman et al. (1973) simultaneously observed 8.5 to 10.0-sec period swell impinging on the coast inside Simpson Lagoon and on the seaward side of one of the Jones Islands (Fig. 19.1). The height of the swell on the barrier island was estimated to be in the range 1.5 to 2.0 m, and 10.0 to 20.0 cm on the lagoon shoreline.

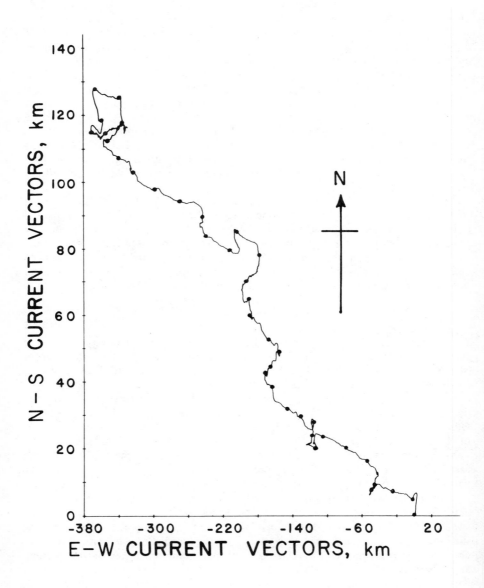

Fig. 19.11 Progressive current vector diagram for Simpson Lagoon (at Oliktok Point), August 11 to September 18, 1972. One day intervals shown.

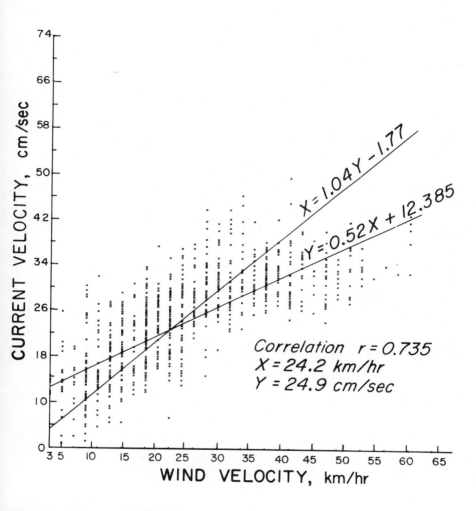

Fig. 19.12 Linear regression and correlation of current velocity (Y) and wind velocity (X). The correlation is significant at the 0.95 level. One hour sample intervals, August 11 to September 18, 1972, Oliktok Point.

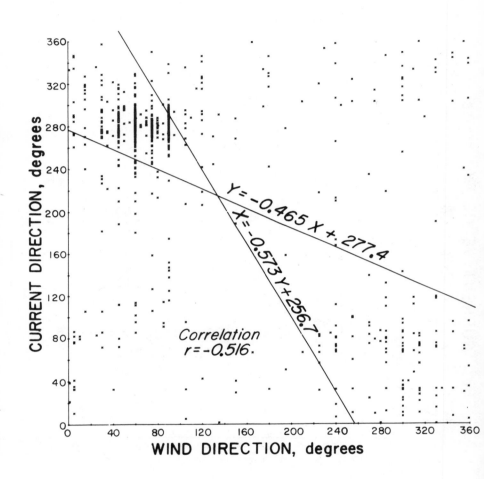

Fig. 19.13 Linear regression and correlation of current direction (Y) and wind direction (X). The correlation is significant at the 0.95 level. One hour sample intervals, August 11 to September 18, 1972, Oliktok Point.

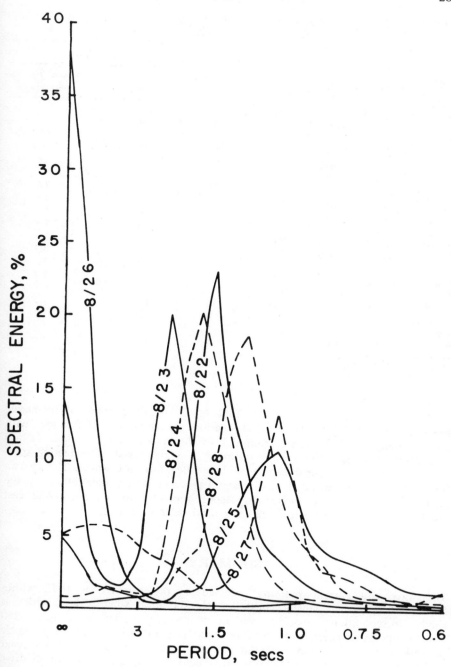

Fig. 19.14 Wave spectra at Oliktok Point, August 1972.

TABLE 19.2 Wave periods for direction winds; also wind velocities and duration
from given observation time. Oliktok Point, Alaska, August 1972

Wind direction	Date	Local time	Wind Velocity (km/hr)	Wind Duration (hrs)	Wave period Significant seconds	Wave period Second highest energy (sec)
N-E	25	0930	8.0	5	1.1	—
	22	1000	12.8	20	1.9	—
		1315	16.1	23	1.9	5.0
		2030	16.1	32	1.9	7.5
	23	0900	17.7	45	1.9	7.5
		1430	28.2	50	2.1	*
	27	1600	12.1	1	1.1	7.5
	28	1015	16.1	5	1.4	3.8
		1930	20.1	15	1.9	*
	29	1530	20.1	35	1.9	7.5
NNE-NE	23	2030	25.7	56	2.1	*
	24	1930	20.1	4	1.9	0.9
		1515	16.1	1	1.1	*
N-W	25	2020	9.6	7	*	1.2
	26	1600	12.1	27	*	0.9
	30	0800	17.6	6	1.9	7.5
	31	2230	32.1	4	2.5	*
S-W	31	1620	20.1	7	2.5	*
S-E	29	1900	12.0	2	1.7	15.0

*Wave periods > 15.0 sec.

The data of Figure 19.14 and Table 19.2 suggest a trend of increasing spectral wave energy with increasing periods of from 1.0 to 3.0 sec, with corresponding wind velocities from 3.5 to 7.8 m/sec. Both Hunkins (1962) and Wiseman et al. (1973) have recorded a similar trend of increasing spectral wave energy and significant period for Beaufort Sea waves. The character of wind-waves generated in Simpson Lagoon is largely determined by the available fetch, which is limited by lagoon geometry to the north and east but is open to the arctic ice-pack in a northwesterly direction.

SUMMARY

The wave and current patterns observed in the summer of 1972 within Simpson Lagoon are largely controlled by the local wind regime, which is similar to conditions recorded over many years at Barter Island some 250 km to the east. These data, together with less extensive observations made in adjacent years, suggest that the patterns recorded in this particular year are typical of summer conditions in this area. Northeasterly winds drive surface (depths in Simpson Lagoon are mostly less than 2 m) currents in a westerly direction, and westerly winds — which occur with increasing frequency as winter approaches — reverse this trend. The westerlies are less fetch-limited and are associated with low pressure centers and surface storms. It is possible that 4 to 5 day periodicities in the wind-drift current records may be correlated with 3 to 7 day period atmospheric weather cycles. The direction of net water transport over the period 11 August through 18 September was towards the WNW with a mean speed of 10.6 cm/sec. There appears to be little time-lag between wind and current reversals, and the significant correlations between wind and current speeds and directions suggest a close coupling between current patterns in Simpson Lagoon and the prevailing meteorological processes. Wind stress is also important in the genesis of short-period waves, the spectral energies of which increase with significant wave period, wind speed and direction. Tidal and inertial currents are subordinate to wind-drift currents in this environment.

The statistical correlations between wind and water transport presented here suggest a means for predicting the direction and rate of movement of contaminants leaked into this arctic coastal environment. Such a capability would be a considerable aid to environmental protection as industrial exploitation increases. Along the Alaskan arctic coast, for example, there are already existing facilities which monitor the local wind and pressure regimes, notably at Point Barrow and Barter Island, and at intermediate DEW-line sites. These real-time weather data provide the basis for a remote and economic means of predicting local wind-drift current patterns, as described, which would be of considerable utility until more precise knowledge of the coastal currents is available.

Acknowledgments

Financial support for this research was provided by the Alaska Sea Grant Program (supported by NOAA Office of Sea Grant), Grant no. 04-3-158-41; EPA Office of Research and Monitoring, Contract no. R-801124; Arctic Institute of North America Grant no. 152; the Prudhoe Bay Environmental Subcommittee; and the State of Alaska. We are indebted to the Naval Research Laboratory (ONR) at Point Barrow, for aerial transport and logistic support, and to Drs. R. D. Muench and T. C. Royer for critical review of the manuscript. Contribution no. 291, Institute of Marine Science, University of Alaska.

REFERENCES

ALEXANDER, V., ET AL.

1974 Environmental studies of an arctic estuarine system: Final report. Report R74-1, Inst. Mar. Sci., Univ. Alaska, Fairbanks, 359 pp.

DOODSON, A. T., and H. D. WARBURG

1941   Admiralty manual of tides. H.M.S.O., London.

DYGAS, J. A., R. W. TUCKER, and D. C. BURRELL

1972   Heavy minerals, sediment transport and shoreline changes at the Barrier Islands and coast between Oliktok Point and Beechey Point. *In* Baseline data study of the Alaska arctic aquatic environment, Report R72-3, Inst. Mar. Sci., Univ. Alaska, Fairbanks.

FEE, E. J.

1969   Digital computer programs for spectral analysis of time series. Special Report No. 6, Center for Great Lakes Studies, University of Western Michigan.

HUNKINS, K.

1962   Waves on the Arctic Ocean. *J. Geophys. Res.* 67: 2477-2489.

HURST, C. K.

1971   Investigation of storm effects in the Mackenzie delta region. Engineering Programs Branch, Dept. Public Works, Ottawa.

KINNEY, P. J., ET AL.

1972   Nearshore and estuarine environments of the Alaskan arctic coast: Parameters for engineering solutions. *In* Proc. 1st Int'l Conf. Port and Ocean Engineering under Arctic Conditions, Vol. 1. Department of Port and Ocean Engineering, Technical University of Norway, Trondheim, pp. 48-72.

MATHEWS, J. B.

1970   Tides at Point Barrow. *The Northern Engineer* 2: 12-13.

REED, J. C.

1974   Symposium on Beaufort Sea coastal and shelf research. Arctic Institute of North America, San Francisco, California.

ROSSBY, C. G.

1937   On the mutual adjustment of pressure and velocity distributions in certain simple current systems. *J. Mar. Res.* 1: 239-263.

1939   Relation between variations in the intensity of the zonal circulation of the atmosphere and the displacements of the semi-permanent centers of action. *J. Mar. Res.* 2: 38-55.

ʃATER, J. E., A. G. RONHOVDE, and L. C. VAN ALLEN

    1971   *Arctic environment and resources.* Arctic Institute of North America, Washington, D. C., 309 pp.

ʃEARBY, H. W., and M. HUNTER

    1971   Climate of the North Slope of Alaska. Tech. Mem. AR-4, National Oceanic and Atmospheric Administration, 54 pp.

ʃISEMAN, W. J., ET AL.

    1973   Alaskan arctic coastal processes and morphology. Technical Report No. 149, Coastal Studies Institute, Louisiana State University, Baton Rouge, Louisiana, 171 pp.

biological features

CHAPTER **20**

Primary productivity of sea-ice algae

R. C. CLASBY [1], V. ALEXANDER [2], *and* R. HORNER [3]

Abstract

A study of algal primary productivity associated with sea ice was carried out in the nearshore Chukchi Sea. Divers applied the carbon-14 method using specially designed incubation chambers for *in situ* work. The seasonal distribution of the "ice bloom" activity was determined, resulting in an annual input estimate of about 5 g C/m^2 by the ice community. Therefore, this ice-related primary production appears responsible for a significant proportion of the total annual photosynthetic carbon input into arctic coastal waters. Species composition and the role of factors related to primary production are also discussed.

INTRODUCTION

The Arctic is in general considered unproductive, whether from a marine or terrestrial point of view. This is in part due to the unfavorable nutrient regime. In contrast with Antarctic regions, there is almost no upwelling in the Arctic Ocean, and the contribution from rivers is small. English's (1961) measurements from Drift Station Alpha confirm this extremely low productivity. In addition to nutrient limitation, light is the second major problem. There is the long dark period during the winter when positive net productivity is not possible; during the summer there is continuous insulation, although at a low solar angle and often with extensive cloud cover. Although the ice cover greatly reduces the amount of light available in the water below, it affords at the same time a means for photosynthetic organ-

[1] *Seward Marine Station, (University of Alaska) Institute of Marine Science, Seward, Alaska 99664.*
[2] *Institute of Marine Science, University of Alaska, Fairbanks, Alaska 99701.*
[3] *14816 Bothell Way, N. E., #432, Seattle, Washington 98155.*

isms to maintain themselves at the surface without the problem of sinking. The environment within the ice represents vertical stability which is greater than any other marine situation, and organisms growing within the ice are exposed to the best light conditions available during the ice-covered period. *In situ* nutrient regeneration probably combines with seawater exchange to provide a relatively favorable nutrient environment.

In turn, the ice organisms seriously attenuate light passing through to the planktonic community below. The density of this ice population may exceed that of most algal blooms in the water. Although these organisms occupy only a band of restricted width at the bottom of the ice, the timing of this population renders it significant in relation to the total algal community. There is also reasonably good evidence for a significant role in the planktonic food chain for organisms originating in the epontic community.

The earliest reports of algae living in sea ice in the Arctic are those of Ehrenberg (1841, 1853) and Dickie (1880), listing diatoms collected during the Arctic Expedition of Sir George Nares. The early work on this algal community was primarily taxonomic (Bursa 1961 a, b; Cleve and Grunow 1880; Nansen 1906). Information during the last 10 years on the physical environment and physiology of the ice algal community includes the reports of Apollonio 1961, 1965; Bunt 1963; Bunt and Lee 1969, 1970; Bunt and Wood 1963; Meguro et al. 1966, 1967. The data presented in this paper are the results of our efforts to quantify the dynamics of this community and assess its contribution to the productivity of the arctic water. To facilitate this study, a special sampler, incubators, and allied techniques were developed for use by SCUBA divers in allowing *in situ* experiments to be run on the ice algal community (Clasby et al. 1973). A more comprehensive treatment of the project may be found in Alexander et al. (1974).

METHODS

A combined sampler-incubation chamber 4 cm long was constructed of 4.8 cm diameter plexiglass core tube lining. One end was closed off with a plexiglass plate fitted with a serum bottle type septum. The top of the sampler was serrated to cut into the ice. A diver removed the septum to remove water from the chamber and screwed the sampler into the underside of the ice to a depth of 2 cm. The septum was then replaced and the chamber inoculated with $_{14}C$-bicarbonate for primary productivity experiments and nitrogen-15 labeled ammonia and nitrate for nitrogen uptake experiments. The inoculating syringe was gently pumped to mix the solutions with the water without disturbing the algae. After 5 to 6 hrs the sampler was retrieved. A heavy metal spatula was used to chip away ice from around the chamber and then sever the top of the ice core. The sample was retained in the chamber by a core cap. Dark uptake experiments were conducted in a chamber that had been darkened, capped, and then suspended from ice pitons. After incubation, the samples were brought to the surface, transferred to 250-ml squat jars, and one drop of 0.4 percent $HgCl_2$ was added as a preservative. Samples of the algal layer for chlorophyll *a*, standing stock counts, and particulate nitrogen were taken with the same device.

Primary productivity samples were allowed to melt at room temperature approximately 1 to 2 hrs, filtered onto Millipore® HA filters, and rinsed with 5-ml 0.005 HCl and 5-ml filtered seawater. The wet filters were then placed in scintillation vials containing a counting solution of 100-g naphthalene, 7-g PPO, 0.3-g POPOP, and 400-g Cab-O-Sil® dissolved in 1-liter dioxane (Schindler 1966) or Aquasol®. Primary productivity samples were counted on a Nuclear Chicago model 6848 liquid scintillation counter. Counting efficiency was in the range of 80 to 90 percent, and all counts were corrected to 100 percent by the channels ratio method. Primary productivity calculations were made using the equations of Strickland and Parsons (1968), modified for liquid scintillation and expression of results on a per square meter basis. Nitrogen-15 samples were analyzed by the method of Dugdale and Dugdale (1965) with sample preparation for mass spectrometry according to Barsdate and Dugdale (1965). The UNESCO-SCOR (1966) method was used for determination of chlorophyll a. Light was measured with a Sekonic underwater light meter and results compared with a Middleton model CN7-156 pyranometer for conversion into langleys.

To measure photosynthetic rates under varying light conditions, rectangular boxes of plexiglass were designed and constructed for suspension beneath the ice from ice pitons. The boxes were fitted with sliding tops so that the ice incubation chambers could be placed in them without bringing the sample to the surface. Perforated metal screening was wrapped around the boxes to simulate various light intensities.

Interstitial water for chemistry was removed from 2-cm thick ice cores by vacuum filtration. This method did not prove to be satisfactory, as water was lost from the core during sampling and some melting did occur during filtration. During 1973, samples for chemistry were taken using the plexiglass corers and the interstitial water drawn off by pipette. We believe that this method resulted in a sample more representative of the chemical environment in which the algae were living. Nitrate, nitrite, phosphate, and silicate analyses were done on a Technicon Autoanalyzer using the methods of Strickland and Parsons (1968). Ammonia was determined by the method of Solorzano (1969) as modified for the autoanalyzer (D. Schell, unpublished). pH and alkalinity were run according to Strickland and Parsons (1968). Salinities were run on a Beckman RS 7A induction salinometer. Samples were analyzed for species composition and standing stock using the Utermohl (1931) inverted microscope technique.

The study site was located on the shore-fast ice, approximately 0.5 km offshore from the Naval Arctic Research Laboratory at Barrow, Alaska (Fig. 20.1). The shore-fast ice during the winter of 1972 was very smooth and covered with snowdrifts. Storms during the period of ice formation in the fall of 1972 produced a very rough shore-fast ice. There was little snowfall during the winter of 1972 to 1973, and no drifts formed. Maximum thickness of the shore-fast ice, 160-170 cm, occurred during late May to early June.

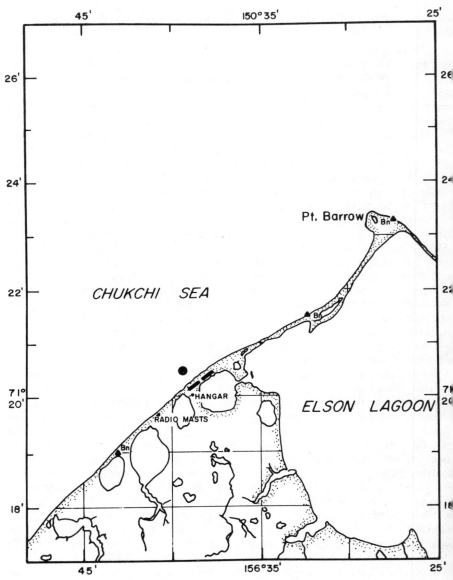

Fig. 20.1 Location of the primary sampling site near the Naval Arctic Research Laboratory Barrow, Alaska.

RESULTS

The first test of the new plexiglass corers for measuring primary productivity of the ice algal community was performed on 7 May 1972. At that time, the algal bloom was already under way and was visible to the unaided eye. Some difficulty was encountered in extracting the ice core, and considerable amounts of sample were lost. These losses gave rise to the use of the spatula method for freeing the ice core. This method proved quite efficient, although at times some samples were still lost.

Results of the primary production and chlorophyll *a* measurements for 1972 are shown in Figure 20.2. The highest rate of production, 4.56 mg C/m^2 hr, occurred on 21 May. The rates decreased sharply over the next 7 days, gradually tapering off to a low of 0.30 mg C/m^2 hr on 8 June. The algal layer was absent after that date. Chlorophyll *a* concentration followed the same general trend; a high of 30.49 mg Chl a/m^2 occurred on 21 May, dropped sharply and then tapered off to a low of 2.96 mg Chl a/m^2 on 8 June. The large variability in the primary productivity data for the first two sets of determinations is probably a reflection of sample loss due to problems encountered during the development of the technique. More practice with the technique seemed to reduce the magnitude of this problem.

Primary productivity measurements for the 1973 season commenced on 6 April with the level being near the limit of detection: 0.28 mg C/m^2 hr.

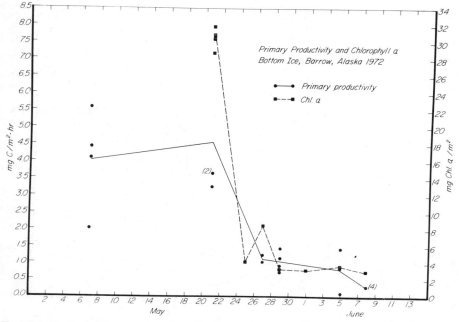

Fig. 20.2 *In situ* **primary productivity and chlorophyll** *a* **in the bottom ice at Barrow during spring of 1972.**

The primary productivity curve for the season was bimodal (Fig. 20.3), the first peak on 23 April at 1.93 mg C/m^2 hr dropping to 0.43 mg C/m^2 hr on 7 May and rising to the highest point, 14.92 mg C/m^2 hr, on 29 May, then decreasing to 10.53 mg C/m^2 hr on 4 June just prior to ice melt. The chlorophyll *a* concentration (Fig. 20.3) of the bottom ice also followed a bimodal curve similar to the primary productivity curve. The highs and lows for the chlorophyll *a* concentrations were 0.87 mg/m^2 on 5 April, 8.26 mg/m^2 on 27 April, 2.90 mg/m^2 on 11 May, 23.00 mg/m^2 on 23 May, and 18.08 mg/m^2 on 11 June. The first peak is possibly an artifact due to patchiness in the algal distribution.

To determine possible triggering mechanisms for the ice algal bloom, we decided to test for the minimum amount of light needed to initiate photo-

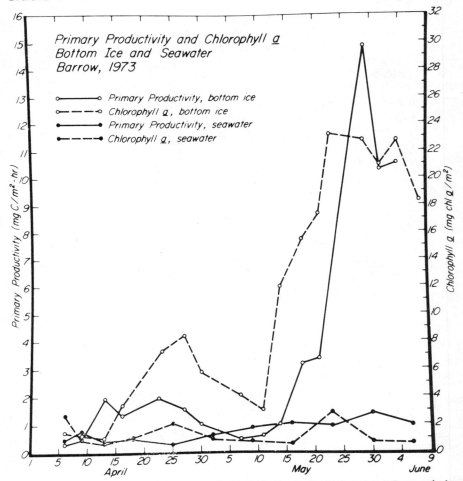

Fig. 20.3 *In situ* **primary productivity and chlorophyll** a **in the bottom ice at Barrow during spring of 1973.**

synthesis in the community. The experiments were run *in situ*. Samples were collected using the incubator corers, capped, injected with $_{14}$C-bicarbonate and placed in the screened boxes. The boxes were suspended and oriented under the ice. The remainder of the experiment was conducted following the standard primary productivity procedures used in this study. The 1972 and 1973 experimental results are shown in Figure 20.4. The 1972 experiment showed a slight increase in the rate of primary production over the range of 0.008 to 0.042 langleys/hr. Above 0.042 langleys/hr, the rate increased sharply. Rates of primary production at light levels below 0.042 langleys/hr were near the limits of detection for the method as outlined by Strickland and Parsons (1968). These results suggest that the minimum light needed to stimulate primary production in this algal community is close to 0.042 langleys/hr.

The 1973 data showed a much higher threshold (0.171 langleys/hr) than found in 1972 and a more gradual increase in productivity per unit light.

Observations of the underside of the ice made by divers indicated that the epontic community was patchy in nature, and that this patchiness was correlated with light attenuation due to snow depth. In 1972, a 14-m transect was run out diagonally from the diving hole through a relatively deep snowdrift. Snow depth, light and chlorophyll *a* were measured at 2-m intervals along this transect. The data showed an inverse relationship between the amount of chlorophyll *a* and snow depth (Fig. 20.5). Not so readily apparent is the relationship between light and the other two parameters.

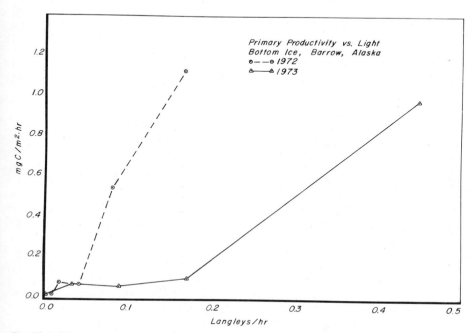

Fig. 20.4 Primary productivity vs light for the bottom ice at Barrow.

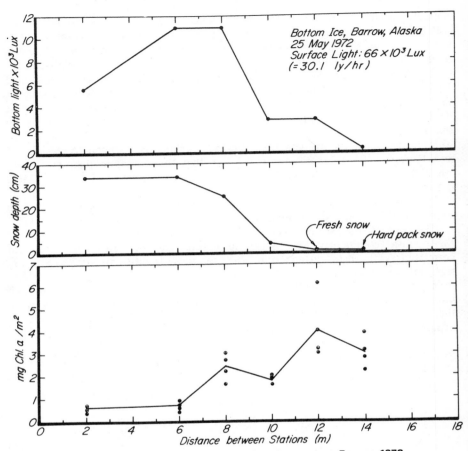

Fig. 20.5 Light, snow depth, and bottom-ice chlorophyll *a* transect, Barrow, 1972.

After the data were obtained, we realized that the light measurements were taken below the algal layer and, therefore, reflect the amount of light reaching the phytoplankton and not that reaching the ice algae. The data do show the strong light-attenuating properties of the ice algal layer. It is apparent that the high albedo and light-attenuating properties of snow may be a governing factor for the distribution and abundance of the epontic community.

Snowdrifts did not form in the area near the diving hole in 1973 because of ice, snow, and wind conditions, so the chlorophyll transect was not repeated. Small-scale patchiness was measured on 11 May 1973 by taking six cores within a 0.5 m² area. Average chlorophyll *a* concentration was 2.90 mg/m² with a standard deviation of ± 1.02 mg/m².

A nitrogen-15 ammonia uptake experiment was performed on 1 June 1972 to determine the rate of uptake by the ice algae. The experiment was run in triplicate and gave a rate of 9.29 μg-atoms NH_4^+ -N/m^2 hr with a standard deviation of $\pm.42$ μg-atoms NH_4^+ -N/m^2 hr. Ammonia uptake were also measured on 18 April and 16 May 1973 and the resultant rates were 6.90 (± 1.49) and 25.00 (± 3.03) μg-atoms NH_4^+ -N/m^2 hr respectively. Nitrate uptake was measured on 18 April and 23 May 1973 and the resultant rates were 5.25 (± 1.80) and 21.43 (± 0.00) μg-atoms NO_3^- -N/m^2 hr respectively. Experiments were run on 16 May 1973 to measure the response of ammonia uptake to varying concentration of ammonia and on 23 May for nitrate uptake at various nitrate concentrations. The initial ammonia concentration was 18.4 μg-atoms NH_4^+ -^{14}N/liter and additions of 3.3, 6.7, 13.3 and 26.7 μg-atoms NH_4^+ -^{15}N/liter were made. Initial nitrate concentration was 16.9 μg-atoms NO_3^- -^{14}N/liter. Additions of nitrogen-15 had to be made at these high levels so that the increase in nitrogen-15 in the algal material could be detected by the mass spectrometer. Figure 20.6 shows the results of the two experiments. Maximum uptake of nitrate was at 1.10 μg-atoms/liter and at 0.71 μg-atoms/liter for ammonia. Results of the ammonia and nitrate uptake experiments are shown in Table 20.1.

Table 20.2 shows the results of the chemical analyses of the bottom ice interstitial water. For 1972, salinities were lower than expected from freeze concentration and were probably a result of dilution by melt water during filtration, although care was taken to prevent this. Concentration of inorganic nutrients was high and approximately an order of magnitude above that of adjacent seawater. The data show an increase in nutrient concentration with time, rather than depletion, possibly a result of increasing air tempera-

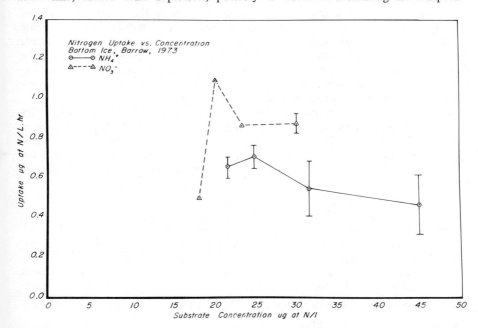

Fig. 20.6 Nitrogen uptake by the bottom-ice algae at varying nitrogen concentrations.

TABLE 20.1 Uptake rates of nitrogen-15 ammonia and nitrate compounds

Sample	Uptake rates (μg-atoms/m^2-hr)			
	1 June 1972	18 April 1973	16 May 1973	23 May 1973
Ammonia	9.29 ± 1.42	2.90 ± 1.49	25.00 ± 3.03	
Nitrate		5.24 ± 1.80		21.43 ± 0.00

tures causing enlargement and increased migration of the interstitial water in the brine cells.

The bottom ice interstitial water samples for 1973 (Table 20.3) also had low salinities near that of the underlying seawater. Inorganic nitrogen in the interstitial water was two to three times higher than in the seawater, although phosphate and silicate were near seawater concentrations. The change in interstitial nitrate with time had a bimodal distribution (Fig. 20.7) similar to the primary productivity curve, with the nitrate maxima occurring before the primary productivity maxima. Interstitial ammonia was somewhat more variable than nitrate, but showed depletion prior to the second primary productivity maximum.

Incoming solar insulation varied from 8 to 56 langleys/hr depending on cloud cover. The mean for 1972 was approximately 25 langleys/hr, although that for 1973 was 50 langleys/hr. Light levels under the ice were on the order of 1 percent of the incoming solar radiation.

Algal cells are present in the ice from the time the ice forms in the fall, but they are few in number and scattered throughout the entire thickness of the ice. Organisms present early in the winter (January-February) include unidentified flagellates and pennate diatoms, primarily species of *Navicula* and *Nitzschia*, with *Nitzschia closterium* being the most abundant. The number of cells present in the ice increases so that by late March a relatively large population consisting of *Navicular* and *Nitzschia* spp., *Gyro-Pleurosigma* spp., and unidentified pennate diatoms is present.

In 1972, the most abundant diatom in the ice in May when productivity was highest was *Nitzschia frigida*. In May, 1973, the most abundant diatom was *Navicula marina*. This diatom was heavily parasitized by at least three species of chytridiaceous fungi. Although the chytrids attacked other diatoms, including *Pleurosigma stuxbergii, N. marina* appeared to be the most heavily infected.

TABLE 20.2 Bottom-ice interstitial water chemistry, Barrow, Alaska, 1972

Date	Salinity ($^o/_{oo}$)	pH	ΣCO_2 (meq/liter)	NH_4^+-N	NO_3^- + NO_2^--N	PO_4^{-3}-P	SiO_3^{-3}-Si
				(μg-atoms/liter)			
5 May	20.35	8.18	1.53	16.0	10.6	1.6	11.0
26 May	34.09	8.25	2.63	44.0	13.6	2.3	14.5
9 June	27.69	8.05	2.21			2.1	30.3

TABLE 20.3 Bottom-ice interstitial water chemistry, Barrow, 1973

Date	Salinity ($^o/_{oo}$)	pH	ΣCO_2 (meq/liter)	NH_4^+-N	$NO_3^- + NO_2^-$-N (µg-atoms/liter)	PO_4^{-3}-P	SiO_3^{-3}-Si
11 Apr	26.06	7.96	2.39	3.5	14.7	0.8	26.4
18 Apr	32.91	8.02	2.45	3.4	13.0	0.7	25.4
7 May	32.30	8.08	2.43	8.0	14.0	9.5	21.3
11 May	–	8.03	2.22	0.1	13.8	–	–
16 May	29.64	–	–	18.4	14.1	3.6	29.4
23 May	–	–	–	1.5	17.8	4.3	36.6
29 May	27.66	8.12	2.07	3.4	9.4	7.0	49.2
4 Jun	32.48	8.04	2.16	3.4	12.1	5.0	31.6

Standing stock in the ice varied from about 6000 cells per liter in middle ice in March to nearly 24 million cells per liter in bottom ice in late May in 1972, and from about 60,000 cells per liter in bottom ice in January to 43 million cells per liter in bottom ice in early June of 1973. Species of the genera *Navicula* and *Nitzschia* were nearly always the most abundant diatoms present. Other genera nearly always present and in large numbers were *Amphiprora, Gomphonema,* and *Pleurosigma.* Cells belonging to the genera *Gyrosigma* and *Pleurosigma* could not always be separated under the inverted microscope and were lumped into a composite "genus" *Gyro-Pleurosigma,* which was also nearly always present.

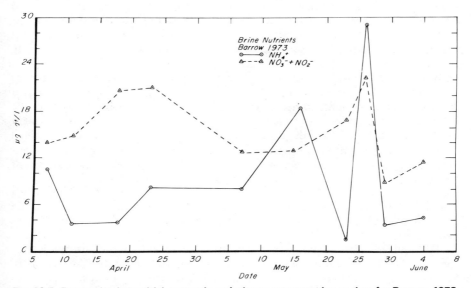

Fig. 20.7 Bottom-ice interstitial ammonia and nitrate concentration vs time for Barrow, 1973.

Ciliates belonging to the order Hypotrichida are common in the ice, as are heliozoans. Among the Metazoans present, nematodes are common. Polychaete larvae, turbellarians and copepods may also be present. The worms and copepods feed on the ice organisms, as evidenced by partially digested ice diatoms in their digestive tracts and fecal pellets. Amphipods, including *Pseudalibrotus litoralis* (Kroyer), *Gammaracanthus loricatus* (Saline) and *Gammarus wilkitzkii* Birula, have been observed and photographed by divers on the brown ice layer, where they apparently feed.

DISCUSSION

The epontic population develops in response to a minimum critical light level during April, with a population peak in May followed by a rapid decline in June as the ice layer containing the algae disintegrates. At the time of maximum photosynthesis in the ice, there is not yet very high activity in the seawater below the ice. In view of the active grazing occurring on the ice undersurface, the food-chain implications of this early primary production mechanism could be very significant. This community is dependent on primary production, and yet it contains heterotrophic components which are active in organic assimilation and nutrient regeneration.

Nutrient supply within the ice appears quite high — presumably several sources are available. Water passing through the ice undoubtedly carries with it nutrients from the entire ice column. With both an active microbial population and some grazers, *in situ* regeneration is probably significant. Exchange with the seawater below undoubtedly occurs, and this "inverted benthos" probably abstracts nutrients from seawater as the currents pass under the ice. This constitutes, then, a highly effective mechanism to extend the growing season at a latitude where the season is otherwise extremely short.

We believe that our primary productivity measurement technique is superior for the particular area in question, although it conceivably would not work as well under other physical ice conditions at the bottom of the ice cover. The values obtained may be as close estimates of primary productivity as possible with present technology. Annual production, obtained by integration of the 1973 curve, is about 5 g C/m^2; although this is not very much for most coastal areas, it represents a significant proportion of the annual primary productivity in the Arctic. The assimilation efficiency of the chlorophyll is not high, with a maximum on 29 May of 0.7 mg C/mg Chl *a* hr. It is interesting to note the difference in light response by the population during the two seasons, possibly a result of different species composition.

The greatest lack of knowledge concerns the extent of the distribution of the ice algal layer throughout the arctic region, and the nature of its distribution, i.e., "patchiness" and continuity. Our study of this single area suggests considerable patchiness, partly related to snow depth.

Variability or patchiness also occurs with regard to species composition and standing stock. This variability occurs between years, possibly as a response to differing ice conditions during different years, and also within a relatively small area in the same year, possibly because of ice conditions and distance from shore. A sample collected approximately 15 km offshore on 17 April 1973 contained several million cells per liter of *Fragilariopsis* spp., while a sample collected at Barrow on 16 April 1973 contained few *Fragilariopsis* spp.

SUMMARY

This study has shown that primary production in the underside of the ice is a significant input of organic matter into arctic coastal waters, and that primary production by such populations reaches 15 mg C/m^2hr, with chlorophyll *a* concentrations as high as 30 mb/m^2. Nutrient concentrations in the ice do not appear to pose a limitation, at least as far as nitrogen is concerned. Light, on the other hand, is an important factor in determining the onset and distribution of the bloom. The diving technique used in this study was successful and, we believe, superior to any approach working from the ice surface.

Conference Discussion

Korringa — One should be very careful in comparing phosphate data for different seas. Where light is abundant, free phosphate levels are often extremely low, but this does not indicate a low fertility rate. One should measure total phosphorous in addition to free phosphorous in studies on productivity and fertility.

Dunbar — There is something philosophically satisfactory in the fact that in the Arctic Ocean, one of the least productive regions of the world, one can measure, locally, phosphate concentrations far higher than elsewhere in the world.

Acknowledgments

We would like to thank Mr. John F. Schindler, former Director, and the staff of the Naval Arctic Research Laboratory, Barrow, Alaska for logistic support, and we particularly thank Mr. Walter Akpik, Jr. for his assistance in the field. Divers were Mr. R. C. Clasby (project technician), Mr. Grant E. M. Matheke, and Mr. G. E. Hall; Mr. Douglas Redburn and Mr. Ken Coyle assisted with the fieldwork in 1972. Financial support was provided by National Science Foundation Grant no. GV29342. Logistic support was provided in part by Office of Naval Research Grant no. NOOO14-67-A-0317-003. Contribution no. 287, Institute of Marine Science, University of Alaska.

REFERENCES

ALEXANDER, V., R. HORNER, and R. C. CLASBY

1974 Metabolism of Arctic sea ice organisms. Report R74-4. Inst. Mar. Sci., Univ. Alaska, Fairbanks, 120 pp.

APOLLONIO, S.

1961 The chlorophyll content of Arctic sea-ice. *Arctic* 14: 197-200.

1965 Chlorophyll in Arctic sea ice. *Arctic* 18: 118-122.

BARSDATE, R. J., and R. C. DUGDALE

1965 Rapid conversion of organic nitrogen to N_2 for mass spectrometry; an automated Dumas procedure. *Anal. Biochem.* 13: 1-5.

BUNT, J. S.

1963 Diatoms of Antarctic sea-ice as agents of primary production. *Nature* 199: 1255-1257.

BUNT, J. S., and C. C. LEE

1969 Observations within and beneath Antarctic sea ice in McMurdo Sound and the Weddell Sea, 1967-1968. Methods and data. Inst. Mar. Sci., Univ. Miami Tech. Rep. 69-1, 32 pp.

1970 Seasonal primary production in Antarctic sea ice at McMurdo Sound in 1967. *J. Mar. Res.* 28: 304-320.

BUNT, J. S., and E. J. F. WOOD

1963 Microalgae and Antarctic sea ice. *Nature* 199: 1254-1255.

BURSA, A.

1961a Phytoplankton of the *Calanus* Expedition in Hudson Bay, 1953 and 1954. *J. Fish. Res. Board Can.* 18: 51-83.

1961b The annual oceanographic cycle at Igloolik in the Canadian Arctic. II. The Phytoplankton. *J. Fish. Res. Board Can.* 18: 563-615.

CLASBY, R. C., R. HORNER, and V. ALEXANDER

1973 An *in situ* method for measuring primary productivity of Arctic sea ice algae. *J. Fish. Res. Board Can.* 30: 835-838.

CLEVE, P. T., and A. GRUNOW

    1880   Beitrage zur Kenntniss der arctischen Diatomeen. *K. svenska Vetensk Akad. Handl.* 17(2): 1-122.

DICKIE, G.

    1880   On the algae found during the Arctic Expedition. *J. Linn. Soc. Bot.* 17: 6-12.

DUGDALE, V. A., and R. C. DUGDALE

    1965   Nitrogen metabolism in lakes. III. Tracer studies of the assimilation of inorganic nitrogen sources. *Limnol. Oceanogr.* 10: 53-57.

EHRENBERG, C. G.

    1841   Einer Nachtrag su dem Vortrage uber Verbreitung und Einfluss des mikroskopischen Lebens in Sud-und-Nord-Amerika. *Monatsber. d. Berl. Akad.* 1841: 202-207.

    1853   Uber neue Anschauungen des kleinsten nordlichen Polarlebens. *Monatsber. d. Berl. Akad.* 1853: 522-529.

ENGLISH, T. S.

    1961   Some biological oceanographic observations in the central North Polar Sea, Drift Station Alpha, 1957-1958. Arctic Institute of North America Scient. Rep. No. 15, 79 pp.

MEGURO, H., K. ITO, and H. FUKUSHIMA

    1966   Diatoms and the ecological conditions of their growth in sea ice in the Arctic Ocean. *Science* 152: 1089-1090.

    1967   Ice flora (bottom type): A mechanism of primary production in polar seas and the growth of diatoms in sea ice. *Arctic* 20: 114-133.

NANSEN, F.

    1906   Protozoa on the ice-floes of the North Polar Sea. *In* F. Nansen (ed.), Norwegian North Polar Expedition, 1893-1896. Scient. Results Norw. N. Polar Exped. 5(16): 1-22.

SCHINDLER, D. W.

    1966   A liquid scintillation method for measuring carbon-14 uptake in photosynthesis. *Nature* 211: 844-845.

SOLÓRZANO, L.

1969 Determination of ammonia in natural water by the phenol-hypochlorite method. *Limnol. Oceanogr.* 14: 799-801.

STRICKLAND, J. D. H., and T. H. PARSONS

1968 A manual of sea water analysis. Bull. Fish. Res. Board Can. 165. 311 pp.

UNESCO-SCOR

1966 Monographs on oceanographic methodology. I. Determination of photosynthetic pigments in sea water. UNESCO, Paris. 69 pp.

UTERMÖHL, H.

1931 Neue Wege in der quantitativen Erfassung des Planktons. *Verh. int. Verein. theor. angew. Limnol.* 5: 567-596.

CHAPTER **21**

Effects of oil on microbial component of an intertidal silt-sediment ecosystem

S. A. NORRELL[1] *and* M. H. JOHNSTON[2]

Abstract

The ability of crude oil to supply oxidizable soluble organic material to bacterial populations in marine sand ecosystems is significantly documented. In this chapter, a review is made of the fundamental relationships reported by others to exist between bacterial biomass and sediment properties. These observations are compared to the distinguishing characteristics of a glacial silt intertidal zone studied during 1973 and 1974 near Port Valdez, terminus of the trans-Alaska pipeline. Bacterial plate-count and respirometry data are used to estimate standing crop and activity response of microbial populations to oil contamination. The presence of various sulfur bacteria is also examined because of their known association with both marine environments and with oil deposits.

INTRODUCTION

Bacteria are often considered to be the basis of food-chains in marine sediments. Zobell and Feltham (1938) showed that certain marine invertebrates used bacteria as food. Perkins (1958) concluded that many mud ingesting feeders were, in fact, feeding upon bacteria adhering to sand grains and to detritus; McIntrye et al. (1970) reported on the role of bacteria as a source of the protein and the energy flowing through a sand ecosystem. In their studies of flatfish nursery ground, McIntrye and Murison (1973) concluded

[1]*Anchorage Senior College, 3221 Providence Drive, University of Alaska, Anchorage, Alaska 99504.*
[2]*Institute of Marine Science, University of Alaska, Fairbanks, Alaska 99701.*

that the meiofauna feed mainly on bacteria and diatoms attached to sand grains and present in the interstitial water. They were able to show that a substantial part of the meiofaunal carbon budget was supplied by bacteria and that a crucial factor in maintenance of the ecosystem was the delivery of soluble organics to the bacteria by movement of water through the sediment. Fenchel (1969) details the interrelationships that exist between various biotic compartments, including bacteria, in oxidized sand in marine ecosystems.

The attachment of bacteria to sand grains and the relative populations of bacteria on the grains and in the interstitial water has been extensively studied by Meadows and Anderson (1968). Their studies suggest that, because most of the bacteria are attached to the sand grains in microcolonies, the sediment acts to hold the bacterial population in place and that movement of fluid through the sediment is necessary for the delivery of nutrients to the bacteria. Furthermore, Steele et al. (1970) report that drainage and sublittoral pumping through sand beaches with mean grain diameter of 250 μ may be the dominant mechanism for supplying oxygen and soluble organic matter to the bacteria adhering to the sand grains.

In a well sorted, wave-washed, and oxidized beach ecosystem, the size of the bacterial population was strongly, but negatively, correlated with mean grain size. Although bacterial biomass was strongly correlated with total carbon and nitrogen, the bacterial carbon was estimated to account for only 1.2 percent of the total carbon and 2.5 percent of the total nitrogen, and no correlation was observed between bacterial biomass and oxidation state (as Eh) of the sediment (Dale 1974). He further concluded that the strong statistical relationships are ultimately traceable to the dominant influence of the waves and tides on the properties of the intertidal sediments, and that a statistically simple relationship may exist between bacterial biomass and sediment properties.

Cummins (1974) presents evidence that, in a freshwater system, the microbes on particulate organic matter not only play a role in reducing the size of the particles, but also serve as the major food source for stream invertebrates. He concluded that the microbial biomass layer is at least as important a food source as the particles themselves, and that development of the bacterial layer is dependent upon the chemical properties of the particles. The final embedding of these particles in sediment appears to be necessary for their eventual conversion to dissolved organic matter and for return to the water ecosystem as dissolved, rather than particulate, nutrients.

The ability of crude oil to supply oxidizable soluble organic material to bacterial populations in marine sand ecosystems had been shown in several studies. Generally, the oil is able to cause an increase in bacterial biomass when added over prolonged periods, but it does not permanently affect the size or composition of the bacterial population when added only once, as would occur from a spill. Laboratory studies using artificially produced sand systems with a mean grain size of 350 μ (Bloom 1970) and 250 μ (Johnston 1970), as well as studies of spills on natural beach systems at San Francisco Bay (Cobet and Guard 1973; Guard and Cobet 1973) and in beach communities affected by the Torrey Canyon spill on the Cornish Coast (Gunkel 1968), are in agreement with the reports of Steele et al. (1970) and Meadows and Anderson (1968). Increased oxygen uptake was shown in columns made of sand (350-μ mean particle diameter) from Nobska Beach, Woods Hole, Massachusetts, when the columns were flushed with seawater, contaminated with Kuwait crude oil, plus the dispersant "Corexit 7664" (Enjay Chemical Company), and with dispersant alone (Bloom 1970). These results suggested that the dispersant, by itself or in combination with oil, had no obvious deleterious effects on either the meiofaunal population or the bacteria. Bleakley and Boaden (1974), however, report that the dispersant Lissapol N produced some morbidity in copepods (Order: Harpacticoida) at as low a concentration as 1 ppm, although an observed decline in morbidity may "reflect the recovery of the surviving meiofauna probably associated with the bacterial degradation of surfactant."

The sediment ecosystem in Valdez Arm is, however, of different physical and physiological characteristics than those reported on above, and this environment has been extensively characterized by Feder (1974). The sediment is composed of fine glacial silt, which is deposited at an annual rate of about 1.67 cm/yr. The mean particle size of only 4 to 16 μ is uniform to a depth of 5 cm. Because there is no effective wave action, there is no apparent sorting of the particles on the surface. The salinity of the interstitial water is always higher than the overlaying tidal waters, often by a factor of at least two, even under conditions of heavy rainfall (up to 2 m of rain during the period from July to October). The study area is traversed by several streams from nearby snowfields, and surface runoff of the rainwater is constant and typical. The sediment contains no dissolved or precipitated sulfide except under heavy algal mats. Iron concentrations of 0.1 to 0.3 ppm show no gradient with depth. There is less than 0.2 percent by weight of organic matter, and the interstitial water has a constant pH of between 7.2 and 7.4. The low organic content and absence of animal remains in the sediment suggest that deposited organisms are either removed by rapid digestion on the surface or are flushed out by the ebb tide. Very little of the dead organic matter becomes embedded in the sediment. The absence of detectable sulfide suggests a highly oxidized and aerobic environment (Fenchel 1969) and is consistent with the presence of detectable dissolved oxygen down to 15 cm (A. S. Naidu, personal communication).

The upper one or two centimeters of the sediment appear to be the biologically active layers. Feder (1974) examined the meiofaunal population of this ecosystem and found that most of the organisms were located in the first centimeter of sediment; concentrations contained in the upper horizon ranged from 1000 to 1700 organisms per 10 cm^2, about 85 percent of which were Nematodes.

A major objective of this study was to estimate the size and activity of the bacterial population in Valdez intertidal sediments and to determine the effect of oil on this population. The presence of various sulfur bacteria was also examined because of their association with both marine environments and with oil deposits. Typically, in highly oxidized marine sediments, these organisms form ecosystems of great complexity and almost universal distribution (Fenchel 1969; Fenchel and Riedl 1970). Sulfur bacteria are almost always associated with oil deposits, although their actual role in the oxidation of oil or in oil synthesis is not known (Guarraia and Ballentine 1972). Kusnetzov (1967) does report an instance in which oil became the source of energy for sulfate-reducing bacteria, with hydrogen sulfide sometimes being produced at the rate of 0.2 mg/liter-24 hrs.

MATERIALS AND METHODS

Sampling and site preparation

Samples for bacterial analysis were collected during the summer months of 1973 and 1974 from an intertidal zone near Port Valdez, Alaska, located on Valdez Arm of Prince William Sound, in conjunction with the studies described by Feder (1974). Specifically, samples were taken from the undisturbed site at Island Flats, an oil seepage site at Old Valdez, Alaska (produced by the burial of an oil tank during the historic 1964 earthquake) and from under algal mats on the Island Flats site.

The 1973 samples were collected to determine the comparative size of the bacterial populations in open sediment, in sediment directly under (decaying) algal mats, and in sediment exposed to continuous oil seepage from the buried oil tank at Old Valdez.

On 18 June 1974, four sets of 25 glass rings, approximately 15 cm in diameter and 4 cm in height, were placed in the sediment of the intertidal study area on Island Flats. These rings, although only slightly immersed in the sediment, were not displaced by daily tides. Each set of 25 rings comprised a separate "site," designated as Control (C), 500-ppm, 2000-ppm and Chronic (CH). The rings in the 500-ppm site each received two equal applications of 500 ppm Prudhoe Bay crude oil on two consecutive days during each low tide series of the summer of 1974 (for a total of 1000 ppm). Similarly, the rings of the 2000-ppm site received 4000 ppm oil. Each ring in the Chronic (CH) Site received 200 ppm oil on each of 5 consecutive days during each low tide series, for a total of 1000 ppm oil. The rings within the Control (C) Site were never oiled.

During each low tide interval throughout the sampling periods, five to six sediment cores were taken to a depth of 2 cm from each site. Not more than one core was taken from each ring, and the sediment within any one ring was sampled only once for bacterial analysis. (The remaining sediment was used for meiofaunal studies). When sampling was conducted outside the sites containing the glass rings, cores were taken randomly from apparently homogeneous areas.

Sampling dates were dependent upon tide levels, and specific samples were taken at least one tide series after the previous addition of oil. For example, sediment samples on 3 July 1974 received only one series of oil application (on 19, 20 June), while those sampled late in the season (15 September 1974) had been oiled on five previous low tide exposures. The samples taken from the Old Valdez site were considered to represent con stantly oiled sediments. The sampling and oil application schedule is shown in Table 21.1.

The individual cores, from replicate rings, were taken using a steel corer approximately 4 cm in diameter, and the contained sediment was immediately expressed into a sterile Whirlpak bag in such a way that no fluid or sediment was lost. The samples were transported and stored on ice or under refrigeration (4° C) for analysis in the laboratory at Fairbanks. In all cases, processing of the cores was begun no later than 48 hrs after sampling. Each of the five replicate cores was analyzed separately and the results averaged for each site.

Total bacterial population

Total aerobic bacterial populations were estimated by standard plate count-dilution methods (Meynell and Meynell 1965; Parkinson et al. 1971) using Tryptone Glucose Extract Agar (Difco) prepared with filtered seawater taken from Port Valdez offshore waters and supplemented with 8 mg/liter cycloheximide. All dilutions were made with filtered, sterile seawater from the same source. In every case except for the initial samples, the sediment was allowed to settle for 5 min prior to dilution. All samples were counted in triplicate after 10 days' incubation at 10° C in the dark. Unless other-

TABLE 21.1 Oil amendment and sampling protocol

Dates on which samples were taken	Dates on which oil was added		
	Control sites	Chronic sites	500 or 2000 ppm sites[2]
6/18	Samples taken for baseline and control data, no previously oiled samples.		
7/3	None	19-23 June[1]	19, 20 June
7/20	None	19-23 June 3-7 July	19, 20 June 3, 4 July
8/5	None	19-23 June 3-7 July 20-24 July	19, 20 June 3, 4 July 20, 21 July
8/26	None	19-23 June 3-7 July 20-24 July 1-5 Aug	19, 20 June 3, 4 July 20, 21 July 2, 3 Aug
9/15	None	19-23 June 3-7 July 20-24 July 1-5 Aug 16-20 Aug	10, 20 June 3, 4 July 20, 21 July 2, 3 Aug 16, 17 Aug

[1] 200 ppm oil added on each day, inclusive of the dates shown.
[2] 500 or 2000 ppm oil added, in equal amounts, on the dates shown.

wise noted, all incubations were aerobic and are reported as total "aerobic" counts.

Sulfur-cycle bacteria

Two methods were used to test for the presence of sulfate-reducing bacteria. One method involved plating appropriately diluted samples from single cores, following the procedure described for total aerobic population counts except that, in this case, the medium consisted of Trypton Glucose Extract Agar (Difco), 24 g/liter; Na_2SO_4, 5 g/liter; $FeSO_4-7H_2O$, 0.09 g/liter; and filtered seawater to 1000 ml (Aaronson 1970). Plates were incubated in BBL Anaerobe Jars using $CO_2 + H_2$ gas packs. After 10 days incubation in the dark at $10°$ C, the plates were examined for blackened colonies (sulfate reducers) and total "facultative counts."

The second method of determining the presence of sulfate-reducing bacteria involved using a liquid medium (Aaronson 1970), consisting of KH_2PO_4, 0.5 g/liter; NH_4Cl, 1.0 g/liter; sodium lactate, 6.0 g/liter; $CaCl_2-6H_2O$, 60.0 mg/liter; $MgSO_4-7H_2O$, 60.0 mg/liter; yeast extract, 1.0 g/liter; $FeSO_4-7H_2O$, 0.1 g/liter; sodium citrate-$2H_2O$, 0.3 g/liter; $(NH_4)_2SO_4$, 7.0 g/liter; and filtered seawater to 1000 ml. The pH of the medium was adjusted to 7.5 before autoclaving. One-gram sediment samples from separate cores were placed in 70-ml sterile glass-stoppered bottles, or 5.0-g sediment samples from single cores were placed in 275-ml sterile glass-stoppered bottles. The bottles were then filled to the brim and stoppered to exclude air. Samples were incubated in the dark at $10°$ C and examined periodically for blackening of the medium.

The production of H_2S by heterotrophic bacteria from organic sources was examined by subculturing heterotrophs on an appropriate medium to detect hydrogen sulfide production. Every colony growing on or breaking the surface of the agar in selected plates used for total counts was picked and inoculated into sterile deeps of a solid medium consisting of Tryptone Glucose Extract Agar, 24 g/liter; L-cystine, 0.1 g/liter; Na_2SO_4, 0.5 g/liter; lead acetate, 0.3 g/liter; and filtered seawater to 1000 ml (Aaronson 1970). Tubes were incubated at $10°$ C in the dark and were examined periodically for areas of black precipitation in the region of growth.

The presence of green photosynthetic sulfur bacteria was determined using the following medium (Larsen 1952; Aaronson 1970): NH_4Cl, 1.0 g/liter; KH_2PO_4, 1.0 g/liter; Na_2S-9H_2O, 1.0 g/liter; $MgCl_2$, 0.5 g/liter; NaCl, 2.0 g/liter; and filtered seawater to 975 ml. The pH was adjusted to 7.0, before autoclaving. After autoclaving and cooling of the medium, 2.0 g $NaHCO_3$ in 25 ml filtered seawater, previously sterilized by Millipore filter, were added. Sterile glass-stoppered bottles were inoculated with sediment, and medium was added as described for use of the liquid sulfate-reducer medium. Bottles were incubated in the light at $10°$ C and examined periodically for green colonies.

The presence of purple photosynthetic sulfur bacteria was determined as for green photosynthetic sulfur bacteria; in this case, however, 2.1 g/liter Na_2S-9H_2O was used, and the pH of the medium was adjusted to 8.0

before autoclaving (Larsen 1952; Aaronson 1970). Bottles were examined periodically for purple colonies.

A modified Winogradsky-type column technique (Fenchel 1969) was used to simultaneously determine the presence of "white sulfur" chemoauto-trophic, sulfur-oxidizing bacteria and of green and purple photosynthetic sulfur bacteria. The solid phase, consisting of $CaSO_4$, 10 g/liter; glucose, 1 g/liter; peptone, 1 g/liter; agar, 15 g/liter; and filtered seawater to 1000 ml, was added to large test tubes to a depth of about 4 cm. The liquid phase and inoculum, consisting of 20 ml of a 10^{-1} dilution of sediment in sterile filtered seawater, were added to the tubes, covering the solidified medium. The tubes were segregated into two sets. In one group, the sediment in the dilution was not allowed to settle before the 20 ml were dispensed over the medium. In the other set, the sediment in the dilution was allowed to settle for 10 min, and only the resultant supernatant was dispensed over the medium. Tubes containing the unsettled dilution of sediment were incubated at $10°$ C in the dark and periodically examined for the presence of white sulfur bacteria. Tubes containing both unsettled and settled dilutions of sediment were incubated at $10°$ C in the light and were periodically exam-ined for the presence of white sulfur bacteria, and later, for the presence of green and purple photosynthetic sulfur bacteria.

The presence of non-sulfur, photosynthetic bacteria (Athiorhodaceae) was determined using a liquid medium (Pratt and Gorham 1970), consisting of NH_4Cl, 1.0 g/liter; KH_2PO_4, 1.96 g/liter; K_2HPO_4-3HO, 3.33 g/liter; $MgCl_2$, 0.2 g/liter; $NaCl$, 2.0 g/liter; Bacto-Yeast Extract, 0.2 g/liter; Bacto-Peptone, 2.9 g/liter; and filtered seawater to 975 ml. After the medium was autoclaved for 20 min at 15 psi and cooled, 5.0 g $NaHCO_3$ in 25 ml filtered seawater, previously sterilized by Millipore filtration, was added. Sterile glass-stoppered bottles were inoculated with sediment, and medium was added as previously described. Bottles were incubated at $10°$ C in the light and were examined periodically for pink, orange, or straw-col-ored pigmentation of the medium, or for pigmented colonies growing on the surface of the sediment.

TABLE 21.2   Preliminary survey of bacterial biomass in sediments from island flats study area and oil seep site from Old Valdez pre-earthquake oil stor-age area

Sampling date:	Bacterial count CFU/cc of sediment (x10³)	
	8/27/73	9/27/73
Control site	70 (34)	46 (37)
Oiled site (surface application on 6/27)[2]	70 (32)	N. D.[3]
Algae covered	189 (43)	6 (16)
Old Valdez seep	N. D.[3]	179 (59)

[1] CFU = Colony forming units/cubic centimeters.

[2] 500 ppm of Prudhoe Bay crude oil was added to the sediment surface within containment structures during the ebbing tide; the oil was added on six successive days.

[3] ND = Not Determined.

A Most Probable Number (MPN) statistical dilution estimate of the number of Athiorhodaceae was made following standard MPN methods. The medium and culture conditions were as described above, except that, for each sample tested, triplicate aliquots of 10.0, 1.0, and 0.1 ml of a 1:1 dilution of sediment to medium were added to separate 16 x 125-mm test tubes. The tubes were then filled to the brim with medium and capped to exclude air. Incubation was as previously described for Athiorhodaceae.

An attempt was made to determine what percentage of colonies developing on total aerobic population plates were Athiorhodaceae. Colonies were picked from plates as previously described for bacteria-producing hydrogen sulfide from organic sources and were inoculated into tubes containing liquid Athiorhodaceae medium. The tubes were incubated as previously described for Athiorhodaceae.

Oxygen uptake by sediment

Oxygen consumption rates of unamended, glucose-supplemented, and oil-supplemented sediment were obtained by incubating homogenized sediment with appropriate supplements in Gilson respirometer flasks. Sediment was prepared by removing visible remains of meiofauna and mixing with filtered seawater. In one experiment, the oxygen consumption rates were determined both before and after addition of glucose and oil and were calculated in such a way that the oxygen uptake obtained under each experimental condition could be compared directly. Specific experimental conditions are given with the results. Various combinations of sediment and sterile seawater were used in attempts to obtain maximum uptake rates. In all cases, triplicate flasks were run, the results averaged, and the oxygen uptake was reported as microliters of oxygen consumed per hour per gram of sediment (μ-liters O_2/hr-g). The temperature coefficient (Q_{10}) was determined by comparing the rate of oxygen uptake at both 10 and 20° C. Unless noted as otherwise, equilibration of flasks prior to uptake determinations was between 12 and 18 hrs at the experimental temperature.

RESULTS

Table 21.2 shows the counts obtained, in colony-forming-units/cubic centimeters of sediment (CFU/cc), for five replicate samples taken in 1973, with the standard deviations of the replicate counts shown in parentheses. The differences in counts of bacteria between the algae-covered site in the August sampling and the oil seep site in the September sampling, when compared to controls, were found to be highly significant (P = less than 0.001); however, no other significant differences were observed, including the counts obtained from sediment that had been lightly oiled for 2 mos (four-tide series) previously. Interestingly, the algae-covered site, which by the September sampling period had been subjected to freezing, no longer supported a larger bacterial population when compared to the bare sediment controls.

The data in Table 21.3 show the estimated size of the aerobic and facultatively anaerobic heterotrophic bacterial population, in colony forming units per gram of sediment, for all samples taken during the 1974 field season. The results of a determination of the effects of sediment settling are also shown (18 June sample). These figures indicate that there is a loss of almost 50 percent of bacteria when the sediment is allowed to settle for 30 min before the dilution series is completed, suggesting that the major proportion of the bacterial populaion may be found on the sediment grains, rather than in the interstitial water. However, at the dilutions used, the amount of sediment present prevented accurate counting, and the sediment was allowed to settle for 15 min in subsequent counting experiments. Also, the ratio of aerobic to facultatively anaerobic organisms is, with the exception of the 3 July sample, relatively constant at about 10:1 (from 8.1:1 to 15.4:1). Interestingly, the addition of oil to surface of the sediment, at all concentrations tested, including the chronic additions, had a small but consistent enriching effect on the size of the bacterial population that could be detected, but not statistically supported, by plate counting methods. In fact, the lack of differences observed in the overall seasonal means is striking; on the other hand, the seasonal change observed in the daily sampling means does suggest a seasonal pattern for the size of the bacterial biomass. The population seems to peak during August, followed by a gradual decline through September.

Sulfur cycle bacteria

Hydrogen sulfide producers: Sulfate-reducing bacteria. When the plating method for determining the presence of sulfate-reducing bacteria was used to estimate the presence of these organisms in Control Site sediment taken on 18 June 1975, no blackened colonies were observed on any of the plates (containing approximately 1000 colonies) after 10 days of incubation at 10° C. When the same sediment samples were enriched for these organisms with selective media, similarly negative results were obtained. Three glass-stoppered bottles containing 275 ml of medium and two containing 70 ml of medium were inoculated with 5.0 and 1.0 g of sediment, respectively. After 7 weeks of incubation at 10° C, all bottles showed heavy turbidity, but none showed blackening due to the production of hydrogen sulfide. At the end of 11 weeks incubation, only one bottle, containing 275 ml of medium and 5.0 g of medium, showed evidence of hydrogen sulfide production.

Hydrogen sulfide producers: Aerobic hydrogen sulfide production. Table 21.4 shows, for 4 of the sampling days, the percent of cultures from each of the sampling sites which were able to produce hydrogen sulfide from an organic source. The mean number of colonies sampled from each series was 66, and the cultures were maintained until no change in the percent of positives was observed, or until approximately 8 weeks, by which time dehydration precluded further use. Between 60 and 97 percent of the colonies sampled produced hydrogen sulfide within 52 days. Although there was no significant difference in H_2S production between the control cultures and the cultures

TABLE 21.3 Heterotrophic bacterial counts on sediment samples from oiled and control island flats sites taken during 1974 sampling season

Sampling dates	Number of previous oil applications	Count condition	Heterotrophic bacterial CFU/gm x 10				Daily sampling means	
			Control	500 ppm	2000 ppm	Chronic	Total	Oiled only
6/18/74	None	Sediment suspended	675					
		Settled 30 min.	361					
7/3/74	1[1]	Aerobic[2]	34	41	41	106	55	63
		Anaerobic	3	10	8	41	15	20
7/20/74	2	Aerobic	245	301	289	315	287	302
		Anaerobic	21	16	27	35	25	26
8/5/74	3	Aerobic	125	37	370	380	228	262
		Anaerobic	6	7	10	40	16	19
8/16/74	4	Aerobic	724	460	247	386	454	364
		Anaerobic	113	40	16	26	49	27
9/15/74	5	Aerobic	87	201	137	194	154	177
		Anaerobic	5	4	9	25	11	13
Overall season means		Aerobic	243	208	216	276	235	233
		Anaerobic	30	15	14	33	23	20

[1] Numbers of times oil was applied to sediment surface prior to sampling. In all cases sampling occurred at the next tide series after the oil application. See Material and Methods, Table 21.1.

[2] Refers to incubation of plates. Anerobic counts were obtained from sulfate-reducing counts and are all colonies that grew, whether or not they produced blackened zones around the colonies (see Methods).

TABLE 21.4 Percent of colonies producing H_2S from organic sources

| Sampling date | Incubation days | Percentage of colonies producing H_2S | | | |
		Control	500 ppm	2000 ppm	Chronic
6/18	14	34			
	31	67			
	42	68			
7/3	14	42	40	41	42
	35	74	56	47	70
	52	74	60	64	74
7/20	10	25	17	37	33
	28	95	94	94	79
	42	97	95	96	79
8/5	14	94	79	72	83
	28	97	84	82	88
	35	97	84	82	88

from the oiled sites, a slight seasonal increase was apparent; the increase is reflected both in the percentage of cultures showing H_2S production and in the time needed for that production to become visible.

Photosynthetic sulfur bacteria: Chromatiaceae (formerly Thiorhodacae) and Chlorobiaceae (formerly Chlorobacteriaceae). The photosynthetic sulfur bacteria were generally found to be present in only small numbers. Only two of three enrichments for the green sulfur bacteria (Chlorobiaceae) prepared with 5.0 g sediment were positive after 7 weeks of incubation. After 11 weeks, all 5-g enrichments showed positive results, but the 1-g enrichments produced only three colonies of bacteria on the surface of the settled sediment. Enrichments for the purple sulfur bacteria (Chromatiaceae) were even less successful, producing only one colony of purple bacteria after 11 weeks enrichment from a total of 13 g of sediment. The long enrichment times precluded examination of late season samples, and only the sediment collected on 18 June was enriched for the photosynthetic sulfur bacteria.

Purple nonsulfur bacteria: Rhodospirillaceae (formerly Athiorhodaceae). In enrichment for the purple nonsulfur bacteria from 18 June sediment samples, all enrichments were positive within 4 weeks, and a Most Probable Number (MPN) analysis was completed on sediment collected at the 20 July sampling period. After 3 weeks of incubation, most of the 10-ml inoculated tubes were positive; after 4 weeks, tubes at all dilutions showed positive results. Table 21.5 shows the MPN analysis for Rhodospirillaceae after 4 weeks incubation of enrichment cultures inoculated with aliquots of 1:1 suspension of sediment in filter-sterilized seawater. No attempt was made to determine if the enrichments were oxygen tolerant or if any were capable of converting sulfide into sulfur. Since the medium contained only organic additives,

TABLE 21.5 Most Probable Number analysis of enrichment cultures for Rhodospirillaceae (formerly Athiorhodaceae)

Sample	Number of tubes positive 10:1.0:0.1			MPN per 100 ml of 1:1 Diln.	Mean MPN 50 g sed	Mean, 95% confidence limits
Control	3	3	2	1100		
	3	3	1	450		
	3	3	2	1100	672	95 – 3140
	3	3	1	460		
	3	3	0	240		
500 ppm	3	2	2	210		
	3	2	2	210		
	3	1	1	460	257	140 – 945
	3	3	3			
	3	2	1	150		
2000 ppm	3	3	2	1100		
	3	3	2	1100		
	2	2	2		668	94 – 3011
	2	1	0	15		
	3	3	1	460		
Chronic	3	3	3			
	3	3	1	460		
	2	2	3		314	47 – 1616
	2	2	0	21		
	3	3	1	460		

the sulfur transitions that may have occurred would be of sulfide contained either in the seawater or in the sediment. However, since these sediments are especially low in sulfides (Feder 1974), it is doubtful that these organisms play any role in the sulfur cycle in Valdez silt sediments.

Micro-aquaria model ecosystems

The ability of the microbial population in Valdez sediments to establish a sulfur cycle system, as described by Fenchel (1969), was verified using Fenchel's micro-aquarium technique. The micro-aquaria were inoculated with both suspended sediment and aliquots of supernate from which sediment had been allowed to settle out. As suggested by the reduction in counts due to the settling of the sediment reported above, the aquaria inoculated with clear aliquots showed much delayed, if any, activity. Only one of five cultures showed any evidence of bacterial activity after 3 mos incubation. However, cultures inoculated with suspended sediment produced changes consistent with those reported by Fenchel, although the changes were consistently delayed. After 4 weeks incubation in the light, the medium throughout the tube turned black; by 6 weeks, a dense band of growth was observed approximately one-third of the way down the liquid part of the column. This band slowly moved down the liquid column, causing the medium above the band to lighten to a cloudy gray color, while the medium below remained black. After 12 weeks, the band had moved to the bottom of the liquid phase, and the entire liquid part of the column had turned light gray. The solidified lower part of each column remained black, except for a thin layer of clear agar at the agar-liquid interface. Identically prepared cultures, when incubated in the dark, followed similar but slower patterns of changes; after the 12 weeks of incubation, the band of bacterial growth had moved only approximately three-fourths of the way down the liquid column. At no time, however, were bands of green or purple pigmented bacteria observed in the liquid column, although some pigmented forms were apparent in the settled sediment. The significance of these transitions is discussed by Fenchel (1969) and in the Discussion section below.

Micro-respirometry

Oxygen uptake was found to be very low. Observed rates were consistent with that which would be expected because of the low biomass present in

TABLE 21.6 Oxygen uptake by sediments enriched *in vitro* with glucose and by *in-situ* surface application of oil

Sample date	Sediment source	Core number	Supplement	Rate of uptake[4] μ-liters O_2/hr-g
6/18/74 (10°C)[1]	Control site	1	None	0.56
			1.8 mg glucose[2]	0.78
		2	None	0.45
			1.8 mg glucose	0.89
		3	None	0.12
			1.8 mg glucose	0.33
		Ave. of	None	0.37
			1.8 mg glucose	0.66
7/20/74 (20°C)	Control site	1	None[3]	0.55
		2	None	1.00
		Ave	—	0.77
	Chronically oiled site	1	None	0.46
		2	None	0.89
		Ave	—	0.67

[1] Temperature at which oxygen uptake rate was determined.
[2] 0.5 ml of 0.02 M glucose added to reaction flasks containing 3.0 g sediment and 3.0 ml sterile seawater after unsupplemented rate was determined. Final glucose concentration was 1.8 mg glucose/flask, or 0.277 mg/ml in reaction mixture.
[3] Reaction mixture of 7/20 samples contained 5.0 g sediment plus 3.0 ml sterile seawater.
[4] Rate determined by averaging triplicate flasks prepared from each core.

the sediments. Initially, uptake rates were determined for 3 g of sediment at 10° C, but values were later determined for 8 g of sediment at 20° C. To increase diffusion of gases into the sediments, the samples were routinely mixed with sterile seawater, to a volume which could safely be used in the manometric flasks. Nevertheless, the rates obtained and reported here approach the lower limits of machine sensitivity and are included only because they are consistent with other data and with expected changes due to sediment amount, temperature effects, and added organic materials.

Oxygen uptake was found to be proportional to the amount of sediment used, over a range of 3 to 8 g of sediment, but the rate was constant when converted to oxygen uptake per hour per gram (μ-liters O_2/hr-g) of sediment. The experiments reported in Table 21.6 and Figure 21.1 indicate that sediments that have been chronically exposed to oil do not show increased O_2 uptake when compared to unoiled sediment control samples (Fig. 21.1a). However, the mixing of glucose with sediment *in vitro* did cause an increase in the rate of oxygen uptake (Fig. 21.1b). The increase in

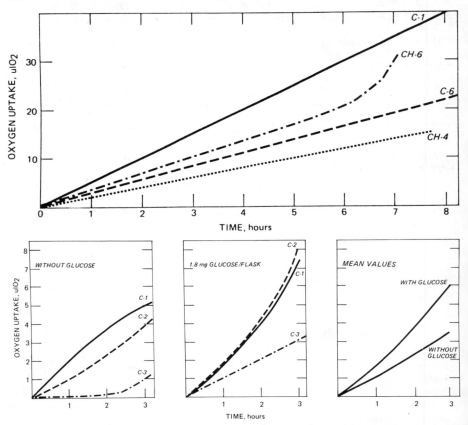

Fig. 21.1 Effect of added organic material on oxygen uptake (a) by unsupplemented sediments and (b) by glucose-supplemented control sediment samples.

the rate of O_2 consumption was observed for each core and, when averaged, showed a doubling of uptake (Table 21.6).

When oil was mixed with sediment *in vitro* (as opposed to surface application *in situ*) and oxygen uptake measured 24 hrs later, a two-fold increase in the rate of uptake was observed. Table 21.7 shows the results of experiments designed to test the response of the sediment microbial population to "mixed-in" oil and glucose. In the experiments reported here, the glucose and oil were added to sediments after approximately 4 hrs incubation in the respirometer and the resulting changes in uptake recorded. When the uptake was measured 2 hrs after mixing glucose and oil with separate sediment samples, the glucose was observed to cause an increase in the rate of oxygen uptake, but the uptake rate of the oil-amended sediment did not change significantly. After 24 hrs, however, the oxygen uptake rate for the oil-amended sediments surpassed the enhanced rates observed for the glucose-amended sediments (Figs. 21.2 and 21.3). As before, similar increases in respiratory activity were observed for all samples.

TABLE 21.7 Oxygen uptake by control site sediments enriched *in vitro* with glucose and oil

| | Experimental conditions | | |
	Series 1: Control sediment	Series 2: Control sediment	Series 3: Control sediment
Reaction mixture[1]			
Sediment (g)	5.0	8.0	5.0
Sterile seawater (ml)	3.0	None	3.0

| | Rates of O_2 uptake, μO_2/hr-g | | | |
	Series 1	Series 2	Series 3	Series 4
Average initial rate, before organic supplement. (N = 6 per series)	1.14	0.88	0.88	0.96
Average glucose-enhanced rate[2] (N = 3 per series)	1.66	1.80	1.76	1.74
Average oil-enhanced rate[3] (N = 3 per series)				
After 2 hrs	1.06	1.00	1.08	1.04
After 24 hrs	2.53	2.07	1.54	2.04

[1] Shows reaction mixture before addition of supplements. All samples were collected on 8/16/74.

[2] Glucose (10 mg/flask: final concentration to 1.25 mg/ml) was added to three reaction vessels and the uptake rate determined for 3 hrs, following a 2-hr equilibration.

[3] The rates shown were determined over two 3-hr periods, beginning at 2 hrs, and again at 24 hrs after the addition of 0.5 ml Prudhoe Bay crude oil to the reaction mixture.

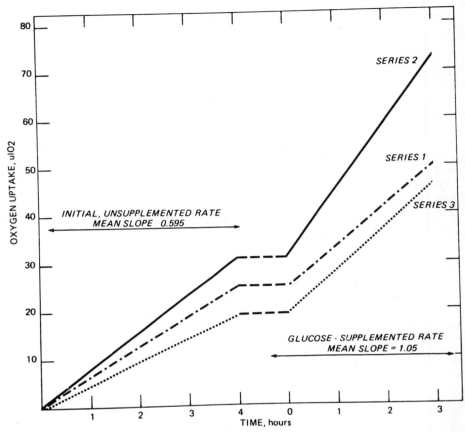

Fig. 21.2 Effect of glucose on oxygen uptake by control sediment conditions as noted in Table 21.7.

Temperature coefficients were calculated for both unamended and glucose-supplemented samples by using data from Tables 21.6, 21.7, and from Figure 21.4. The Q_{10} for unsupplemented and supplemented sediments was 2.405 and 2.636, respectively.

DISCUSSION

The dominant distinguishing characteristics of the glacial silt intertidal ecosystem studied at Port Valdez, Alaska, as compared to sediment intertidal systems studied by others, appear to be comparatively small mean grain size of the sediment particles, the lack of significant wave action, and less contribution of organic material from external sources.

Although the relatively small sediment grain size of 4 to 16 μ provides ample surface area for the establishment of bacterial microcolonies (Dale

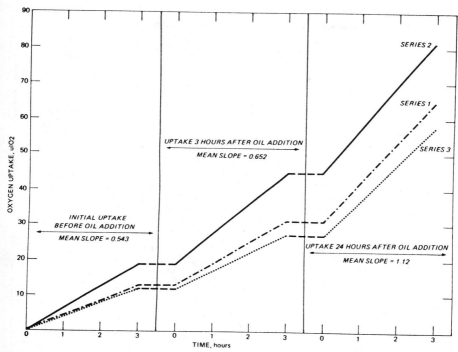

Fig. 21.3 Effect of added oil on oxygen uptake by control sediment conditions as noted in Table 21.7.

1974), the compactness and close packing of the particles greatly reduce the size of the interstitial spaces through which nutrients might be delivered to the sediment-bound bacteria. Although the total surface area on the particles might be larger than in sand systems, the small physical size precludes the growth of more than a few bacterial cells on any one sediment grain. The smaller interstitial spaces would be expected to clog rapidly, providing little or no opportunity for migration of microorganisms or for the movement of nutrients through the sediment. Clearly, the bulk of any organic material that might be deposited on the sediment, whether soluble or particulate, will be removed at the surface, probably by tidal action. Such features as the low mean temperatures and expected reduced metabolic activity ($Q_{10\text{-}20°}=2.5$), fairly rapid tidal surface flushing, almost continuous surface washing by precipitation, and the compactness of the sediment itself preclude rapid digestion of organic material at the surface by bacteria or extensive movement of dissolved organics into the sediment for retention and slower digestion. The relatively gentle tidal changes (stability of the test rings) would also prevent physical burial of detritus. It is apparent that either no significant enrichment of the bacterial population occurs or, if it does, it is removed from the surface by ebb tide along with any deposited organic material. Certainly not enough bacteria or nutrients penetrate the sediment to utilize what little oxygen is present.

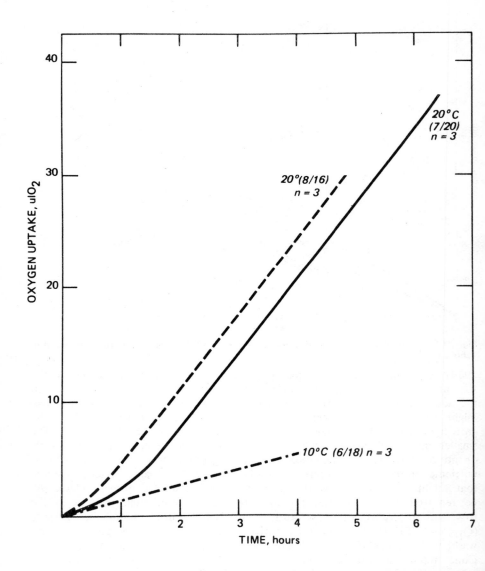

Fig. 21.4 Effect of reaction temperatures on oxygen uptake by unsupplemented sediment samples from control sites.

The low bacterial counts, the failure of the population to respond to *in situ* addition of organics, and the low number of anaerobic sulfur reducers support the conclusion that the sediment at Valdez intertidal zones do not support even a modest microbial biomass, and would not respond to surface applied organics, such as would occur following an oil spill. On the other hand, the *in vitro* experiments reported here suggest that the bacterial population will respond to added organic material if the nutrients can be delivered to the organisms by careful mixing with sediment.

The increased uptake of oxygen and the succession of changes in microaquaria support the hypothesis that this ecosystem is organically poor, but that it will respond when organic nutrients are made available to the microorganisms by mixing either glucose or oil with the sediment. The microaquaria technique of Fenchel (1969) demonstrated that a relatively normal succession of bacterial types, with expected modification due to the paucity of photosynthetic and chemoautotrophic forms, will occur under appropriate environmental conditions.

The initial blackening of the medium in the microaquaria is due to the production of hydrogen sulfide from organic sources by the large proportion of the population that are able to digest sulfur-containing amino acids. As the chemoautotrophic and small numbers of photosynthetic bacteria develop, the hydrogen sulfide is oxidized, resulting in a lightening of the medium and, as the redox discontinuity point moves down through the liquid phase, the chemoautotrophic bacterial growth band similarly moves, resulting in eventual removal of hydrogen sulfide from the entire liquid phase. The modest growth of pigmented organisms in the sediment at the bottom of the liquid phase represents response of the Rhodospirillaceae (Athiorhodaceae) to the organic components of the medium and the anaerobic conditions that exist below the redox discontinuity level. In contrast, the failure to detect even modest numbers of sulfide-oxidizing bacteria (chemoautotrophic or photoautotrophic) is consistent with the reported low sulfide content and aerobic conditions of the sediment.

It should be noted, however, that the procedures used in this report do not measure dynamics in the bacterial populations, but, instead, measure the standing crops at any given time. The plate-count and respirometry data measure only the numbers or activity of the bacterial population, but they give no indication of the turnover of the biomass (see Strickland 1971, and Gray and Williams 1971, for a discussion of this relationship) or of changes in the proportional distribution of certain species within the population. It is not unreasonable to expect changes in the species represented or increased turnover due to grazing without noting changes in the size of the standing crop. There are, indeed, some indications that this might be the case. For example, the percentage of bacterial colonies able to produce hydrogen sulfide from organic sources was observed to change seasonally, from a low of 34 percent from the 18 June sample to a high of 94 percent from the 5 August sample (both after 14 days incubation). Furthermore, the application of the oil resulted in a slight reduction in the sulfide producers, when compared to the control sediments (Table 21.4). However, during these same sampling periods, no change was observed in the total heterotrophic bacterial counts (Table 21.3). H. M. Feder (personal communication) also reports a statistically significant increase in two species of sediment harpacticoid copepoids (*Halectinosoma gothiceps* and *Heterolaophonte* sp.) thought to feed on bacteria, in sediments taken from oiled sites but not in control site sediment samples.

We conclude from these data that the surface addition of oil to the Valdez intertidal areas will not materially affect the sediment ecosystem with similar physical and biological properties. We anticipate that tidal action, together with digestion of oil in the water column, will be found as the major factor in biological removal of oil. However, if significant areas of the intertidal zone cannot support biological removal of the oil, increased amounts of oil (at least that which would have been digested on the sediment) will be expected to be found in the water column.

In contrast, certain areas of the Valdez intertidal zones that do not have similar physical and biological characteristics might be expected to show effects of oil contamination. H. M. Feder and D. G. Shaw (personal communication) report that clams in salt marsh areas are rapidly destroyed by even small additions of oil, and the movement of oil hydrocarbons into more permeable sediments is to be anticipated. Pierce et al. (1975) demonstrated enrichment of a population of hydrocarbon-degrading bacteria in Rhode Island beach sediments subjected to a spill of No. 6 Residual Fuel Oil. The concentration of petroleum hydrocarbons was shown to rapidly decline during the enrichment period, but the rate of hydrocarbon removal remained constant throughout the summer and declined to less than 1 mg of hydrocarbon per gram of dry sediment per day during the winter months. Measurable movement of hydrocarbons through the sediment was observed, presumably due to the effects of tidal and wave action on the interstitial fluids.

REFERENCES

AARONSON, S.

    1970   *Experimental microbial ecology.* Academic Press, New York.

BLEAKLEY, R. J., and P. J. S. BOADEN

    1974   Effects of an oil spill remover on beach meiofauna. *Ann. Inst. Oceanogr.* 50(1): 51-58.

BLOOM, S. A.

    1970   An oil dispersant's effect on the microflora on a beach sand. *J. Mar. Biol. Assn. UK* 50: 919-923.

COBET, A. B., and H. E. GUARD

1973   Effect on bunker fuel on the beach bacterial flora. Proc. Joint Conf. on Prevention and Control of Oil Spills. Amer. Petrol. Inst., New York, pp. 815-819.

CUMMINS, K.

1974   Structure and function of stream ecosystems. *BioScience* 24(11): 631-641.

DALE, N. G.

1974   Bacteria in intertidal sediments: Factors related to their distribution. *Limnol. Oceanogr.* 19: 509-518.

FEDER, H. M.

1974   The sediment environment of Port Valdez and Galena Bay, Alaska, and the effect of oil on this ecosystem. Interim Progress Report 1973 to 1974. Project No. R800944 (18080HFP).

FENCHEL, T. M.

1969   The ecology of marine microbenthos. *Ophelia* 6: 1-183.

FENCHEL, T. M., and R. J. RIEDL

1970   The sulfide system: A new biotic community underneath the oxidized layer of marine and sand bottoms. *Mar. Biol.* 7(3): 255-268.

GRAY, T. R. G., and S. T. WILLIAMS

1971   Microbial productivity in soil. In *Microbes and biological productivity, Symposium* 21, Society for General Microbiology, edited by D. E. Hughes and A. H. Rose. Cambridge University Press, pp. 255-286.

GUARD, H. E., and A. B. COBET

1973   The fate of a bunker fuel in beach sand. Proc. Joint Conf. on Prevention and Control of Oil Spills. Amer. Petrol. Inst., New York, pp. 827-834.

GUARRAIA, L. J., and R. K. BALLENTINE (Eds.)

1972   The aquatic environment: Microbial transformations and water management implications. Symposium sponsored by the Environmental Protection Agency Office of Water Program Operations.

GUNKEL, W.

1968 Bacteriological investigations of oil polluted sediments from the Cornish coast following the Torrey Canyon disaster. In *Biological effects of oil pollution on littoral communities*, edited by J. D. Carthy and D. R. Arthur. Proc. Symp., Feb 1968, at Pembroke, Wales. Field Studies Council, London, pp. 151-158.

JOHNSTON, R.

1970 The decomposition of crude oil residues in sand columns. *J. Mar. Biol. Assn. UK* 50: 925-937.

KUSNETZOV, S. I.

1967 The role of microorganisms in the transformation and degradation of oil deposits. Proc. World Petroleum Congress. Elsevier Ed. 8: 171-181.

LARSEN, H.

1952 On the culture and general physiology of the green sulfur bacteria. *J. Bacteriol.* 64: 187-196.

McINTYRE, A. D., A. L. S. MUNRO, and J. H. STEELE

1970 Energy flow in a sand ecosystem. In *Marine food chains*, edited by J. H. Steele. Edinburgh, pp. 19-31.

McINTYRE, A. D., and D. J. MURISON

1973 The meiofauna of a flatfish nursery ground. *J. Mar. Biol. Assn. UK* 53: 93-118.

MEADOWS, P. S., and J. G. ANDERSON

1968 Microorganisms attached to marine sand grains. *J. Mar. Biol. Assn. UK* 48: 161-175.

MEYNELL, G. G., and E. MEYNELL

1965 Theory and practice in experimental bacteriology. Cambridge Univ. Press, London, UK, 347 pp.

PARKINSON, D., R. G. GRAY, J. HOLDING, and H. M. NAGEL-DE-BOOIS

1971 Heterotrophic microflora. In *Methods of study in quantitative soil ecology: Population production and energy flow*. IBP Handbook No. 18. Blackwell Scientific Publications, Oxford, pp. 34-50.

PERKINS, E. J.

   1958   The food relationships of the microbenthos, with particular
          reference to that found at Whitstable, Kent. *Annals and
          Magazine of Natural History*, series 3. 1: 64-74.

PIERCE, R. H., A. M. CUNDELL, and R. W. TRAXLER

   1975   Persistence and biodegradation of spilled residual fuel oil on
          an estuarine beach. *Appl. Microbiol.* 29: 646-652.

PRATT, D. C., and E. GORHAM

   1970   Occurrence of Athiorhodaceae in woodland, swamp, and pond
          soils. *Ecology* 51(2): 346-349.

STEELE, J. H., A. L. S. MUNRO, and G. S. GIESE

   1970   Environmental factors controlling the epipsammic flora on
          beach and sublittoral sands. *J. Mar. Biol. Assn. UK* 50:
          907-918.

STRICKLAND, J. D. H.

   1971   Microbial activity in aquatic environments. In *Microbes and
          biological productivity, Symposium* 21, Society for General
          Microbiology, edited by D. E. Hughes and A. H. Rose, Cam-
          bridge University Press, pp. 231-254.

ZoBELL, C. F., and C. B. FELTHAM

   1938   Bacteria as food for certain marine invertebrates. *J. Mar. Res.*
          1: 312.

CHAPTER 22

Shallow-water benthic fauna of Prudhoe Bay

H. M. FEDER *and* D. SCHAMEL[1]

Abstract

The investigation described in this chapter was designed to provide preliminary background biological information for nearshore invertebrate benthos in Prudhoe Bay in the summer and thereby to develop a data base suitable for the initiation of a long-term monitoring program for the area. The study, in which both infaunal and slow-moving epifaunal species were considered, was conducted in conjunction with a causeway construction project by the Atlantic Richfield Company.

INTRODUCTION

Limited data on the biology of the Alaska arctic coastal marine environment are currently available (see Feder et al., in press). The only seasonal data are the primary productivity and phytoplankton studies of Horner (1969, 1972, 1973); Matheke (1973); Alexander (1974), at Point Barrow and in the Colville Delta; and the benthic studies of MacGinitie (1955) at Point Barrow. Data from summer studies of the phytoplankton and benthos from the Colville Delta area are available in the literature (Crane 1974; Kinney et al. 1971, 1972; and Alexander et al. 1974). Some shallow-water summer benthic samples from the western Beaufort Sea are also described by Carey et al. (1974). Information from biological explorations along the Canadian arctic coast should be valuable for comparison with Alaskan studies, since many arctic species are circumpolar in distribution (Ellis 1960; Feder et al., 1976). Data on phytoplankton, primary productivity, zooplankton, and hydrography are available for Prudhoe Bay and nearby lagoon areas (Coyle 1974; Horner et al. 1974). However, there is no published information on the benthos for this area, although some qualitative data were obtained near

[1] *Institute of Marine Science, University of Alaska, Fairbanks, Alaska 99701.*

Prudhoe Bay during exploratory dives in conjunction with phytoplankton research (R. Horner, G. Matheke, and S. Maynard, unpublished).

Benthic invertebrate organisms can be useful as indicator species for a disturbed area (Mileikovsky 1970) because many of them tend to remain in place, react to long-term perturbations (Mileikovsky 1970; Perkins 1974), and generally reflect the composition of the substratum (Lie 1968). The importance of the benthic fauna as a monitoring tool in the Arctic may be magnified by the fact that species with direct development are more prevalent here than in temperate regions (Thorson 1950; Ellis 1960; Chia 1970; Mileikovsky 1971). Many arctic species have pelagic larvae, but the planktonic stage is greatly abbreviated (Mileikovsky 1971). A possible consequence of direct development and a short pelagic larval stage is the establishment of "local" populations that recruit new individuals largely from the local stock of immediately adjacent areas. Although local perturbations may affect arctic benthic communities for a relatively short period of time, disruption of these communities over a broad area would require a significantly longer time for recovery. Replacement of adults from adjacent areas would be slow in the latter case and, with the slow growth and development to sexual maturity typical of many arctic forms (Ellis 1960; Thorson 1950), the re-establishment of a benthic assemblage of species would require many more years here than in temperate waters (Chia 1970; also see Mileikovsky 1970 for review on marine larvae and pollution problems). The recolonization of arctic benthic communities will also depend on additional factors, including the larval development patterns of the species affected, the distance to the nearest upcurrent source of these species, the velocity of the currents, and the condition of the local substratum on the arrival of the larvae (Mileikovsky 1970) or adults.

Insufficient long-term information about the environment and the basic biology and recruitment of species in that environment can lead to the erroneous interpretation of drastic changes in types and density of species that might occur if an area becomes altered (Nelson-Smith 1973; Pearson 1971, 1972; Rosenberg 1973, for general discussions of benthic biological investigations in industrialized marine areas). Populations of marine species fluctuate over a time span of a few to 30 years; such fluctuations are typically unexplainable because long-term physical, chemical, and biological data are seldom gathered (Lewis 1970 and personal communication). Additionally, the presence or absence of benthic species can be in part determined by the nature of the substrate. Specifically, the close relationships of benthic faunal assemblages to particular sediment characteristics have been shown for some areas (Jones 1950; Sanders 1968). Furthermore, the ability of larval forms of benthic species to select or reject a substratum on the basis of physical and chemical properties has been determined experimentally (Wilson 1953). Thus, changes in the substrate character may be reflected by changes in resident fauna. However, such changes can be properly interpreted only if the biota and associated substrata are investigated over a reasonable time base prior to and after disturbance of the particular area (see Pearson 1970 and Rosenberg 1973 for such an approach in monitoring areas affected by industrial activity).

METHODS

The study was conducted on the west side of Prudhoe Bay, Alaska (70° 23'N, 148° 31'W) from 15 to 25 August 1974 (Fig. 22.1, Table 22.1). Tidal fluctuations in the area average 15 cm. Two parallel offshore transects, each 1500 m in length, were established 170 m on either side of the projected new causeway (Fig. 22.1). Sampling stations were established onshore and at 200, 500, 800, 1100 and 1400 m from shore along each transect (12 stations); all offshore stations were marked temporarily with anchored crab-pot floats. Two additional transects were established 1.5 km on either side of the new causeway site; only stations in deeper water were sampled on these transects due to time constraints. Stations were also established adjacent to the existing causeway on the east side of the bay, and near three barrier islands (Argo, Gull, and Niakuk) (Fig. 22.1, Table 22.1).

Samples were taken primarily by way of divers operating from a 5-m river skiff. Three Fager core (Fager et al. 1966) replicates (each 0.0069 m^2) and at least two airlift replicates (each 0.25 m^2) were taken at each station. Both methods provided sampling to a depth of approximately 4 cm. Geolog-

TABLE 22.1 Stations sampled in Prudhoe Bay during August 1974

Station	Distance from shore (m)	Depth (m)	Bottom
A. Stations located on two transects on either side of the projected causeway:			
East transect			
2	200	0.9	sand-gravel-mud
4	500	0.9	sandy-mud
6	800	1.4	muddy-sand
8	1100	1.7	mud
10	1400	1.8	mud
West transect			
26	200	0.9	sand-gravel-mud
28	500	1.1	sandy-mud
30	800	1.5	muddy-sand
32	1100	1.7	mud
34	1400	1.8	mud
B. Stations located on additional transects:			
37	200	0.9	sandy-mud
38	500	0.6	sandy-mud
39	800	0.9	sandy-mud
40	1100	1.2	sandy-mud
41	1400	1.4	sandy-mud
C. Stations occupied at selected localities:			
43	nearshore	2	sandy-mud
44	nearshore	2	sandy-mud
Gull Island	nearshore	–	sandy-mud
Argo Island	nearshore	–	sandy-mud
Niakuk Island	nearshore	–	sandy-mud

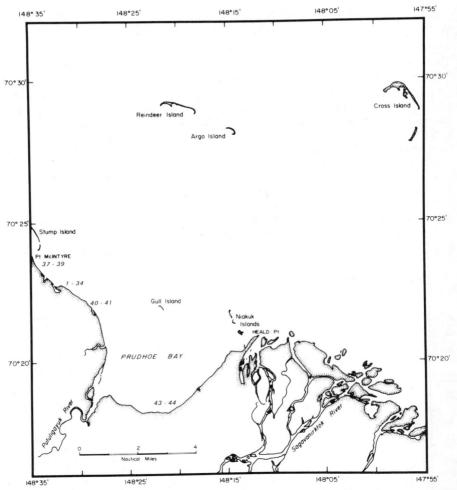

Fig. 22.1 Station locations occupied in Prudhoe Bay for the present study. The new causeway sampling sites are represented by stations 1-34. The old causeway sites are represented by stations 43 and 44.

ical and hydrocarbon samples were taken at all stations (see Feder et al. 1975, 1976). Sampling was also accomplished on selected stations with a small sea-sled and minnow traps baited with fish scraps. Fager core samples were taken near the three barrier islands. A helicopter was used for transportation of the divers and their equipment to the offshore islands. Sampling was accomplished here by swimming to arbitrarily selected study sites. Shore stations on the mainland were examined for organisms *in situ* at random locations. Qualitative examination of the narrow intertidal zone was made at the bases of all transects. All biological material was immediately transferred to plastic bags in the field and preserved in 10 percent hexamine-buffered formalin. Samples were screened through 1-mm mesh Nitex screen in the laboratory. Species identifications, counts, and biomass determinations were made at the Marine Sorting Center, University of Alaska, Fairbanks.

Species diversity was determined by the Gleason and Shannon-Wiener indices (Lie 1968). The former index is a ratio of total number of species to total number of individuals, and it does not weight the contribution of each species to total diversity. The Shannon-Wiener Index is a step-wise summation of the ratio of numbers of individuals of each species to total number of individuals. The latter method weights the contribution of each species to total diversity. In both methods, index values are positively correlated with diversity. Since these indices are based on different calculations, their numerical values are not directly comparable. However, they are measures of the same phenomena, and trends should be similar.

RESULTS AND DISCUSSION

Quantitative studies

No macrofaunal marine invertebrates were found on or within the sediment of the beach or along the narrow intertidal zone.

Fager cores were satisfactory for sampling the sedentary polychaetous annelids, but many of the motile species, inclusive of infaunal crustaceans, readily avoided the coring tube or escaped through the top. The airlift was satisfactory for sampling most infaunal burrowing polychaetes, crustaceans, and slow-moving epifaunal species. The larger sedentary polychaetes were not sampled quantitatively by this technique.

Thirty-eight invertebrate species representing eight phyla (Table 22.2) were collected from the subtidal stations near the new causeway. Polychaetes (13 species) and amphipod crustaceans (13 species) were the dominant groups.

Three of the four transects occupied (encompassing stations 2, 4, 6, 8, 10, 26, 28, 30, 32, 34, 40 and 41; see Fig. 22.1) showed similar physical and faunal trends with increasing distance from shore. A sediment transition was observed along these transects, varying from sand (with little organic detritus) at the inner stations to muddy sand (with abundant organic detritus) at the outer stations. Species diversity, number, and biomass all tended to increase with increasing distance from shore (Figs. 22.2 to 22.5, Table 22.3).

Polychaetes followed the same basic trend along the transects as that noted above (Figs. 22.6 and 22.7). A polychaete biomass discrepancy between samples taken by Fager cores and airlifts at stations 32 and 34 (Fig. 22.6) is apparent. This difference is even more striking when it is noted that the sampling method with the greatest surface coverage (the airlift) resulted in a lower total polychaete biomass. A similar polychaete biomass discrepancy occurred to a lesser extent at stations 8 and 10 (Fig. 22.6). The airlift sampler was apparently unable to pick up many individuals of the large polychaete *Ampharete vega* before they moved deeper into their tubes, as indicated by the many empty *Ampharete* tubes in these samples.

TABLE 22.2 A list of species collected at the 15 stations in the vicinity of the new causeway area, Prudhoe Bay. Most probable types of larval development for these species (P=pelagic; D=direct).

Taxon	Species	Common name	Type larval development
Phylum Porifera			
Order Haplosclerida	*Haliclona rufescens*	sponge	P
Phylum Platyhelminthes			
Class Turbellaria	unknown	flatworm	D
Phylum Nemertea			
	unknown	proboscis worm	P
Phylum Priapulida			
	Halicryptus spinulosus		D
Phylum Mollusca			
Class Gastropoda Family Trochidae	*Margarites helicina*	snail	D
Class Pelecypoda Family Hiatellidae	*Cyrtodaria kurriana*	Northern propeller clam	P
Unknown family	unknown		
Phylum Annelida			
Class Polychaeta			
Family Phyllodocidae	*Eteone longa*	bristle worm	P
Family Nereidae	*Nereis zonata*	bristle worm	P
Family Orbiniidae	*Phylo* sp.	bristle worm	P
Family Spionidae	*Spio mimus*	bristle worm	P
	Scolecolepides arctius	bristle worm	P
	Prionospio cirrifera	bristle worm	P
	Pygospio elegans	bristle worm	D
Family Cirratulidae	*Cirratulus cirratus*	bristle worm	D
	Chaetozone setosa	bristle worm	P
Family Capitellidae	*Capitella capitata*	bristle worm	D
Family Ampharetidae	*Ampharete vega*	bristle worm	D
Family Sabellidae	*Chone duneri*	bristle worm	D
Family Sphaerodoridae	*Sphaerodoronsis minuta*	bristle worm	P

TABLE 22.2 (continued)

Taxon	Species	Common name	Type larval development
Phylum Arthropoda			
Class Oligochaeta			
Unknown family	unknown	—	D
Class Crustacea			
Subclass Ostracoda			
unknown order	unknown	—	P
Subclass Copepoda	unknown	—	P
Subclass Malacostraca			
Order Mysidacea	*Mysis* sp.	opossum shrimp	D
Order Cumacea	*Diastylis sulcata*	—	D
Order Isopoda	*Saduria entomon*	pill bug	D
Order Amphipoda	*Pontoporeia affinis*	sand flea	D
	Pontoporeia femorata	sand flea	D
	Pseudalibrotus sp.	sand flea	D
	Onisimus sp.	sand flea	D
	Monoculopsis longicornis	sand flea	D
	Oediceros saginatus	sand flea	D
	Paroediceros lynceus	sand flea	D
	Gammaracanthus loricatus	sand flea	D
	Apherusa megalops	sand flea	D
	Gammarus zaddachi	sand flea	D
	Aceroides latipes	sand flea	D
Phylum Chordata			
Class Ascidiacea	unknown	sea squirt	P

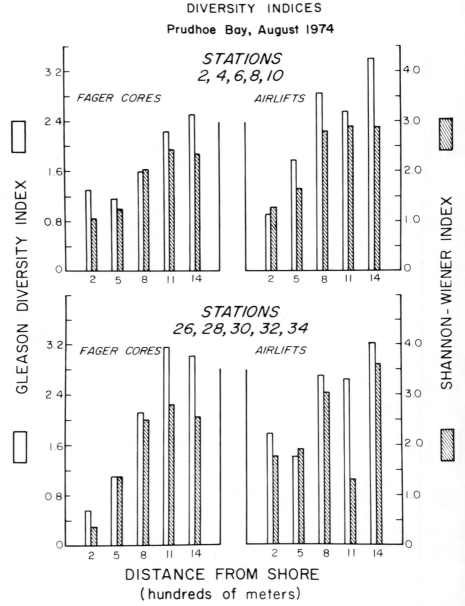

Fig. 22.2 Diversity indices for biological stations 2-34, adjacent to causeway site, sampled by Fager cores and airlifts.

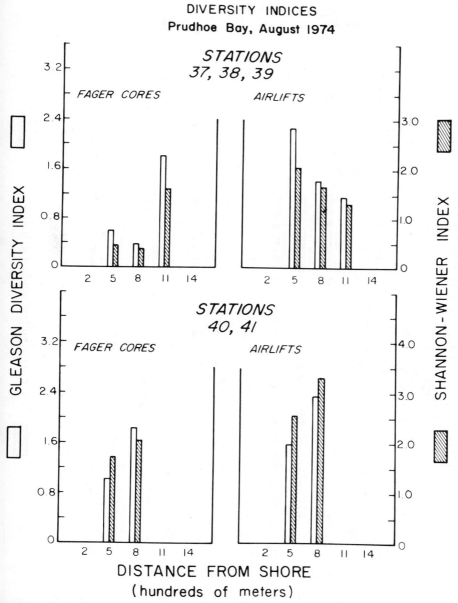

Fig. 22.3 Diversity indices for biological stations 37-41, sampled by Fager cores and airlifts.

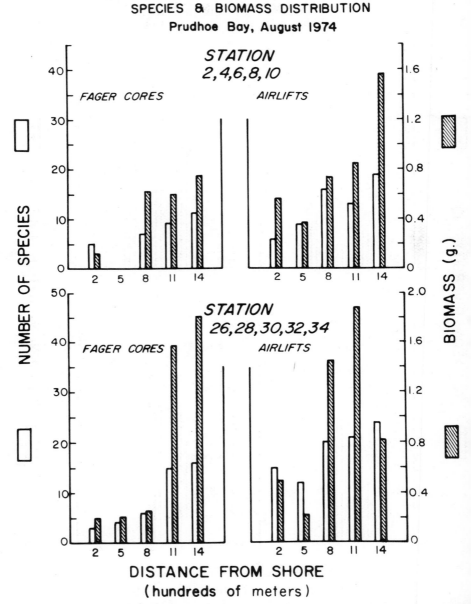

Fig. 22.4 Species and biomass at biological stations 2-34, adjacent to causeway site, sampled by Fager cores and airlifts.

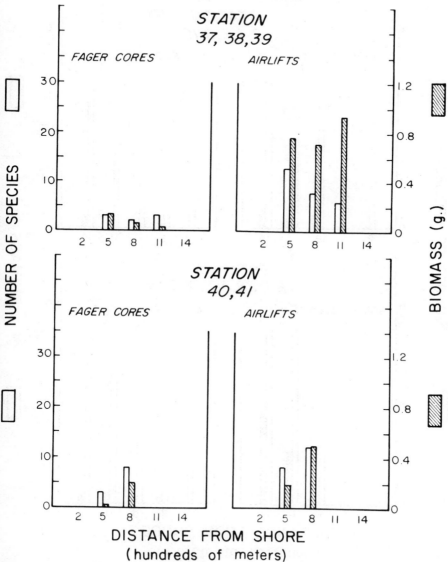

Fig. 22.5 Species and biomass at biological stations 37-41, sampled by Fager cores and airlifts.

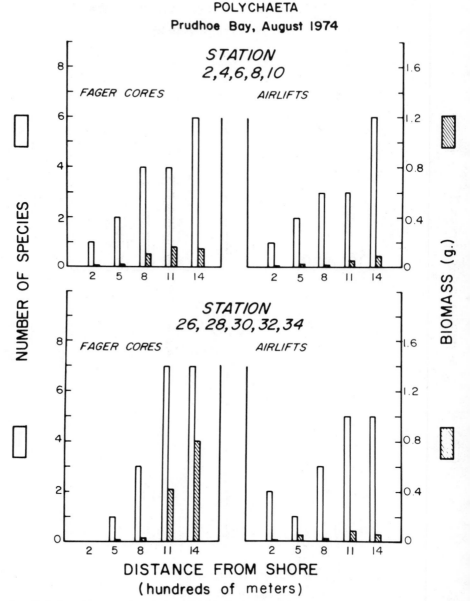

Fig. 22.6 Distribution of polychaetous annelids at biological stations 2-34, adjacent to causeway site, sampled by Fager cores and airlifts.

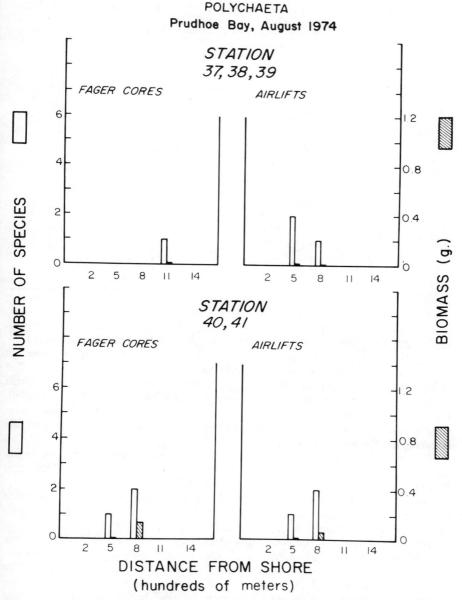

Fig. 22.7 Distribution of polychaetous annelids at biological stations 37-41, sampled by Fager cores and airlifts.

TABLE 22.3 Number of species collected at stations occupied in Prudhoe Bay during August 1974

Station	Number of species	Distance from shore (m)
2	9	200
4	9	500
6	18	800
8	16	1100
10	22	1400
26	11	200
28	9	500
30	17	800
32	24	1100
34	25	1400
37	14	500
38	8	800
39	8	1100
40	9	500
41	14	800
43	1	90
44	1	90

Few empty tubes appeared in the Fager core samples of both stations. Inadequate sampling of this one species by the airlift apparently occurred.

Amphipods tended to increase in species, but there was a decrease in biomass outward from shore along the transects (Figs. 22.8 and 22.9). The change in biomass can be attributed primarily to the distribution of a single species, *Pontoporeia affinis*, which greatly decreased in abundance seaward. Even so, it represented a large proportion of the total amphipod biomass at all stations. At the outer stations, *Pontoporeia affinis* seemed to be partially replaced by *Pontoporeia femorata*.

The clam *Cyrtodaria kurriana* occurred only at the deeper stations, beginning at 800 m from shore and in at least 1.4 m of water (Figs. 22.10 and 22.11).

The isopod *Saduria entomon* showed no major pattern of distribution in either numbers or biomass at any of the stations, but it occurred primarily at stations deeper than 1.4 m.

The fourth transect, encompassing stations 37, 38 and 39, differed from the other three. It physically and faunistically resembled the shallower portions of the other transects. Stations in the fourth transect were quite shallow (0.6 to 1.0 m) and subject to ice scour and wave action. All three stations were poor in detritus. This may be due to the action of ice and wind or to the distance of the stations from nearby rivers which carry tundra debris to the coast. Currents run from east to west along the Beaufort Sea coast (Burrell et al. 1973), and the nearest supply of detritus east of the transect is the Putuligayuk River, about 6 km away (Fig. 22.1). The other three transects are closer to this river. In the shallow fourth

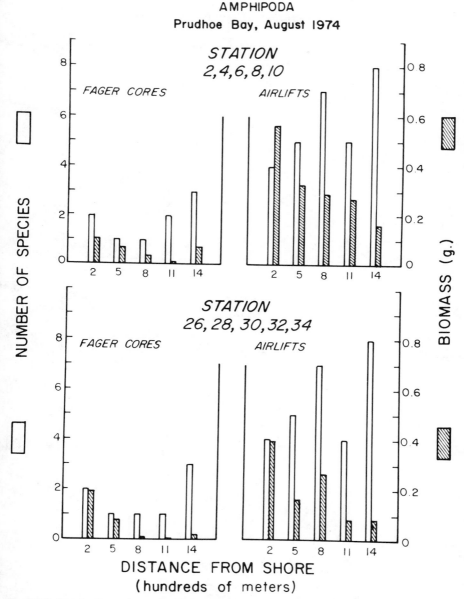

Fig. 22.8 Distribution of amphipod crustaceans at biological stations 2-34, adjacent to causeway site, sampled by Fager cores and airlifts.

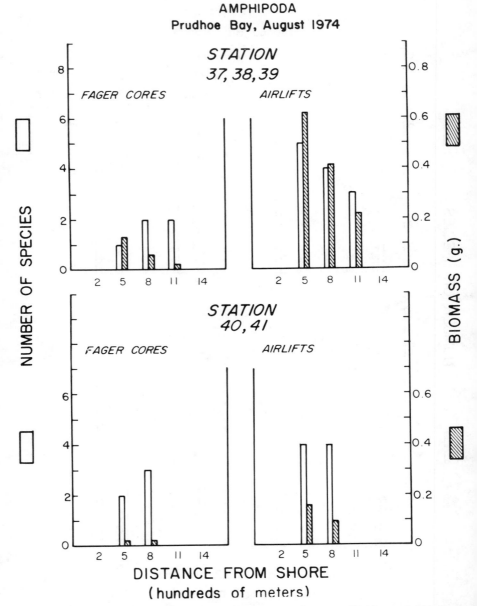

Fig. 22.9 Distribution of amphipod crustaceans at biological stations 37-41, sampled by Fager cores and airlifts.

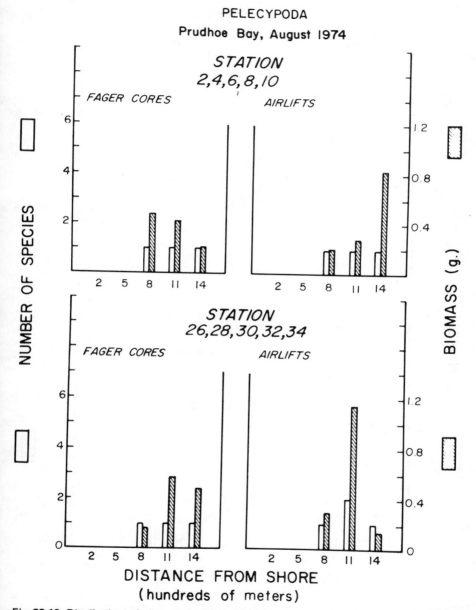

Fig. 22.10 Distribution of the pelecypod mollusk *Cyrtodaria kurriana* at biological stations 2-34, adjacent to causeway site, sampled by Fager cores and airlifts.

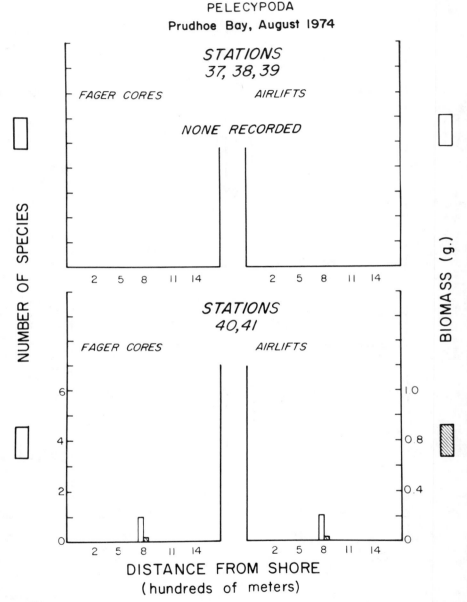

Fig. 22.11 Distribution of the pelecypod mollusk *Cyrtodaria kurriana* at biological stations 37-41, sampled by Fager cores and airlifts.

transect, the tube-dwelling polychaete *Ampharete vega* and the clam *Cyrtodaria kurriana* were not present. However, at the stations in deep water (stations 6, 8, 10, 30, 32, 34 and 41) along other transects, the two species occurred together in large numbers. *Pontoporeia affinis* was the most common amphipod throughout this transect, whereas it was most prevalent nearshore in the other transects. The amphipod *Pseudalibrotus* sp. was found at all stations in this transect, but it occurred only at the shallow stations in the other transects.

Numerous polychaetes collected during this study were bearing eggs. However, many of the amphipods and cumaceans had reproduced prior to our collection, as indicated by the presence of tiny individuals of *Onisimus* sp., *Pseudalibrotus* sp., *Pontoporeia femorata, Pontoporeia affinis,* and *Diastylis sulcata.* One isopod (*Saduria entomon*) was carrying eggs, and another had an inflated but empty brood pouch. This reproductive pattern generally agrees with MacGinitie (1955), who found reproduction of invertebrate species at Point Barrow, Alaska (325 km west of Prudhoe Bay), to extend from early summer through October. Summer production of the pelagic larvae of benthic species (Redburn 1974) and the young of *Saduria entomon* (Bray 1962; Crane 1974) have been recorded in the Arctic.

TABLE 22.4   A list of the Biologically Important Species (BIS) at the 15 stations in new causeway area, Prudhoe Bay, Alaska, August 1974. (See Feder et al. 1973 for BIS criteria.)

Phylum Porifera

Haliclona rufescens

Phylum Nemertea

unknown

Phylum Mollusca

Margarites helicina
Cyrtodaria kurriana

Phylum Annelida

Eteone longa
Scolecolepides arctius
Pygospio elegans
Cirratulus cirratus
Ampharete vega
Chone duneri

Phylum Arthropoda

Mysis sp.
Diastylis sulcata
Saduria entomon
Pontoporeia affinis
Pontoporeia femorata
Pseudalibrotus sp.
Onisimus sp.
Monoculopsis longicornis
Oediceros saginatus

Only 15 (42 percent) of the total species (Table 22.2) and six (32 percent) of the Biologically Important Species (Table 22.4; see Feder et al.
1973 for criteria to determine Biologically Important Species) collected in
our study are known to release pelagic larvae. Thorson (1950) stated that
70 percent of the worldwide benthic species have pelagic larvae, of which
regional extremes of near zero are found in polar regions and 90 to 95
percent in the tropics.

Most of the polychaetous annelids and a number of the amphipods collected are sessile, tube-dwelling species. The polychaetes are primarily deposit
feeders. The feeding methods used by most of the amphipods are unknown.
However, based on collections made in baited traps, at least some of the
amphipod species (*Pseudalibrotus* sp., *Onisimus* sp., and *Gammaracanthus
loricatus*) can function as scavengers. Only three suspension feeding species
were identified. The remaining species appear to be primarily scavengers or
deposit feeders (Table 22.5).

Qualitative studies

Sea-sled samples. The following species were collected at the old causeway
site (station 43; see Figs. 22.1 and 22.12 for location of sample sites):
unknown cnidarian (anemone), *Mysis* sp., *Pseudalibrotus* sp., *Onisimus* sp.,
Gammaracanthus loricatus, Pontoporeia affinis, Monoculopsis longicornis, and
Oediceros saginatus. With the exception of the cnidarian, all of the species
taken were epifaunal and motile. Very minor changes in water depth
occurred over the sampling sites with depths ranging from 15 to 30 cm.
Numbers of *Mysis* sp. increased outward from shore and decreased away
from the causeway, as indicated by the intensive sampling of 24 August (see
Fig. 22.13). Some differences in mysid distribution were apparent in the less
intensive sampling of 25 August. These data indicate that numbers of mysids
fluctuate in space and time and that intensive sampling will be necessary if
these crustaceans are to be used in monitoring studies.

Limited sampling at the new causeway site resulted in the following array
of species: *Mysis* sp., *Monoculopsis longicornis, Saduria entomon, Pontoporeia
affinis, Diastylis sulcata, Gammaracanthus loricatus, Oediceros saginatus,
Pseudalibrotus* sp., and unknown copepods.

Fish-trap samples. At the old causeway site, only one species, the amphipod
Pseudalibrotus sp., was collected. The number of individuals collected are shown
in Table 22.6. Great variability between the days and methods of sampling is
evident.

At the new causeway site, four species were collected: *Saduria entomon,
Onisimus* sp., *Pseudalibrotus* sp., *Gammaracanthus loricatus.* The trends noted
for the three species of amphipods are shown in Table 22.7. The tendency
for *Onisimus* to increase in density with increased water depth and for
Pseudalibrotus to decrease in density with increased water depth were observed in the Fager and airlift samples (Feder et al. 1975, 1976). Table 22.8
shows the trends noted for the isopod *Saduria entomon.* Adults of both
sexes were most numerous 500 m from shore; juveniles tended to increase

TABLE 22.5 Feeding methods used by invertebrate species collected at the 15 stations at the new causeway area, Prudhoe Bay. Phylum: P=Porifera, F=Platyhelminthes, N=Nemertea, L=Priapulida, R=Arthropoda, C=Chordata, A=Annelida (based on Feder et al. 1973; G. J. Mueller and H. M. Feder, unpublished data).

Species	Phylum	Deposit feeder	Suspension feeder	Scavenger	Predator	Unknown
Haliclona refescens	P		X			
Flatworms (1 species?)	F			X?		
Proboscis worms (2 species?)	N				X	
Halicryptus spinulosus	L					X
Margarites sp.	M	X?				
Cyrtodaria kurriana	M		X			
Eteone longa	A					
Nereis zonata	A	X		X	X	
Phylo sp.	A	X			X	
Spio mimus	A	X				
Scolecolepides arctius	A	X				
Prionospio cirrifera	A	X				
Pygospio elegans	A	X				
Cirratulus cirratus	A	X				
Chaetozone setosa	A	X				
Capitella capitata	A	X				
Ampharete vega	A	X				
Chone duneri	A		X			
Sphaerodoropsis minuta	A	X				
Oligochaeta (1 species)	A	X				
Ostracod (1 species)	R					X

TABLE 22.5 (continued)

Species	Phylum	Deposit feeder	Suspension feeder	Scavenger	Predator	Unknown
Copepod (1 species)	R					X
Mysis sp.	R					X
Diastylis sulcata	R	X				
Saduria entomon	R			X		
Pontoporeia affinis	R					X
Pontoporeia femorata	R					X
Pseudalibrotus sp.	R			X		
Onisimus sp.	R			X		
Monoculopsis longicornis	R					X
Oediceros saginatus	R					X
Paroediceros lynceus	R					X
Gammaracanthus loricatus	R			?		
Apherusa megalops	R					X
Gammarus zaddachi	R					X
Aceroides latipes	R					X
Tunicate (1 species)	C		X			

TABLE 22.6 Fish samples taken at the old causeway site. Numbers of *Pseudalibrotus* sp. collected

Station	Date	Trap covered with gunny sack (no. individuals)	Trap not covered with gunny sack (no. individuals)
43	August 24	269	123
	August 25	143	242
44	August 24	177	—

TABLE 22.7 Fish samples taken at the new causeway site

	Distance from shore (number of individuals/trap)			
Species	300 m	500 m	800 m	1100 m
Gammaracanthus loricatus	1	0	0	0
Onisimus sp.	2	1	20	29
Pseudalibrotus sp.	18	14	4	9

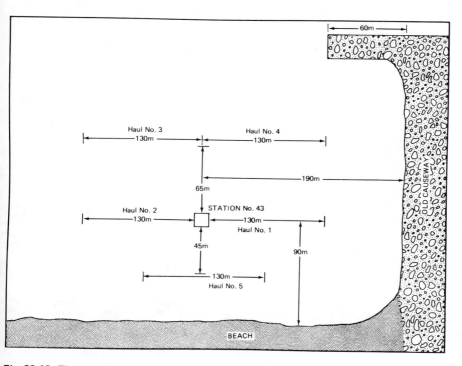

Fig. 22.12 The sampling scheme for sea-sled activities at station 43, adjacent to the old causeway.

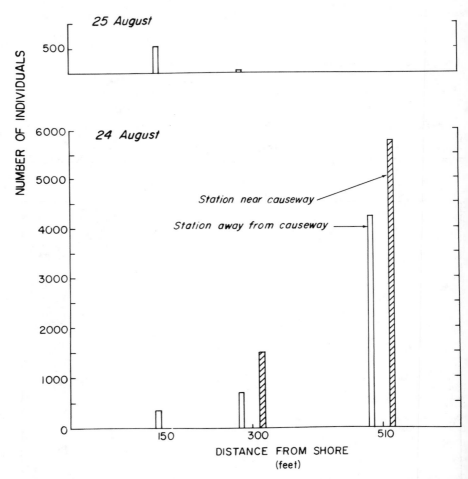

NUMBERS & DISTRIBUTION OF *MYSIS* SP.

Prudhoe Bay, 1974

STATION 43

Fig. 22.13 Distribution and numbers of *Mysis* sp. at Biological Station 43, adjacent to the old causeway at Prudhoe Bay, 1974. Sampling by sea-sled. (See Fig. 22.12 for location of sample tracks).

TABLE 22.8   Fish trap samples taken at the new causeway site. Numbers of *Saduria entomon* collected

Age/Sex	Distance from shore (numbers of individuals/trap)			
	300 m	*500 m*	*800 m*	*1100 m*
Adult/male	0	10	4	4
Adult/female	0	11	1	2
Juveniles (telson length <7 mm)	0	0	1	6

with increasing water depth. Additional sampling is essential to document trends.

Barrier island samples. Two Fager cores were taken at Gull, Niakuk, and Argo islands. At Gull Island, three species were collected: *Saduria entomon, Prionospio cirrifera,* and *Chone duneri.* All samples were taken in 0.6 m of water. The numbers of individuals of each species were low. The two polychaetes taken (*Prionospio* and *Chone*) were previously found only at the deeper stations (at least 1.4 m of water) at the new causeway site. At Niakuk Island, three species were collected: *Saduria entomon, Pontoporeia affinis,* and *Monoculopsis longicornis.* All samples were taken in 0.6 to 0.9 m of water. The species collected here were previously found at all stations sampled at the new causeway site. The latter two species do not characterize any particular water depth.

At Argo Island, eleven species were collected: an unidentified oligochaete, an unidentified bryozoan, an unidentified tunicate, *Ampharete vega, Halicryptus spinulosa, Castalia* sp., *Terebellides stroemi, Chaetozone setosa, Sphaerodoropsis minuta, Cirratulus cirratus,* and *Haploscoloplos elongata.* Two cores were taken on the lee side of the island, at depths of 2 and 3 m. The core from shallower water contained only the oligochaete, the bryozoan, and the tunicate. The few species and low numbers of individuals taken may be the result of a single sample collected in a locally impoverished area or may reflect effects of ice scouring. More intensive sampling is indicated to determine the quantitative nature of the distribution of the biota here. The core from 3 m contained some species not noted at any other station occupied in Prudhoe Bay: *Castalia* sp. (polychaete), *Terebellides stroemi* (polychaete), and *Haploscoloplos elongata* (polychaete). *Castalia* sp. is reported from the deeper subtidal stations in the Canadian Arctic (Ellis and Wilce 1961). Most of the other species taken here were also found at the deeper stations of the new causeway transects.

GENERAL DISCUSSION

Most of the species collected in Prudhoe Bay are known taxonomically, and biological information, based on studies elsewhere, is available for some of the common species (Crane 1974; MacGinitie 1955). The presence of at least some ubiquitous species along the Beaufort Sea coast that are relatively well known biologically may facilitate the development of effective monitoring schemes during long-term industrial activity here.

Two nearshore arctic phenomena must be considered in the development of an effective environmental monitoring scheme for Prudhoe Bay: the existence of ice-scour zones and the establishment of "local" populations of species. In the present study, species diversity, numbers, and biomass tended to increase seaward. These general trends are widespread in arctic waters and are thought to be mainly wave and ice related phenomena (Crane and Cooney 1973; Ellis and Wilce 1961; Sparks and Pereyra 1966). Much of the shallow marine area investigated in the present study is ice-stressed for 8 to 9 months of the year and shows a typical low species diversity (see Dunbar 1968 for discussion of arctic marine ecosystems). This low diversity was especially obvious at the nearshore stations on the causeway transects. Nine species were found both at stations 4 (0.9 m in depth) and 28 (1.1 m in depth). The number of species at the outer stations on these transects increased to 22 and 25 at stations 10 and 34 respectively (Table 22.3). Those organisms within the zone of ice scour must withstand wave and ice action in spring, summer, and fall, as well as high salinities accompanied by low levels of dissolved oxygen in winter (Crane 1974). The present study involved transects with sampling stations within, as well as below, the zone of ice scour. This sampling regime should permit assessment of any differential effects of perturbations upon the benthic assemblages of these two zones.

The monitoring of widespread perturbations may be facilitated by the existence of "local" populations. Such populations recruit new individuals largely from the local stock or immediately adjacent areas. Many of the benthic species collected in our study are motile, and their young undergo direct development. Thus, their replacement over a wide area would depend on non-pelagic individuals migrating from peripheral areas. On the other hand, species with pelagic larvae may be able to colonize a perturbed area more readily. Such a recruitment pattern may result in modification of the species and age-class composition of that area. Restricted reproductive periods might also hamper repopulation of an area, particularly if the effects of the disturbance last through the settling period.

The apparent broad distribution of the more common shallow-water marine invertebrate species along the arctic coast suggests a widely dispersed brood stock available for repopulation of stressed areas there. The sampling scheme developed for the investigation in Prudhoe Bay reported here should permit long-term assessment of many of the species of the shallow benthos in the nearshore waters of Prudhoe Bay. Such biological studies can then be integrated with simultaneous examination of the sediment composition and hydrocarbons at stations from specific transects in Prudhoe Bay (Feder et al. 1975, 1976). The results should provide a meaningful basis for critical evaluation of the benthic populations at these stations in space and time, under naturally or unnaturally stressed conditions.

Fishes collected in Prudhoe Bay during the same period as the benthic investigation reported here were feeding heavily on the common crustaceans collected by our sampling gear (Furniss 1975). Amphipods, isopods, and mysids were the major benthic species used by the least cisco (*Coregonus sardinella*), Arctic char (*Salvelinus alpinus*), Arctic cisco (*Coregonus autumnalis*), fourhorn sculpin (*Myoxocephalus quadricornis*), broad whitefish (*Coregonus nasus*), Arctic grayling (*Thymallus arcticus*), and Arctic flounder (*Liopsetta glacialis*). Thus, the shallow-water marine benthos is an important food source for fishes migratory through and resident to Prudhoe Bay.

Acknowledgments

This work was supported by the Atlantic Richfield Company and by the NOAA Office of Sea Grant, Department of Commerce (contract 03-3-158-41). The assistance of the staff of the ARCO Base at Deadhorse, Prudhoe Bay, is appreciated; we specifically commend Angus Gavin for his enthusiastic support of all aspects of the field work. All diving activities were organized and accomplished by Rick Rosenthal of Dames and Moore, Anchorage, and Lou Barr of the National Marine Fisheries Service, Auke Bay Biological Laboratory, Auke Bay, Alaska. We thank John Hilsinger of the University of Alaska for his assistance in the field and George Mueller, Director of the Marine Sorting Center of the University of Alaska, for his assistance with equipment design, taxonomic determinations, and discussions. Contribution no. 286, Institute of Marine Science, University of Alaska.

REFERENCES

ALEXANDER, V.

1974 Primary productivity regimes of the nearshore Beaufort Sea, with reference to the potential role of ice biota. *In* Proc. Symposium on Beaufort Sea Coastal and Shelf Research, edited by J. C. Reed and J. E. Sater. The Arctic Institute of North America, Arlington, Virginia, pp. 609-632.

ALEXANDER, V., ET AL.

1974 Environmental studies of an arctic estuarine system. Final Report R74-1, Inst. Mar. Sci., Univ. Alaska, Fairbanks. (Sea Grant Report 73-16), 538 pp.

BRAY, J. R.

1962 Zoogeography and systematics of isopoda of the Beaufort Sea. M.S. Thesis, McGill University, 138 pp.

BURRELL, D. C., J. A. DYGAS, and R. W. TUCKER

1973 Beach morphology and sedimentology of Simpson Lagoon. *In* Environmental studies of an arctic estuarine ecosystem, edited by V. Alexander et al. Final Report R74-1, Inst. Mar. Sci., Univ. Alaska, Fairbanks, pp. 45-144.

CAREY, A. G., R. E. RUFF, J. G. CASTILLO, and J. J. DICKINSON

1974 Benthic ecology of the western Beaufort Sea continental margin: Preliminary results. *In* Proc. Symposium on Beaufort Sea Coastal and Shelf Research Program, edited by J. C. Reed and J. E. Sater. The Arctic Institute of North America, Arlington, Virginia, pp. 665-680.

CHIA, F. S.

1970 Reproduction of arctic marine invertebrates. *Mar. Poll. Bull.* 1: 78-79.

COYLE, K. O.

1974 The ecology of the phytoplankton of Prudhoe Bay, Alaska, 1974. Alaska Dept. Fish and Game, 16 pp.

CRANE, J. J.

1974 Ecological studies of the benthic fauna in an arctic estuary. M.S. Thesis, Univ. Alaska, Fairbanks, 105 pp.

CRANE, J. J., and R. T. COONEY

1973 The nearshore benthos. *In* Environmental studies of an arctic estuarine ecosystem, edited by V. Alexander et al. Final Report R74-1, Inst. Mar. Sci., Univ. Alaska, Fairbanks, pp. 411-466.

DUNBAR, M. J.

1968 *Ecological development in polar regions: A study in evolution.* Prentice Hall, Englewood Cliffs, New Jersey.

ELLIS, D. V.

1960 Marine infaunal benthos in arctic North America. Tech. Paper No. 5, Arctic Institute of North America, 53 pp. Montreal, Canada.

ELLIS, D. V., and R. T. WILCE

1961 Arctic and subarctic examples of intertidal zonation. *Arctic* 14(4): 224-235.

FAGER, E. W., ET AL.

1966 Equipment for use in ecological studies using Scuba. *Limnol. Oceanogr.* 11: 503-509.

FEDER, H. M., G. J. MUELLER, M. H. DICK, and D. B. HAWKINS

1973 Preliminary benthos survey. In *Environmental studies of Port Valdez*, edited by D. W. Hood, W. E. Shiels, and E. J. Kelley. Occas. Publ. No. 3, Inst. Mar. Sci., Univ. Alaska, Fairbanks, pp. 305-391.

FEDER, H. M., D. G. SHAW, and A. S. NAIDU

1975 Environmental study of the marine environment in Prudhoe Bay, Alaska. Interim report, Inst. Mar. Sci., Univ. Alaska, Fairbanks, 90 pp.

FEDER, H. M., D. G. SHAW, and A. S. NAIDU

1976 The nearshore marine environment in Prudhoe Bay, Alaska. *In* The arctic coastal environment of Alaska, Vol. 1. Sea Grant report 76-3. Report 76-1, Inst. Mar. Sci., Univ. Alaska, Fairbanks, 161 pp.

FURNISS, R. A.

1975 Fisheries investigations at Prudhoe Bay, Alaska. Alaska Dept. Fish and Game, 16 pp.

HORNER, R.

1969 Phytoplankton studies in the coastal waters near Barrow, Alaska. Ph.D. Thesis, Univ. Washington, Seattle, 261 pp.

1972 Ecological studies on arctic sea ice organisms. Tech. report R72-17, Inst. Mar. Sci., Univ. Alaska, Fairbanks, 179 pp.

1973 Studies on organisms found in arctic sea ice. Final report to the Arctic Institute of North America. Inst. Mar. Sci., Univ. Alaska, Fairbanks.

HORNER, R., K. COYLE, and D. REDBURN

1974 Biology of the plankton near Prudhoe Bay, Alaska. Final report to NOAA-Sea Grant. Inst. Mar. Sci., Univ. Alaska, Fairbanks.

JONES, N. S.

1950 Marine bottom communities. *Biol. Rev.* 25: 283-313.

KINNEY, P., ET AL.

1971 Baseline data study of the Alaskan arctic aquatic environment: Eight month progress report. Tech. report R71-4 (1970), Inst. Mar. Sci., Univ. Alaska, Fairbanks, 176 pp.

KINNEY, P., ET AL.

    1972   Baseline data study of the Alaskan arctic aquatic environment. Tech. report R72-3, Inst. Mar. Sci., Univ. Alaska, Fairbanks, 271 pp.

LEWIS, J. R.

    1970   Problems and approaches to baseline studies in coastal communities. Proc. FAO Technical Conf. Marine Pollution and its Effect on Living Resources and Fishing. FIR:MP 70/E-22, 7 pp.

LIE, U.

    1968   A quantitative study of benthic infauna in Puget Sound. Washington, USA in 1963--1964. Fisk. Dir. Skr. (Ser. Havunders.) 14: 223-356.

MacGINITIE, G. E.

    1955   Distribution and ecology of the marine invertebrates of Point Barrow, Alaska. Smithsonian Misc. Coll. 128: 1-201.

MATHEKE, G. E. M.

    1973   The ecology of the benthic microalgae in the sublittoral zone of the Chukchi Sea near Barrow, Alaska. M.S. Thesis, Univ. Alaska, Fairbanks, 114 pp.

MILEIKOVSKY, S. A.

    1970   The influence of pollution on pelagic larvae of bottom invertebrates in marine nearshore and estuarine waters. *Mar. Biol.* 6: 350-356.

    1971   Types of larval development in marine invertebrates, their distribution and ecological significance: A re-evaluation. *Mar. Biol.* 10: 193-213.

NELSON-SMITH, A.

    1973   *Oil pollution and marine ecology.* Paul Elek (Scientific Books) Ltd., London, 260 pp.

PEARSON, T. H.

    1971   The benthic ecology of Loch Linnhe and Loch Eil, a sea loch system on the west coast of Scotland. Part 3. The effect on the benthic fauna of the introduction of pulp mill effluent. *J. Exp. Mar. Biol. Ecol.* 6: 211-233.

1972   The effect of industrial effluent from pulp and paper mills on the marine benthic environment. *Proc. R. Soc. Lond. B.* 180: 469-485.

PERKINS, E. J.

1974   *The biology of estuarine and coastal waters.* Academic Press, New York, 678 pp.

REDBURN, D. R.

1974   The ecology of the inshore marine zooplankton of the Chukchi Sea near Point Barrow, Alaska. M.S. Thesis. Univ. Alaska, Fairbanks, 172 pp.

ROSENBERG, R.

1973   Succession in benthic macrofauna in a Swedish fjord subsequent to the closure of a sulphite pulp mill. *Oikos* 24: 244-258.

SANDERS, H. L.

1968   Marine benthic diversity: a comparative study. *Amer. Naturalist* 102: 243-282.

SPARKS, A. K., and W. T. PEREYRA

1966   Benthic invertebrates of the southeastern Chukchi Sea. *In* Environment of the Cape Thompson region, Alaska, edited by N. J. Wilimovsky. USAEC, Oak Ridge, Tennessee, pp. 817-838.

THORSON, G.

1950   Reproduction and larval ecology of marine bottom invertebrates. *Biol. Rev.* 25: 1-45.

WILSON, D. P.

1953   The settlement of *Ophelia bicornis* Savigny larvae. *J. Mar. Biol. Assn UK* 31: 413-438.

Assessment of the Arctic Marine Environment: Selected Topics
Copyright © 1976 by Institute of Marine Science, University of Alaska, Fairbanks

CHAPTER **23**

Fish use of nearshore coastal waters in the western Arctic: Emphasis on anadromous species

P. C. CRAIG *and* P. MCCART [1]

Abstract

An overview of fish utilization of nearshore habitats is presented for the Beaufort Sea coastal region between the Colville River (Alaska) and the Mackenzie River (NWT). Movements and life histories of anadromous species, principally the Arctic char and Arctic cisco, are emphasized.

Most of the 28 fish species caught in nearshore areas are freshwater or anadromous rather than marine. Areas of greatest species diversity are centered in the deltas of the largest drainages, most notably the Mackenzie delta. Potential sources of disturbance to fish in nearshore habitats are discussed.

INTRODUCTION

The status of fish resources in arctic coastal waters has recently received particular attention because of interest in offshore petroleum development. This chapter presents an overview of our current knowledge of fish utilization of nearshore areas along the Beaufort Sea between the Colville and Mackenzie rivers. The purposes of this paper are the following: to serve as a bibliographic reference to reports issued by various governmental agencies and private consulting firms, to describe coastal areas from the fisheries viewpoint, and to examine the movements and life histories of the Arctic char

---

[1] *Aquatic Environments Ltd., 1235A 40 Avenue N.E., Calgary, Alberta, Canada T2E 6M9.*

(*Salvelinus alpinus*) and Arctic cisco (*Soregonus autumnalis*). These anadromous species are among the two most abundant and widely distributed fishes in the study area. The Arctic char is a major sports fish, and both species are important in native domestic fisheries.

SPECIES OCCURRING IN NEARSHORE AREAS

About 28 fish species have been recorded in nearshore habitats between the Colville and Mackenzie rivers (Fig. 23.1) (Winslow and Roguski 1970; McRoy et al. 1971; Roguski and Komarek 1972; Kendel et al. 1975; Percy et al. 1974; Ward and Craig 1974; Galbraith and Fraser 1974; Slaney 1974; Furniss 1975; Griffiths et al. 1975). As shown in Figure 23.2, the areas of greatest diversity are the deltas of the largest North Slope rivers — the Colville and, most notably, the Mackenzie. Once beyond the Mackenzie's influence, however, the number of species drops dramatically, with fewer species recorded off portions of the Arctic Wildlife Range coast. This may be in part due to the limited effort so far expended in the latter areas.

The distribution pattern shown in Figure 23.2 applies particularly to the anadromous and freshwater species. As studies of coastal fish populations

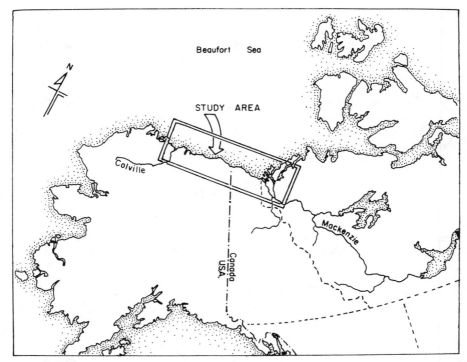

Fig. 23.1 Map of the study area showing the relative sizes of the two largest drainage, the Colville and Mackenzie rivers.

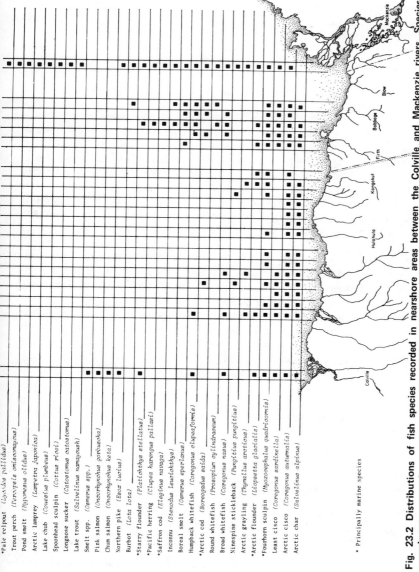

*Pale eelpout *(Lycodes pallidus)*
Trout perch *(Percopsis omiscomaycus)*
Pond smelt *(Hypomesus olidus)*
Arctic lamprey *(Lampetra japonica)*
Lake chub *(Couesius plumbeus)*
Spoonhead sculpin *(Cottus ricei)*
Longnose sucker *(Catostomus catostomus)*
Lake trout *(Salvelinus namaycush)*
Smelt spp. *(Osmerus spp.)*
Pink salmon *(Oncorhynchus gorbuscha)*
Chum salmon *(Oncorhynchus keta)*
Northern pike *(Esox lucius)*
Burbot *(Lota lota)*
*Starry flounder *(Platichthys stellatus)*
*Pacific herring *(Clupea harengus pallasi)*
*Saffron cod *(Eleginus navaga)*
Inconnu *(Stenodus leucichthys)*
Boreal smelt *(Osmerus eperlanus)*
Humpback whitefish *(Coregonus clupeaformis)*
*Arctic cod *(Boreogadus saida)*
Round whitefish *(Prosopium cylindraceum)*
Broad whitefish *(Coregonus nasus)*
Ninespine stickleback *(Pungitius pungitius)*
Arctic grayling *(Thymallus arcticus)*
*Arctic flounder *(Liopsetta glacialis)*
*Fourhorn sculpin *(Myoxocephalus quadricornis)*
Least cisco *(Coregonus sardinella)*
Arctic cisco *(Coregonus autumnalis)*
Arctic char *(Salvelinus alpinus)*

* Principally marine species

Fig. 23.2 Distributions of fish species recorded in nearshore areas between the Colville and Mackenzie rivers. Species records are approximate, since sampling efforts varied throughout the area. Most samples shown here were taken in nearshore, brackish water areas less than 10 m in depth. Marine species collected off Herschel Island are not included (McAllister 1962).

continue, it is likely that more marine species will be collected. McAllister (1962), for example, found a variety of marine species off Herschel Island.[2]

The species occurring within the study can be classified in three broad categories:

Freshwater species, with a limited tolerance for saline water, are found occasionally in estuaries when salinities are low. Examples include the grayling, longnose sucker, round whitefish, and pike.

Anadromous species are tolerant of saline water and undertake a seaward migration on one or more occasions during their life. Downstream migration takes place in the spring, and the fish return to fresh water each autumn to spawn or overwinter. Examples include the Arctic char, Arctic cisco, and the least cisco.

Marine species remain in saline waters throughout the year, though some may make forays into brackish or fresh waters during the ice-free season. Examples include the fourhorn sculpin, Arctic flounder, and the Arctic cod.

Most fish caught by gillnet in nearshore areas during the ice-free season are freshwater or anadromous rather than marine fish. The latter category accounted for only 7 to 28 percent of samples collected during several extensive studies (Roguski and Komarek 1972; Kendel et al. 1975; Griffiths et al. 1975; Furniss 1975). Of the principally marine fishes frequenting the study area, a single species – the fourhorn sculpin – is by far the most abundant.

For the anadromous species, nearshore waters may be loosely characterized as areas containing relatively few fry or young (small) juveniles but large numbers of older (larger) juveniles and mature fish. This is especially true for the region west of Herschel Island.

The numbers of anadromous fish present in nearshore areas during the ice-free period are seasonally variable. At one location, Nunaluk Lagoon off the Firth River, Arctic char were most common during the first half of the season while the largest catches of Arctic cisco were taken after mid-July (Fig. 23.3). The marine fourhorn sculpin was most abundant just prior to freeze-up.

Anadromous species use nearshore habitats as feeding grounds or as migratory pathways to feeding locations. All known spawning and overwintering sites are located in fresh water, although data concerning possible overwintering sites in the lower portions of deltas or other nearshore areas are sparse.

PRINCIPAL ANADROMOUS SPECIES: ARCTIC CHAR

Life history

Within the *Salvelinus alpinus* complex, there are two apparent species which have been designated the Eastern Arctic form and the Western Arctic-Bering

---

[2]*A more recent study by Griffiths et al. (in press) shows that additional species, almost all of which are marine fishes, may be present at some coastal locations. These investigators identified eight species in nearshore waters at Barter Island (Alaska) which are not included in Figure 23.2. Small numbers of the following fishes were caught: least cisco, Arctic flounder, capelin, false sea sculpin, and the fry of slender and stout eelblennies and an unidentified snailfish. Arctic cod fry were more common.*

Sea form of the Arctic char (McPhail 1961). The Mackenzie River is, for practical purposes, the dividing line between the distributions of anadromous populations of the two forms in North America. Within the study area, all anadromous populations which have been so far investigated are of the Western Arctic-Bering Sea form.

Where it does occur west of the Mackenzie River, the Eastern Arctic form is entirely freshwater-resident and is confined to a few lakes in the Richardson and Brooks mountain ranges (McCart and Craig 1971; Ward and Craig 1974; McCart et al. 1974; Craig and Wells 1975).

The following is an outline of the life history of anadromous char in the Beaufort Sea drainage with an emphasis on the marine phase of their life cycle. This information is drawn from several reports (McCart et al. 1972; Yoshihara 1973; Bain 1974; Glova and McCart 1974; Griffiths et al. 1975; Craig and McCart 1975) and from the authors' unpublished data. Although this paper describes the life cycle of anadromous char, the most prevalent life history pattern for this species in the study area, the species is ecologically flexible, and other less common life history patterns have been

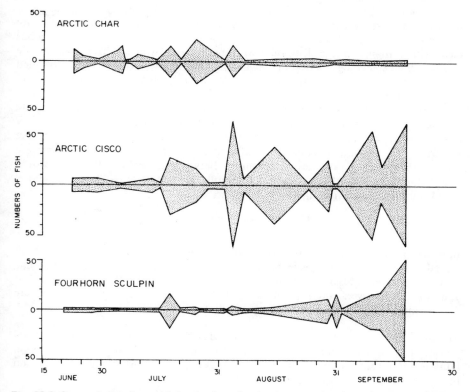

Fig. 23.3 Seasonal abundance of Arctic char, Arctic cisco and fourhorn sculpin in Nunaluk Lagoon off the Firth River, 1974. Samples were taken by gill-net at two-week intervals (standard 24-hr sets), starting at spring break-up and finishing just prior to the formation of surface ice (Griffiths et al. 1975).

documented. For example, several isolated populations of dwarf, non-anadromous char of the Western Arctic-Bering Sea form have been discovered inhabiting perennial springs, and another isolated population is known to inhabit a lake within the Canning River drainage (McCart and Craig 1971; McCart and Bain 1974; Bain 1974; Ward and Craig 1974).

The Arctic char, an abundant fish throughout the study area, spawns in most of the larger Beaufort Sea drainages having perennial springs (Fig. 23.4). From mid-summer to early fall, anadromous char move up the rivers onto the spawning grounds. Spawning takes place over an extended period in late summer and autumn (mid-August through November), with peak spawning occurring in September and October. Juvenile char remain in fresh water for a variable number of years before migrating seaward for the first time (as smolts), usually between the ages of 3 and 5 years. These immature migrants feed in nearshore areas of the Beaufort Sea during the short arctic summer and return to overwinter in the rivers. No char are known to overwinter in the ocean. Immature char may migrate to the ocean several times before reaching sexual maturity at 6 to 8 yrs.

Mature char, unlike salmon, do not die after spawning, and they may spawn more than once during their lifetime. The evidence, however, suggests that consecutive spawning is unusual and that most individuals spawn only every second year. Thus, at any given time, a population of anadromous char will contain mature fish of two types: those which will spawn in the approaching spawning period (spawners), and those which have spawned before but will not spawn in the approaching spawning period (mature non-spawners).

There is some question as to whether maturing Arctic char migrate seaward in the summer of the year in which they spawn or whether they remain in fresh water for this entire period. If the latter is the case, the maturing char would remain in fresh water continuously for approximately 20 months prior to and after spawning. Between spawnings, the char would spend approximately 1 to 3 summer months feeding in coastal waters and 9 to 11 months overwintering in fresh water.

Movements

Information about char movements in marine waters has come from fish tagged in several drainages: Sagavanirktok River (Furniss 1975), Canning River (Craig, in press), Firth River (Glova and McCart 1974), and the Rat and Big Fish rivers (Jessop et al. 1973; Jessop et al. 1974). There are several important findings from these studies:

Timing. The timing of movements is not clearly documented, but the general pattern is that most anadromous char enter the ocean during spring break-up (June) and return to the rivers by mid-August (Fig. 23.3). Some char remain in nearshore waters until freeze-up.

Coastal movements. Once char leave fresh water, the fish disperse along the coastline, in some cases covering large distances (Fig. 23.5). One fish traveled

367

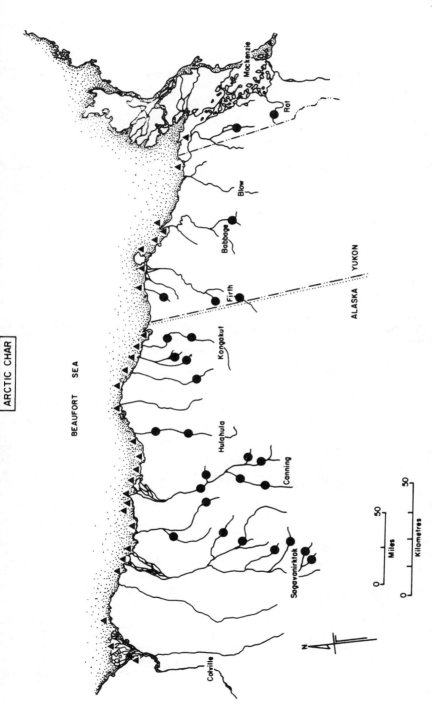

Fig. 23.4 Nearshore distribution (triangles) and known spawning grounds (circles) of anadromous Arctic char.

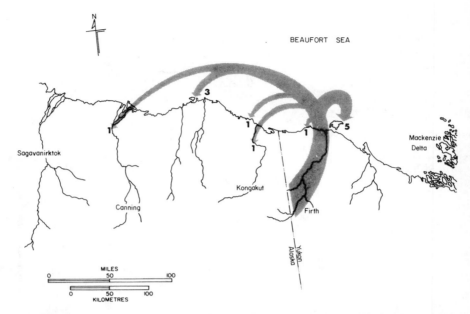

Fig. 23.5 Coastal movements of Arctic char from the Firth River (Glova and McCart 1974; Griffiths et al. 1975). These fish (N = 643) were tagged in the upper Firth River in 1972, and recaptures were made between 1972 and 1974 (13 recaptures in the Firth itself are not shown). The number of recaptures at each site is indicated.

from the Firth to the Canning River (a coastal distance of 250 km) and another from the Sagavanirktok River to Elson Lagoon near Barrow (300 km).

Interdrainage exchange. Although spawners appear to home to their natal stream to spawn, some non-spawning char may wander into non-natal drainages to overwinter. Six fish were tagged in one river and recaptured well upstream in another, as listed below:

Tagging location	Recapture location
1. Sagavanirktok River	Colville River
2. Sagavanirktok River	Canning River
3. Canning River	Sagavanirktok River
4. Firth River	Canning River
5. Firth River	Kongakut River
6. Rat River	Big Fish River

Interdrainage exchange is thought to be the exception rather than the rule, although this may reflect some sampling bias since the majority of recapture efforts were conducted in the same streams where the fish were originally tagged.

Composition of char in nearshore areas. From the foregoing, it can be seen that char from each of the major Beaufort Sea drainages utilize a large portion of the study area. Conversely, the char caught at any particular coastal location have probably originated from several different drainages. Tagging studies have shown, for example, that the char caught in domestic fisheries at Barter Island originated in the Sagavanirktok, Canning, and Firth rivers.

Size distribution

Most char caught in nearshore habitats were medium to large fish ranging in size from 340 to 520 mm (Fig. 23.6). Few fish under 300 mm were

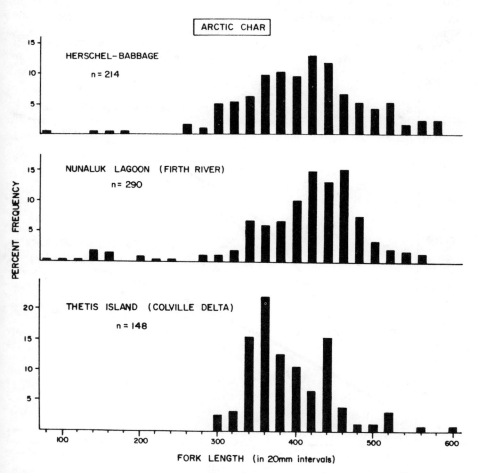

Fig. 23.6 Comparison of length frequencies of Arctic char from the Herschel Island-Babbage River region (Kendel et al. 1975), Nunaluk Lagoon (Griffiths et al. 1975), and Thetis Island (Winslow and Roguski 1970).

collected. The latter correspond in size to juveniles in fresh water, age 0 to 3 or 4, which have not yet smolted. Thus, while char less than 300 mm may be abundant in the deltas of the larger rivers, (Yoshihara 1973), they are not common in marine waters.

Age, growth, and maturity

Arctic char taken in nearshore areas generally range from 3 to 12 yrs in age, but fish aged 5 to 8 predominate. A length-age relationship for one char population (Firth River) is shown in Figure 23.7. The two lines on the figure represent two distinct growth patterns within a single population. It is clear that sea-run anadromous char are much larger at each age interval than either presmolts or resident fish which have matured without migrating to the ocean. This size difference, which is typical of char populations in the study area, demonstrates one apparent advantage of anadromy.

Sexual maturity is reached between the age of 6 to 8 yrs for most anadromous char. As previously mentioned, these fish may spawn more than

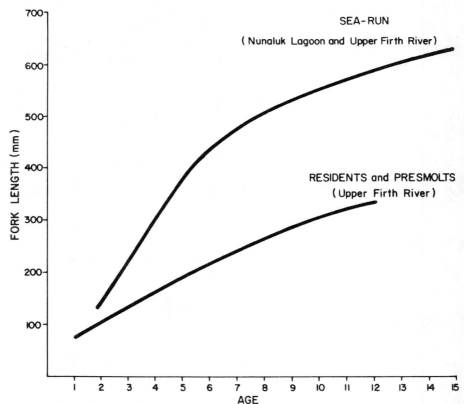

Fig. 23.7 Age-length relationships for sea-run (anadromous) Arctic char with stream residents (including presmolts) from the Firth River (redrawn from Glova and McCart 1974; Griffiths et al. 1975).

once during their life span but generally not in consecutive years. Those fish which are mature but will not spawn in a given year, migrate to the ocean to feed during the summer months. Many spawners, on the other hand, apparently do not leave the rivers during the summer of the year they will spawn. Thus, most (but not all) of the char caught in nearshore areas are either immature fish or mature non-spawners.

Sex ratio

Some members of anadromous char populations never migrate to the ocean (Fig. 23.7). This tendency to remain in fresh water is much more prevalent among males than females. In fact, virtually all of the mature, freshwater residents that have been examined in the Sagavanirktok, Canning, Firth, and Babbage rivers have been males (McCart et al. 1972; Glova and McCart 1974; Bain 1974; Craig, in press), although some mature, non-anadromous females have been found in Fish Creek, Y. T. (P. C. Craig and P. McCart, unpublished data).

A consequence of this pattern is that female char are significantly more abundant in nearshore areas than are males (Table 23.1). Samples of anadromous char taken in fresh water in the vicinity of spawning and overwintering areas indicate a comparable difference in the male/female ratio (McCart et al. 1972; Glova and McCart 1974; Bain 1974).

Foods

Arctic char in nearshore areas feed on a variety of benthic and pelagic organisms (Griffiths et al. 1975; Furniss 1975). At Nunaluk Lagoon the overall sample of ingested foods was 23.4 percent insects (mostly chironomid larvae and dipteran pupae), 29.8 percent crustaceans (amphipods, isopods, and copepods), 31.9 percent fish (five species, mostly fourhorn sculpin), and 14.9 percent unidentified materials. Craig (in press) found that the most

TABLE 23.1 Sex ratios of Arctic char from coastal locations. Asterisk indicates that for values of chi square, $p < 0.05$.

Coastal location	Arctic char		Reference
	N	% female	
Herschel Island to Babbage River	191	63.4*	Kendel et al. 1975
Nunaluk Lagoon	267	59.5*	Griffiths et al. 1975
Arctic Wildlife Range Coast	173	65.3*	Roguski and Komarek 1972
Shaviovik River to Canning River	93	75.0*	Craig, in press
Prudhoe Bay	127	63.0*	Furniss 1975
Thetis Island	148	56.1	Winslow and Roguski 1970
	X̄	62.6	

commonly eaten fish in the Canning River region were juvenile Arctic cod (*Boreogadus saida*).

Although the relatively rapid growth of char would suggest that they feed voraciously when they enter coastal waters, Griffiths et al. (1975) observed that the stomachs of char caught in nearshore areas were often empty (32.5 percent N=206) and that those stomachs containing food averaged only 24.8 percent in fullness.

ARCTIC CISCO

Life history

The Arctic cisco (*Coregonus autumnalis*) ranks as one of the most abundant and widespread fishes in nearshore areas between the Colville and Mackenzie rivers (Fig. 23.8). This species plays an important role in domestic fisheries located at Barter Island and the Colville and Mackenzie river deltas (Furniss 1974; Kogl and Schell 1974; Hatfield et al. 1972; Stein et al. 1973). Recent studies of the Arctic cisco include: Kogl 1972; Hatfield et al. 1972; Roguski and Komarek 1972; Alt and Kogl 1973; Stein et al. 1973; Craig and Mann 1974; Kendel et al. 1975; Furniss 1975; Griffiths et al. 1975.

One striking difference between this species and the Arctic char is that the Arctic cisco apparently uses only the two largest drainages in the region, the Colville and Mackenzie rivers, as spawning and, probably, overwintering areas (Fig. 23.8). Spawning migrations into these drainages differ. In the Mackenzie River, spawners begin migrating upstream in July, approximately two months earlier than in the Colville. They also migrate much further upstream in the Mackenzie and have been captured in large numbers as far upstream as the Great Bear River, 725 km from the river mouth. In the Colville, Arctic cisco spawn only in the lower reaches of the river. The species is a fall spawner, but the spawning periods and locations are not definitely known.

The overall pattern of fry and juvenile movements is also largely unknown. Fry were caught in the Colville delta in the vicinity of suspected spawning grounds. It is thought that in the Mackenzie drainage, fry move from the spawning areas in major tributaries (Peel, Arctic Red, Great Bear, and Mountain rivers) downstream to the middle and lower reaches of the Mackenzie. At least some juveniles, especially the older ones, leave these regions and move out into brackish or marine coastal waters during the summer months. Mature fish (principally non-spawners) also forage along the coastline. In the fall, the juvenile and mature non-spawners return to the Mackenzie or Colville drainages and overwinter in the delta regions. It appears that mature fish in the Mackenzie, which have migrated upstream to spawning areas on the major tributaries, return to the delta region to overwinter when spawning is complete.

It is likely that some of the observed differences between the Mackenzie and Colville populations of Arctic cisco reflect the great size difference between the drainages. The Mackenzie River (4320 km) is more than six times the length of the Colville River (680 km). Due to the greater dis-

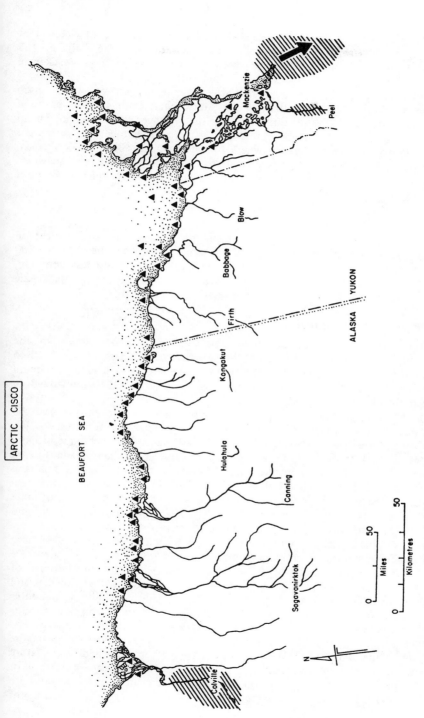

Fig. 23.8 Nearshore distribution (triangle) and approximate spawning grounds (hatched areas) of Arctic cisco. Spawning is thought to occur in the lower reaches of the Colville River and several tributaries of the Mackenzie River (Peel, Arctic Red, Great Bear, and Mountain rivers).

ARCTIC CISCO

BEAUFORT SEA

ALASKA YUKON

Peel

Mackenzie

Blow

Babbage

Firth

Kongakut

Hulahula

Canning

Sagavanirktok

Colville

Miles

Kilometres

50

0

50

N

tances that Arctic cisco travel in the Mackenzie, it might be expected that the spawning run in this river would be earlier than in the Colville. The degree of overlap and interchange between the Colville and Mackenzie "stocks" of Arctic cisco is not known at present.

At this point, it can be seen that the overall pattern of Arctic cisco movements is quite similar to that of the Arctic char. Both species spawn in fresh water in the fall and early winter; the younger juveniles remain in fresh water (or brackish delta waters) for a variable number of years; the older juveniles and mature non-spawners forage along the Beaufort Sea coastline during the summer and return to fresh water in the fall; and spawners do not venture far from their spawning streams, or they remain in fresh water altogether.

Size distribution

As is the case of the Arctic char, most of the Arctic cisco caught in nearshore areas are large fish (Fig. 23.9). Smaller fish in the 200 to 300 mm size category (corresponding to juveniles age 3 to 5) are less common, and fish less than 200 mm (age 0 to 3) are rare except in the Mackenzie delta (Percy et al. 1974; V. Poulin, personal communication).

Age, growth, and maturity

Until the fish reach a size of about 350 mm, which corresponds to the attainment of sexual maturity (age 7 to 9), the growth rate of Arctic cisco is relatively fast. Thereafter, the fish grow more slowly, reaching maximum sizes of 430 to 450 mm and maximum ages of 17 to 21 (summarized in Craig and Mann 1974).

Although the growth rates of Arctic cisco are similar throughout the study area, a wide range of longevities has been recorded for this species. Part of this discrepancy may be due to the two different techniques (scales or otoliths) that have been used to age the fish. Craig and Mann (1974) compared these techniques on the same sample of fish and found that the methods gave similar results through age 10 (Fig. 23.10), but, as in the case of many other northern fish populations, scales appeared to underestimate the ages of older fish (e.g., McCart et al. 1972; Mann 1974; Craig and Poulin 1975). The maximum ages determined were 14 for scales and 21 for otoliths. For this reason, otolith-based ages are recommended for Arctic cisco as well as for other northern populations of fishes.

There is a great variability (7 to 91 percent) in the proportions of immature Arctic cisco present in samples taken between the Colville and Mackenzie rivers (Craig and Mann 1974). Part of this variability is undoubtedly due to the difficulty that field crews have when attempting to distinguish between immature fish, potential spawners, and mature non-spawners. As indicated in Figure 23.11, it is likely that most Arctic cisco in nearshore areas are either immature fish which have never spawned or mature non-spawners in the interval between spawning years. The egg sizes of most females caught in Nunaluk Lagoon throughout the open-water season are substantially smaller than the estimated size of mature eggs at the time of

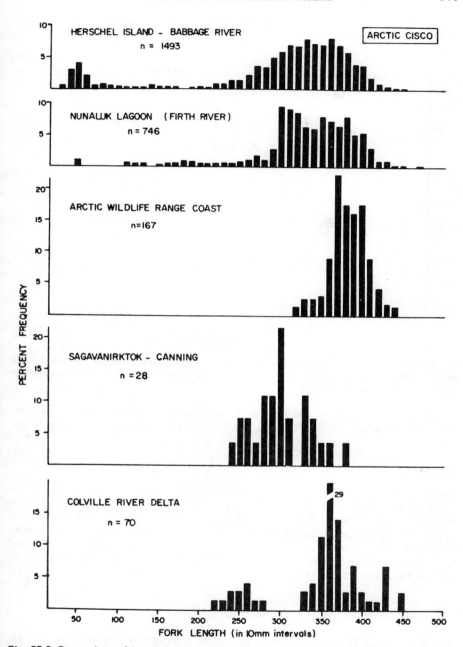

Fig. 23.9 Comparison of length frequencies of Arctic cisco from the Herschel Island-Babbage River region (Kendel et al. 1975), Nunaluk Lagoon (Griffiths et al. 1975), Arctic Wildlife Range coast (Roguski and Komarek 1972), Sagavanirktok River-Canning River region (Craig and Mann 1974), and the Colville River delta (Kogl and Schell 1974).

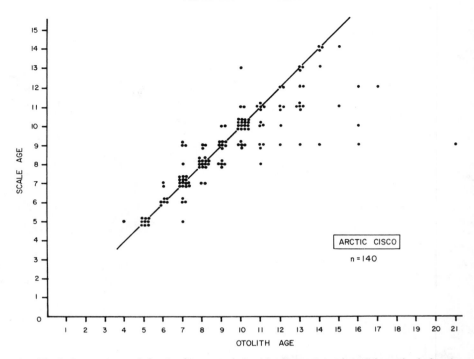

Fig. 23.10 Comparison of Arctic cisco ages derived by the scale and otolith methods for the same sample of 140 fish (Craig and Mann 1974). The straight line indicates a 1:1 relationship between the two aging techniques.

spawning; further, it is unlikely that the Nunaluk females were spawners, since the spawning run in the Mackenzie drainage was already underway by mid-July.

Foods

Arctic cisco in nearshore areas feed predominantly on crustaceans and insects (Berg 1949; Craig and Mann 1974; Griffiths et al. 1975; Furniss 1975). Major food items are copepods, amphipods, mysids, and dipteran adults and larvae (primarily chironomids). As in the case of Arctic char, the stomachs of Arctic cisco caught in nearshore waters were often empty (25 percent, N=344), and those stomachs containing food averaged only 39.9 percent in fullness (Griffiths et al. 1975).

OTHER SPECIES

The least cisco (*Coregonus sardinella*), also a common anadromous species, shares several general characteristics with Arctic cisco. Aspects of the life history of *C. sardinella* are described by: Cohen 1954; Wohlschlag 1954, 1956; Hatfield et al. 1972; Kogl 1972, Alt and Kogl 1973; Stein et al.

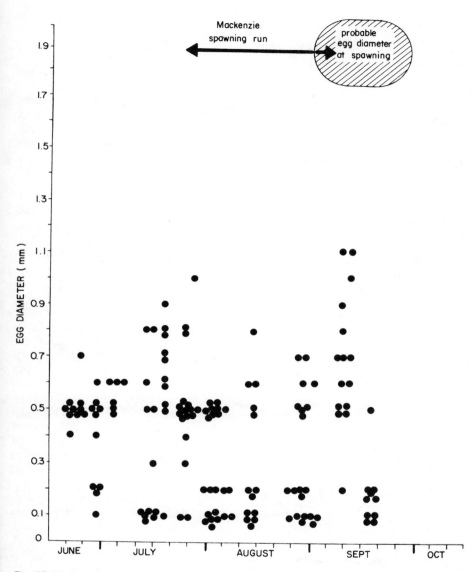

Fig. 23.11 Seasonal patterns of egg sizes in female Arctic cisco at Nunaluk Lagoon (Griffiths et al. 1975). Also indicated are the timing of the upstream spawning run in the Arctic Red River, Mackenzie drainage (Stein et al. 1973), and the estimated size of Arctic cisco eggs at the time of spawning (estimated from the size of least cisco eggs—Mann 1974).

1973; Mann 1974; Kendel et al. 1975; Percy et al. 1974; Furniss 1975; Griffiths et al. 1975. Of these reports, the study by Mann (1974) is one of the more extensive, providing a detailed comparison of anadromous and freshwater resident populations.

Within the study area, only the Colville and Mackenzie rivers serve as spawning areas for the anadromous populations, and from these centers the fish disperse along the coastline. Although least cisco are abundant in certain areas such as Prudhoe Bay (Furniss 1975) and the Mackenzie delta (Kendel et al. 1975; Percy et al. 1974; Galbraith and Fraser 1974), they apparently do not extend as far as the Arctic cisco into the central region between the Colville and Mackenzie rivers (Fig. 23.12). Few least cisco have been collected between the Canning and Malcolm rivers (Roguski and Komarek 1972; Craig and Mann 1974).

Other whitefishes have been recorded in nearshore waters (e.g., humpback and broad whitefish). In Alaska, these species venture little further into coastal waters than the deltas of the largest rivers (Winslow and Roguski 1970; Ward and Craig 1974; Furniss 1975). However, they are more frequently encountered in nearshore waters in the vicinity of the Mackenzie delta (Kendel et al. 1975). Even the freshwater grayling is occasionally caught in nearshore waters when salinities are low (Roguski and Komerek 1972; Griffiths et al. 1975).

Finally, a neglected marine fish, the fourhorn sculpin (*Myoxocephalus quadricornis*), is as widely distributed in the study area as either the Arctic char or Arctic cisco (Fig. 23.12). Like these latter species, the fourhorn sculpin utilizes nearshore habitats as rearing grounds. Fourhorn sculpin spawn in these areas as well, and their fry are often the most abundant and sometimes the only small fish to be found. The available life history data for this species are presented in Andriyashev (1954), McAllister (1961), Westin (1969), Percy et al. (1974) and Griffiths et al. (1975). This is a slow growing fish which matures at ages 5 to 8. Maximum sizes and ages of *M. quadricornis* taken off the Yukon coast were 340 mm total length and age 14, respectively. Small fourhorn sculpin feed on amphipods and copepods, but the larger fish prefer the giant isopod, *Saduria (Mesidotea) entomon*.

Other sources of information on arctic fishes occurring in the study area, not specifically mentioned herein, include studies of Arctic cod (Quast 1974), pond smelt (de Graaf 1974), Arctic flounder (Griffiths et al. 1975), fish in the Mackenzie delta (Mann 1975; de Graaf and Machniak, in press), taxonomic keys (Willimovsky 1958; Morrow 1974), and texts by McPhail and Lindsey (1970) and Scott and Crossman (1973).

DISCUSSION

It is apparent that most of the fishes inhabiting nearshore habitats in the study area are anadromous species. Anadromy, with its alternating freshwater and marine phases, is very common in northern regions. McCart (1970) has discussed the advantages and disadvantages of the anadromous pattern. The

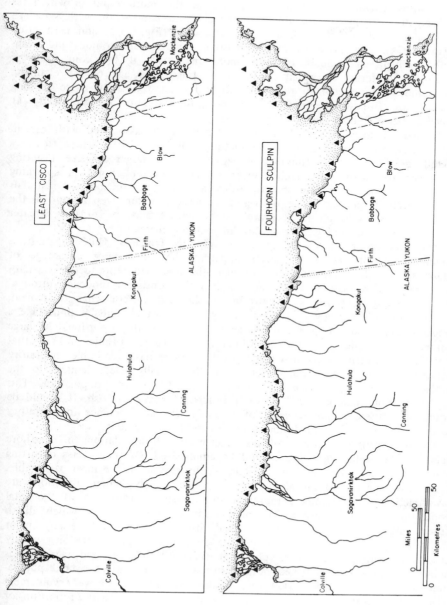

Fig. 23.12 Nearshore distributions of least cisco and fourhorn sculpin.

major advantage applicable to anadromous species within the study area is that which results from the opportunity to utilize the greater resources of the marine environment. This is reflected in the more rapid growth rates and greater lengths of anadromous individuals in comparison with same-age freshwater resident fish such as the Arctic char (Fig. 23.7) and least cisco (Mann 1974). The difference in size between same-age anadromous and freshwater resident individuals is even more impressive if the comparison is expressed in terms of weight rather than length. At age 6, for example, anadromous Arctic char in the Firth River are approximately twice as long as same-age fish which have not yet been to sea, but they are nearly 11 times as heavy (Glova and McCart 1974).

In species such as the Arctic char, larger size is associated with certain other advantages related to reproduction (McCart 1970). These features include greater fecundity (number of eggs produced); larger egg size, resulting in larger fry with presumably better survival; greater choice in spawning sites, because large fish are able to utilize a wider range of bottom materials and tolerate a greater range of water velocities; better preparation of the spawning nest with increased gravel permeabilities and, presumably, greater egg-to-fry survival; and greater success in securing mates.

The first four points apply exclusively to females, while the last is common to both sexes. It seems apparent that the major advantage of anadromy accrues to populations through the increased reproductive capacity of females. This may, in large part, explain the consistent tendency throughout the study area for a proportion of the male Arctic char to remain, mature, and die entirely in fresh water, while almost all females undertake a seaward migration. A practical consequence of the disproportionately large number of females among those which migrate to sea (Table 23.1) is that any fishery which is concentrated on the larger, more desirable, migratory segment of Arctic char populations will differentially select females to the probable detriment of the reproductive capacity of the population. This applies to both marine and freshwater fisheries for migrant fish. It should be noted that there is no evidence of a similar sexual difference in migratory tendency among other anadromous species in the study area.

In considering the effects of disturbance to fish populations in nearshore habitats, the following points should be emphasized. Of the species occurring along the coast, the anadromous one would appear to be most vulnerable. Freshwater species appear alongshore only occasionally, when conditions are favorable, and the numbers represent only a small proportion of the total populations of these species. The marine species that occur are widely distributed, and it is unlikely that localized disturbance would have any major effect on populations as a whole. The anadromous species are also widely distributed during the marine phase of their life history, and, again, it is unlikely that a localized nearshore disturbance would affect overall population size of any species. However, anadromous species are very vulnerable during the freshwater phase of their life cycle, when a large proportion of any population may be crowded into a very small area utilized for spawning and overwintering. In such localities, a catastrophic short-term disturbance, or a less severe one continuing over a longer period, could result in a marked reduction in population levels. Examples of such critical areas include the many spawning and overwintering areas of Arctic char (Fig. 23.4) and the Mackenzie delta, which is a major overwintering area for many anadromous species (Mann 1975). A reduction in populations utilizing the latter area could affect the abundance of ciscoes along hundreds of miles of coast, including both sides of the international border. One of the tasks of the fisheries biologist must be to identify such potentially critical areas and ensure the protection of the fish populations which utilize them.

Although anadromous species are most vulnerable to disturbance in critical areas in fresh water, their populations are to some extent adapted to withstand at least short-term fluctuations in the environment, natural or man-made (e.g., Johnson 1972). As a result of the great longevity, long period of maturity, and the habit of repeat spawning, there is great variation in the ages of fish spawning in any single year. This overlap means that populations are not dependent on the survival of any single year class. The failure of a year class (e.g., through sedimentation of spawning beds) will not necessarily result in any significant reduction in overall population size. An additional factor ensuring population survival is the fact that for many anadromous species, there are major differences in the migration patterns of various life history stages, so that an entire population may not be concentrated in a single locality during the course of the year. This reduces the possibility that an entire population can be destroyed by a single, localized event. Among Arctic char, for example, juveniles and non-spawning mature fish may overwinter in areas distinct from that of the spawning population, either elsewhere in the same drainage (McCart et al. 1972; Yoshihara 1973) or in some other drainage.

We can confidently expect that pressure on fish stocks in northern waters will increase as access is improved and as demand for fish products increases. If fish populations are to be managed on a rational basis, much more of their ecology must be known. With respect to anadromous fishes, detailed studies have only begun. Much more remains to be discovered, including detailed data on the numbers, movements and growth of various species as well as systematic information defining the characteristics of various stocks and the ecological distinctions between them.

Acknowledgments

We thank the many field biologists who have assisted us in our studies. Funding was provided by Alaskan Arctic Gas Study Company, Anchorage, Alaska, and Canadian Arctic Gas Study Limited, Calgary, Alberta, through Northern Engineering Services Company Limited, Calgary, Alberta.

Conference Discussion

Barber — A layer or lens of freshwater has been observed seaward of the Colville River during spring. Can you tell us about the distribution of freshwater and anadromous species in this layer?

Craig — In general, the freshwater and anadromous fishes in inshore areas inhabit waters which fluctuate widely in salinity concentrations, but it is not specifically known how the freshwater plumes of the larger rivers affect the distributions of fishes in arctic coastal waters.

Norton — On the Sagavanirktok River in Alaska, we have watched with alarm the extensive gravel removal and winter water removal on the lower river that have proceeded since the late 1960s. Our concern stems from the fact that we are not finding recruitment of younger fish to the arctic char population of breeding age. It is precisely these sub-adult fish whose overwintering areas remain a mystery, and the Alaska Department of Fish and Game suspects that water and gravel removal have already destroyed overwintering habitat and populations on the Sagavanirktok delta. Do you find this suspicion credible?

Craig — We now know that water withdrawal operations in the Sagavanirktok delta have indeed killed overwintering fish. Attention should now be directed at identifying what proportion of the population has, in fact, been affected by these operations, and, further, how a change in age-structure will affect the Sagavanirktok char population.

Unidentified — I fully agree with your point that reading of scales is less reliable for age determination than otolith readings. The disadvantage of the latter method is that one has to sacrifice the fish. For tagging schemes, we therefore prefer to read the rings in the fin rays, especially in freshwater fishes.

REFERENCES

ALT, K. T., and D. R. KOGL

    1973   Notes on the whitefish of the Colville River, Alaska. *J. Fish. Res. Board Can.* 30(4): 554-556.

ANDRIYASHEV, A. P.

    1954   Fishes of the northern seas of the USSR Izdatel'stvo Akad. Nauk. SSR, Moskva-Leningrad. Translation from Russian, Israel Program for Scientific Translations. Jerusalem, 1964. 617 pp.

BAIN, L. H.

    1974   Life histories and systematics of Arctic char (*Salvelinus alpinus*, Linn.) in the Babbage River system, Yukon Territory. Canadian Arctic Gas Study Limited, Calgary, Alberta, Biological Report Series 18(1): 156 pp. Also: M.Sc. Thesis, Dept. Biol., Univ. Calgary, Calgary, Alberta.

BERG, L. S.

1949 Freshwater fishes of the USSR and adjacent countries (translation). *Zool. Inst. Akad. Nauk.* 27, 39, 30.

COHEN, D. M.

1954 Age and growth studies on two species of whitefishes from Point Barrow, Alaska. *Stanford Ichthyol. Bull.* 4(3): 168-187.

CRAIG, P. C.

(In press) Ecological studies of anadromous and resident populations of Arctic char in the Canning River and adjacent coastal waters of the Beaufort Sea, Alaska. Canadian Arctic Gas Study Limited, Calgary, Alberta, Biological Report Species.

CRAIG, P. C., and G. J. MANN

1974 Life history and distribution of the Arctic cisco (*Coregonus autumnalis*) along the Beaufort Sea coastline in Alaska and the Yukon Territory. Canadian Arctic Gas Study Limited, Calgary, Alberta, Biological Report Series 17(1): 47 pp.

CRAIG, P. C., and P. McCART

1975 Classification of stream types in Beaufort Sea drainages between Prudhoe Bay, Alaska and the Mackenzie Delta, N.W.T. *Arctic and Alpine Research* 7(2): 183-198.

CRAIG, P. C., and V. A. POULIN

1975 Movements and growth of grayling (*Thymallus arcticus*) and juvenile Arctic char (*Salvelinus alpinus*) in a small Arctic stream, Alaska. *J. Fish. Res. Board Can.* 32: 689-697.

CRAIG, P. C., and J. WELLS

1975 Fisheries investigations in the Chandalar River region, northeast Alaska. Canadian Arctic Gas Study Limited, Calgary, Alberta, Biological Report Series 34(1): 114 pp.

DE GRAAF, D.

1974 The life history of the pond smelt *Hypomesus olidus*, Pallas (Osmeridae) in a small unnamed lake in the northern Yukon Territory. Canadian Arctic Gas Study Limited, Calgary, Alberta, Biological Report Series 18(2): 89 pp. Also: M.Sc. Thesis, Dept. Biol., Univ. Calgary, Calgary, Alberta.

DE GRAAF, D., and K. MACHNIAK

(In press) Fisheries investigations in the MacKenzie Delta in the vicinity of the Cross Delta Alternative Pipeline Route. Canadian Arctic Gas Study Limited, Calgary, Alberta, Biological Report Series.

FURNISS, R. A.

1975 Inventory and cataloging of Arctic area waters. Div. of Sports Fish, Alaska Dept. Fish and Game. Annual Report 16: 1-47.

GALBRAITH, D. F., and D. C. FRASER

1974 Distribution and food habits of fish in the Eastern Coastal Beaufort Sea. Fisheries and Marine Service, Environment Canada. Interim Report of Beaufort Sea Project Study B1 (Eastern). 48 pp.

GLOVA, G., and P. McCART

1974 Life history of Arctic char (*Salvelinus alpinus*) in the Firth River, Yukon Territory. Canadian Arctic Gas Study Limited, Calgary, Alberta, Biological Report Series 20(3): 51 pp.

GRIFFITHS, W., P. C. CRAIG, G. WALDER, and G. MANN

1975 Fisheries investigations in a coastal region of the Beaufort Sea (Nunaluk Lagoon, Y. T.). Canadian Arctic Gas Study Limited, Calgary, Alberta, Biological Report Series 34(2): 171 pp.

GRIFFITHS, W., P. C. CRAIG, and J. den BESTE

(In press) Fisheries investigations in a coastal region of the Beaufort Sea. Part 2 (Kaktovik Lagoon, Alaska). Canadian Arctic Gas Study Limited, Calgary, Alberta, Biological Report Series.

HATFIELD, C. T., J. N. STEIN, M. R. FALK, and C. S. JESSOP

1972 Fish resources of the Mackenzie River valley. Fisheries Service, Environment Canada, Interim Report I, Vols. 1 and 2, 247 pp.

JESSOP, C. S., K. T. J. CHANG-KUE, J. W. LILLEY, and R. J. PERCY

1974 A further evaluation of the fish resources of the Mackenzie River valley as related to pipeline development. Fisheries and Marine Service, Dept. of the Environment. Environmental-Social Program, Northern Pipelines Report No. 74-7. Information Canada Cat. No. R-72-13674. 95 pp.

JESSOP, C. S., T. R. PORTER, M. BLOUW, and R. SOPUCK

1973 Fish resources of the Mackenzie River valley: an intensive study of the fish resources of two main stem tributaries. Fisheries Service, Dept. of the Environment. Environmental-Social Program, Northern Pipelines. 148 pp.

JOHNSON, L.

1972   Keller Lake: characteristics of a culturally unstressed salmonid community. *J. Fish. Res. Board Can.* 29: 731-740.

KENDEL, R. E., R. A. C. JOHNSTON, V. LOBSIGER, and M. D. KOZAK

1975   Fishes of the Yukon coast. Environment Canada. Beaufort Sea Project, Tech. Report No. 6, 114 pp.

KOGL, D. R.

1972   Monitoring and evaluation of Arctic waters with emphasis on the North Slope drainages. Div. of Sport Fish, Alaska Dept. of Fish and Game. Job No. G-III-A, Project F-9-3. Annual Report 12: 23-61.

KOGL, D. R., and D. SCHELL

1974   Colville River delta fisheries research. *In* Environmental Studies of an Arctic Estuarine System. Final Report R74-1, Inst. Mar. Sci., Univ. Alaska, Fairbanks, pp. 467-489.

MANN, G. J.

1974   Life history types of the least cisco (*Coregonus sardinella*, Vallenciennes) in the Yukon Territory, North Slope, and eastern Mackenzie River delta drainages. Canadian Arctic Gas Study Limited, Calgary, Alberta, Biological Report Series 18(3): 132 pp. Also M. Sc. Thesis, Dept. Zoology, Univ. Alberta, Edmonton, Alberta.

1975   Winter fisheries surveys in the Mackenzie Delta. Canadian Arctic Gas Study Limited, Calgary, Alberta, Biological Report Series 34(3): 54 pp.

McALLISTER, D. E.

1961   The origin and status of the deepwater sculpin, *Myoxocephalus thompsonii*, a nearctic glacial relict. *Nat. Museum Canada Bull.* 172: 44-65.

1962   Fishes of the 1960 "*Salvelinus*" Program from western Arctic Canada. *Bull. Nat. Museum Canada* 185: 17-39.

McCART, P.

1970   A polymorphic population of *Oncorhynchus nerka* at Babine Lake, B. C. involving anadromous (sockeye) and non-anadromous (kokanee) forms. Ph.D. Thesis, Univ. British Columbia, 135 pp.

McCART, P., and H. BAIN

1974 An isolated population of Arctic char (*Salvelinus alpinus*) inhabiting a warm mineral spring above a waterfall at Cache Creek, N. W. T. *J. Fish. Res. Board Can.* 31(8): 1408-1414.

McCART, P., and P. CRAIG

1971 Meristic differences between anadromous and freshwater-resident Arctic char (*Salvelinus alpinus*) in the Sagavanirktok River drainage, Alaska. *J. Fish. Res. Board Can.* 28: 115-118.

McCART, P., P. CRAIG, and H. BAIN

1972 Report on fisheries investigations in the Sagavanirktok River and neighbouring drainages. Report to Alyeska Service Company, Bellevue, Washington. 170 pp.

McCART, P., W. GRIFFITHS, C. GOSSEN, H. BAIN, and D. TRIPP

1974 Catalogue of lakes and streams in Canada along routes of the proposed Arctic Gas pipeline from the Alaskan/Canadian border to the 60th parallel. Canadian Arctic Gas Study Limited, Calgary, Alberta, Biological Report Series 16: 251 pp.

McPHAIL, J. D.

1961 A systematic study of the *Salvelinus alpinus* complex in North America. *J. Fish. Res. Board Can.* 18: 793-816.

McPHAIL, J. D., and C. C. LINDSEY

1970 The freshwater fishes of north-western Canada and Alaska. *Fish. Res. Board Can. Bull.* 173: 381 pp.

McROY, C. P., G. J. MUELLER, J. CRANE, and S. W. STOKER

1971 Nearshore marine biological results — Colville area. *In* Baseline Data Study of the Alaska Arctic Aquatic Environments. Eighth Month Progress, 1970. Tech. Report R71-4. Inst. Mar. Sci., Univ. Alaska, Fairbanks, 176 pp.

MORROW, J. E.

1974 *Illustrated keys to the freshwater fishes of Alaska.* Alaska Northwest Publishing Co., Anchorage, Alaska, 78 pp.

PERCY, R., W. EDDY, and D. MUNRO

1974 Anadromous and freshwater fish of the outer Mackenzie Delta. Fisheries and Marine Service, Environment Canada. Interim Report of Beaufort Sea Project Study B2, 51 pp.

QUAST, J. C.

1974 Density distribution of juvenile Arctic cod, *Boreogadus saida*, in the eastern Chukchi Sea in the fall of 1970. *Fish. Bull. U. S.* 72(4): 1094-1105.

ROGUSKI, E. A., and E. KOMAREK

1972 Monitoring and evaluation of arctic waters with emphasis on the North Slope drainages. Div. of Sport Fish, Alaska Dept. Fish and Game. Job No. G-III-A, Project F-9-3. Annual Report 12: 1-22.

SCOTT, W. B., and E. J. CROSSMAN

1973 Freshwater fishes of Canada. *Fish. Res. Board Can. Bull.* 184: 966 p.

SLANEY, F. F., and COMPANY

1974 Environmental Program, Mackenzie Delta, N. W. T., Canada. Aquatic Resources. Vol. 6. F. F. Slaney and Co. Ltd., Vancouver, Canada.

STEIN, J. N., C. S. JESSOP, T. R. PORTER, and K. T. J. CHANG-KUE

1973 Fish resources of the Mackenzie River valley. Fisheries Service, Dept. of the Environment. Environmental-Social Program, Northern Pipeline. Interim Report 2: 260 pp.

WARD, D., and P. CRAIG

1974 Catalogue of streams, lakes and coastal areas in Alaska along routes of the proposed gas pipeline from Prudhoe Bay to the Alaskan/Canadian border. Canadian Arctic Gas Study Limited, Calgary, Alberta, Biological Report Series 19: 381 pp.

WESTIN, L.

1969 The mode of fertilization, parental behaviour and time of egg development in fourhorn sculpin, *Myoxocephalus quadricornis* (L.). Fish. Board of Sweden, Inst. Freshwater Res., Drottningholm Report 49: 175-182.

WILIMOVSKY, N. J.

1958 *Provisional key to the fishes of Alaska.* Fisheries Research Laboratory, U. S. Fish and Wildlife Service, Juneau, Alaska. 113 pp.

WINSLOW, P. C., and E. A. ROGUSKI

1970   Monitoring and evaluation of arctic waters with emphasis on the North Slope drainages. Div. of Sport Fish, Alaska Dept. Fish and Game. Proj. F-9-2. Progress Report 11: 279-301.

WOHLSCHLAG, D. E.

1954   Growth peculiarities of the cisco, *Coregonus sardinella* (Val.) in the vicinity of Point Barrow, Alaska. *Stanford Icthyol. Bull.* 4: 189-209.

1956   Information from studies of marked fishes in the Alaskan arctic. *Copeia* 4: 237-242.

YOSHIHARA, H. T.

1973   Monitoring and evaluation of arctic waters with emphasis on the North Slope drainages. Div. of Sports Fish, Alaska Dept. Fish and Game. Job No. G-III-A. Project F-9-5. Annual Report 14: 1-83.

Assessment of the Arctic Marine Environment: Selected Topics
Copyright © 1976 by Institute of Marine Science, University of Alaska, Fairbanks

CHAPTER 24

Development and potential yield of arctic fisheries

C. ATKINSON[1]

Abstract

There has been little advance in the technology and management of fisheries in the arctic seas. Development of the oil fields along the Beaufort Sea has added new dimension to the Alaskan Arctic, however. With improved transportation and communication, the once-isolated villages are rapidly disappearing. High rates of employment and substantial funds made available under the Native Claims Act have given a new incentive to leaders of the native groups to manage their resources in order to assure a continuing income in the future. Arctic fisheries are one of the resources that could be developed in this time of rapidly changing conditions. There is a growing need for assessment of the arctic fisheries, their potential yield, and the problems associated with their development.

INTRODUCTION

The Arctic Ocean and adjacent seas are one of the few regions in the world whose fisheries have not yet been explored and developed. At first glance, the area seems to hold little promise of supporting an active fishery. The environment is severe, the fishing season short, the villages isolated, and the costs of transportation and production extremely high. Further, because of the low incidence of solar light and the extensive ice cover during most of the year, the biological productivity of arctic waters has been considered by many to be relatively low.

Quite contrary to these impressions, however, are the records of the very active whaling industry that developed in the Alaskan Arctic during the

---

[1] *8000 Crest Drive, N.E., Seattle, Washington 98155.*

latter part of the 19th century. Beginning in 1848, when the whaler *Superior* sailed through the Bering Strait into the Arctic Ocean, commercial whaling proved to be a profitable venture. Between 1848 and 1885, over 3000 American ships and 90,000 men sailed into the Arctic in search of the Pacific right and bowhead whales. Then, slowly, as the whale populations and the demand for whale products declined, commercial whaling in the Arctic became less and less profitable, closing out by 1915 (Arctic Environmental Information and Data Center 1975). The number of actively-feeding whales that must have been present to support this size of an operation and the tons of planktonic food that these whales must have consumed during their residence in the Arctic Ocean is a good indication of the biological productivity of these waters.

Due to recent developments, economic conditions in the Arctic are also changing very rapidly. First, discovery and development of the oil fields at Prudhoe Bay have opened up the North Slope of Alaska with new means of transportation and communication, and a new source of cheaper fuel. Second, the current fuel crises have prompted state and federal agencies to develop plans for new offshore explorations for oil. To assess the possible effects of these developments on the other resources and the various coastal communities, the Bureau of Land Management of the U. S. Department of Interior, through the National Marine Fisheries Service, has launched a multi-million dollar "Offshore Continental Shelf" program; much of the information and data collected in these studies will contribute directly to the development of arctic fisheries. Finally, the Alaska Native Claims Settlement Act of 1971 has given new responsibility to the native populations of the Arctic and provided a monetary base for the native associations to develop and manage their resources. Success, such as with the development of very productive salmon fisheries in the Kotzebue area, has aroused new interest in the potential of arctic fisheries.

The area

The area considered as arctic in this report is limited to the Arctic Ocean, north of the Greenland Channel and Bering Strait, including the Kara, Laptev, East Siberian, Chukchi, and Beaufort seas. Although lying within the Arctic Circle, the Greenland, Barents, Pechora, and White seas have been excluded. The species of fish found in the latter area, the vessels fishing it, and the countries involved are the same as those in the northeastern Atlantic, and it is impossible to separate much of the catch and other data.

The ocean

There are three recognizable water masses in the Arctic Ocean: a surface layer of dilute, well-mixed water of arctic origin at the top (0-50 m), overlaying a more saline, warmer layer of water from the Bering Sea (50 to 150 m) and generally moving in a clockwise direction; a mid-layer of high salinity water with temperatures above $0°C$, of Atlantic origin and moving in a counterclockwise direction; and a deep layer of high salinity water with temperatures below $0°C$ (Kinney et al. 1970). Although fishing is restricted to the surface layer, the mid-layer is important as a source of nutrients supplied by up-welling and mixing.

About 20 to 25 percent of the water enters the Arctic Ocean through the shallow (about 60 m) Bering Strait. This water mass is relatively warm (10 to 15°C in August), originating in the westward flowing Alaska Stream south of the Aleutian Islands. A part of this stream passes through the Aleutians, flows northward along the western coast of Alaska where it is diluted by fresh water from the Yukon, Kuskokwim and other rivers, and passes through the Bering Strait into the Arctic Basin (Coachman and Aagaard 1974). This northward flow of water carries with it the eggs and young of many fish and invertebrates normally found in the Bering Sea (Pruter and Alverson 1962; Alverson and Wilimovsky 1966).

The remaining 75 to 80 percent of the water entering the Arctic Basin comes from the Atlantic Ocean, passing along the eastern side of the Greenland Channel between Greenland and the Svalbard Islands.

Almost all of the cold water flowing out of the Arctic Basin passes along the eastern shore of Greenland through the deep Greenland Channel; the amount of water leaving the Arctic through the Bering Strait is thought to be negligible.

The arctic waters are known to be very rich in both phytoplankton and zooplankton for short periods during the summer. For example, English (1966) obtained values for plankton volume from the Chukchi Sea of 10 to 70 times greater than those found off San Diego by Bieri (1962). The intense blooms, however, last for only two or three weeks at the most.

In winter, the standing crop of phytoplankton in water is generally very low due to the low light intensity. However, recent studies by McRoy and Goering (1974) show that communities of algae and other microorganisms grow in the soft, pulpy undersurface of the ice and become a major contributor to the productivity of arctic waters during the winter months.

Further, specimens of Polar cod (*Arctogadus glacialis*), collected from under the ice at Station Charlie in December and January of 1959 and 1960, had large deposits of fat in their bodies and were actively feeding upon small crustaceans at the time, indicating some abundance of zooplankton under the polar ice cap, even in mid-winter (Walters 1961).

There is other evidence of high biological productivity in arctic waters. For example, Swartz (1966) estimated that seabirds such as murres, kittiwakes, puffins, and gulls near Cape Thompson on the Chukchi Sea consumed 13,100 MT of food (mainly Arctic cod and sand lance) over the 3-month ice-free, nesting period from May through September; this is for only one limited area along the arctic coast. We also know that the Arctic cod form a major food item in the diet of the ringed seal during two or three winter months. Although the evidence is fragmentary, there must be large concentrations of fish to support such populations of seabirds and mammals in the Arctic.

THE FISHERIES

At the present time, there are about 16 families and 84 species of marine, eurohyaline, and anadromous fish known from the Arctic Basin — far less than the 300 species reported by Wilimovsky (1974) for the Bering Sea. Of these, however, only the following 7 families and about 20 species are considered to be of commercial importance.

Sturgeon (Acipenseridae)

Sturgeon are found in commercial numbers only in the rivers of European Russia and northwestern Siberia; none have been reported from the Alaskan Arctic. The total catch is reported to be between 1,000–2,000 MT/yr with about 70 percent of the catch coming from the Ob River (Andriyashev 1954).

Herring (Clupeidae)

Sizeable schools of large herring (8-9 inches in length) have been reported from the Chukchi Sea (Alverson and Wilimovsky 1966); Walters (1955) reports schools of herring from Point Barrow and the Mackenzie River delta, and from the Laptev Sea (central Siberian Arctic). The Atlantic form has been reported from the Kara Sea. Although not numerous, these schools could contribute to small, local fisheries.

There is no information available on whether or not these schools of herring are from local populations or are transients from the Bering or the Barents and White seas. If local spawning occurs, then the value of the resource would be enhanced considerably through the export of herring roe (*kazenoko*) to Japan.

Salmon, char, and whitefish (Salmonidae)

The present commercial fisheries along the northern coast of Canada, Alaska, and the USSR centers around this group of fish. Table 24.1, based on information from various sources and for various years, provides only a minimum estimate of the catches.

TABLE 24.1 Estimated annual minimum catches of Salmonidae group fish by USSR and Alaskan fisheries

Group	Estimated minimum catch (MT)	
	Siberia	Alaska
Salmon	Present	1,800
Arctic char	100	Present
Whitefish and cisco	10,500	40
Sheefish	800	100

In the last five years, a very profitable chum salmon fishery has developed in the Kotzebue area (just north of the Bering Strait). The salmon are taken by gillnet and were formerly delivered directly to a freezer ship anchored offshore for export to Japan. At the present time, however, the fish are iced and shipped by air for processing in Anchorage, Seward, or even Seattle. These salmon are of excellent quality and have a high market demand.

Chum salmon are also found in small numbers along the arctic coast from the Lena to the Mackenzie rivers, pink salmon are seen as far north as the Colville River, and king salmon were recently reported from the Mackenzie River.

Depending upon the availability of gas or oil from the Arctic as an inexpensive source of heat, it might prove feasible to increase these northern runs of salmon by artificial propagation (i.e., collecting and incubating the eggs using recirculated heated water, and releasing the young at the time of maximum plankton production in the arctic seas).

The Arctic char (*Salvelinus alpinus*) has been reported as the second most abundant species at Prudhoe Bay (Arctic Environmental and Information Data Center 1974) and as the most common species taken by seine along the beaches in the Point Hope area (Alverson and Wilimovsky 1966). Both the Alpine char and the Dolly Varden (a closely related, more southern form) are serious predators of salmon and for many years were the object of various federal, territorial, state and industrial predator control programs (Atkinson 1954).

The Arctic char is now being fished commercially in the Canadian Arctic, canned as "Arctic Char — Steak Style" in a standard No. 1/2 flat tin (220 g), and marketed both domestically and for export. Efforts by an Alaskan company in the early 1970s to develop an export market for smoked or frozen char in Japan did not materialize. The Alpine char are widely distributed throughout the arctic USSR but are only taken in significant numbers from the Kara and the neighboring Barents and Pechora seas. The Alpine char is an excellent food fish, and as shown by the experience of the Canadians, can be processed and readily sold in the domestic and foreign markets; it could form a valuable part of the production from small, multi-species, local fisheries.

Also, in the last four or five years, a small whitefish fishery has developed in the Colville River, taking three species of whitefish and cisco by gillnet in the summer and selling them in the local markets. Catch records from the Alaska Department of Fish and Game are given in Table 24.2.

Using average weights of 5.1 lbs. for the broad whitefish, 1.0 lbs. for the Arctic cisco and 0.9 lbs. for the least cisco, this small, isolated fishery is

TABLE 24.2   Annual summer catches of whitefish and cisco taken by gill-net from the Colville River, northwest Alaska.

Year	Numbers of fish taken		
	Broad whitefish	Arctic cisco	Least cisco
1971	3,815	38,016	22,713
1972	1,850	37,333	13,283
1973	2,161	71,960	25,408
Average	2,942	49,103	20,468

producing more than 80,000 lbs. (36.4 MT) of fish per year and provides an excellent example of the potential of developing other whitefish and cisco fisheries in the Arctic.

The USSR has exploited the whitefish and cisco stocks of fish along the northern coast of Siberia for at least 40 to 50 yrs. In addition to the eurohyaline species mentioned here, extensive lake fisheries have been developed in Siberia, especially in Lake Baikal.

Perhaps the sheefish is the most spectacular of the arctic fishes, attaining a maximum size of 40 kg (88 lbs.) (Nikol'sky 1971). These very large whitefish occur all along the northern coast of Siberia, but in Alaska they are confined to a limited area of the northern Bering Sea (Kuskokwim and Yukon rivers) and Kotzebue Sound (Kobuk and Noatak rivers). Sheefish are taken mostly by gillnet through the ice in winter, naturally frozen without cleaning, and shipped to local Alaskan markets. A lesser amount is taken during the summer months by gillnet or hook-and-line. Alt (1967) has made a study of the taxonomy and ecology of the sheefish in Alaska, but there are still gaps in our knowledge of the life history of these fish, and little is known of the amount of fishing pressure that these stocks will support.

The sheefish is an excellent food fish, but there is a need to improve the processing of the fish for market. For example, sheefish at the present time are usually sold frozen in-the-round, too large for the normal family and requiring cleaning at home after thawing. It would be a more attractive product if the whole frozen fish were sawn into steaks, the viscera punched out, and the slices film-wrapped for the consumer market. Sheefish can be prepared to resemble swordfish, now generally banned from the U. S. market by high mercury levels; thus, we would already have a built-in domestic demand for sheefish.

Capelin or candlefish (Osmeridae)

Concentrations of capelin, a small northern smelt, are common along the arctic coast of Canada, Alaska, and the USSR, in waters of -1.7 to 1.3°C. (Alverson and Wilimovsky 1966; Walters 1955; J. P. Doyle, personal communication 1975). They spawn on gravelly beaches during the summer months (the end of July at Point Barrow), mostly as 1-yr-old fish, but a few live to spawn in the second and rarely the third year (Andriyashev 1954). Although capelin may reach a size of 9 inches (22 cm), they usually are found at a size of 5.5 to 7 inches in length.

Capelin are a popular food fish in Japan (known as *shishyamo*) and to satisfy the demand, Japanese companies import substantial quantities of frozen capelin from Norway, Iceland, and Canada. Trial shipments from Alaska (Bristol Bay) have proven acceptable in the Japanese markets, and if the economics can be satisfied, a viable export market can be developed.

Capelin are a good quality fish, and the world catch of these fish is increasing rapidly (from 275 MT in 1965 to 1775 MT in 1973). If this trend continues, it is only a matter of time before a demand for these fish will develop in the United States.

Sand lance (Ammodytidae)

The sand lance is a small fish commonly found in arctic waters and ranking second in abundance to the Arctic cod in the Chukchi Sea (Quast 1974). Although little is known of the life history of this species, closely related forms attain a maximum length of 4 inches at 4 yrs of age. The sand lance matures at 3 yrs of age, spawning in the winter (under the ice) on sandy bottom in relatively shallow water (Andriyashev 1954). The sand lance is not usually used for human consumption, but in recent years this fish has been used for fish meal in Europe. With the development of the oil fields along the northern coast of Alaska and the possibility of cheaper fuel, the development of a sand-lance fishery for meal might prove to be economically feasible.

The sand lance forms an important food item for birds and other fish in the Arctic, and the possible effect of mass removal of these fish upon the ecosystem should be considered before developing an intensive fishery for this species.

Cod (Gadidae)

The Arctic cod (*Boreogadus saida*) is by far the most abundant fish in the arctic waters. The fish are circumpolar in distribution and thrive in waters with temperatures as low as -1.5°C. The Arctic cod matures in 4 yrs at a size of about 7.5 inches, spawning under the ice in January and February. They attain a maximum size of about 12.5 inches (Andriyashev 1954). Quast (1974), in his study of the Arctic cod, points out that, "Juvenile Arctic cod and the Pacific sand lance....were virtually the only fish trawled in the surface and mid-depths at night". Cod were found at every station sampled and were 10 times more abundant than the sand lance.

The Polar cod (*Arctogadus glacialis*) has been reported by Walters (1961) to be in some abundance under the ice in the Arctic Ocean from November to January. The size of the fish taken at Ice Station Charlie ranged from 4.3 to 8.3 inches in length and, when taken, were actively feeding upon small crustaceans.

The Saffron cod (*Eleginus gracilis*) has been reported from the Chukchi, Beaufort and Kara seas (Walters 1955; Andriyashev 1954), but so far as is known, this species is not present in commercial abundance. Alverson and Wilimovsky (1966) note that the Saffron cod is found only in the warmer waters of the Arctic, possibly indicating that this species is at the extremity of its range, or that it is a transient from the warmer waters of the Bering or the Norwegian-Barents seas.

Although the Arctic cod is abundant in northern waters and commonly taken by trawl fisheries in the adjacent Greenland, Barents and White seas (348,000 MT in 1972; 82,000 MT in 1973), there is no commercial fisheries for this species in the Arctic Ocean and seas. The cod are taken to some extent by the Eskimos, but according to Andriyashev (1954), the meat is of low palatability and the fish are usually fed to the dogs. If this is true, then the commercial use of these fish would probably be limited to fish paste products such as luncheon meats and a fish base for fish cheese (*kamoboko*), Fish Protein Concentrate (FPC), and fish meal. Again, to be economically feasible, a commercial fishery would be dependent upon the development of a new, year-round method of fishing under the ice, a source of cheap fuel, and low shipping costs.

Flatfish (Pleuronectidae)

Considerable study has been made of the flatfishes in the arctic region, especially in the Chukchi Sea. Of the six species studied by Alverson and Wilimovsky (1966), three are extensively fished by the Japanese and the Russians in the Bering Sea. Two of these three species are common, but by no means abundant, in the Chukchi Sea: the yellowfin sole (*Limanda aspera*) in the shallower, warmer waters, and the Bering flounder (*Hippoglossoides robustus*) in the deeper, colder water.

These two species offer little promise for commercial development. First, the catch from the Chukchi Sea (i.e., 59 half-hour hauls at 48 stations) totaled only 252 flounders and 31 sole, or an average of 8.5 and 1 fish, respectively, for a 1-hr trawl. Converted to weight, the entire catch totaled only about 400 lbs., compared with about 1000 lbs. per hour of fishing for trawlers operating off the coasts of Oregon and Washington.

Second, the flatfish in the Chukchi Sea grow very slowly — the maximum age and size is 11 yrs and 9 inches for the Bering flounder and 6 yrs and 8 inches for the yellowfin sole. Both the average and maximum size are below the acceptable size in United States markets (Alverson and Wilimovsky 1966).

Invertebrates

From limited studies made by Sparks and Pereyra (1966) and Alverson and Wilimovsky (1966), clams (*Macoma calcarea*), scallops (*Chlamys islandica*), and a crangonid shrimp are the three species that might support a commercial fishery of any size. In addition, there are local populations of Tanner and king crab, probably adequate to support small local fisheries. Here, again, there is every indication that these populations of invertebrates may originate in the Bering Sea and are transported by the currents into the Chukchi Sea.

CONCLUSIONS

With few exceptions, the fisheries of the Arctic are a biological unknown. Although regularly taken and eaten by the Eskimos, fish have never been an important item in their diet. The few studies that have been made generally show arctic waters to be low in the production of fish. Except for the local winter ice fishery by hand-line, the fishing season is limited to only two and a half or three months of the year (June/July to September). Low temperatures produce very slow growth in many varieties of fish, which in turn, affect the processing and marketing of the catch. For example, even the largest (and oldest) flatfish taken during the surveys of the Chukchi Sea were below the size acceptable in United States markets. The very heavy predation by birds and marine mammals alone would be sufficient to keep the stocks of fish at low levels of production. Finally, slow growth and poor recruitment together would make the arctic fisheries difficult to maintain and sensitive to manage.

With recent economic development along the North Slope of Alaska, however, the fishery potential of the arctic seas should be re-examined. The Arctic cod, although smaller than other related cods, is found in some abundance and is commercially important in the northern European countries. Capelin, which are marketable in Japan, spawn in numbers along the arctic beaches. In the last three or four years, a profitable salmon fishery has developed in the Kotzebue area; perhaps the abundance of salmon could be extended into the more northern rivers by modern culture techniques, using waste gas from the oil fields to heat the water for hatching the eggs. Tanner crab, shrimp and scallops are present in some areas of the Arctic in numbers sufficient to support small local fisheries.

Unfortunately, there has been little advance in the technology and management of arctic fisheries. We need to develop new methods to observe and study the distribution, behavior, and other habits of fish under the ice. We need to have new types of fishing gear that can fish under the ice during the winter months. We need to study the predator-prey relationship between birds/marine mammals and fish, but more important — we need to establish principles for managing such an ecological system to provide maximum use and enjoyment by man.

Conference Discussion

Korringa — Capelin are now used as food on a large scale in Norway for rearing salmon and steelhead to marketable size in pens installed in fjords. The requirements are a sheltered cove with seawater moved by the tides under ice-free conditions. Success in this type project is a question of economy. For fish feed (i.e., capelin), one can afford to pay 5 to 6 percent of the price per kilogram of the end product (salmon and steelhead). This industry is booming in Norway. The capelin are purchased fresh or deep-frozen, and the main inventory item of the farms is a freezer house.

Craig — Have you looked at the increasing sport-fishing pressure on arctic fishes, particularly impact on the anadromous species such as Arctic char?

Atkinson — No, I have only studied the commercial fisheries, and I am not aware of an extensive marine sportfishery in the Arctic. It is a good point, though, and we should consider the potential of the salmon and whitefish as sports fish in any program for the development of arctic fisheries.

Korringa — It is worthy to note that the limit for mercury in swordfish, the Silzy limit, arose out of a casual comment in a symposium in Hawaii that a concentration of 0.5 ppm was about right. This value was then used by the U. S. as the allowable limit. I do not consider this to be a very sensible limit.

Acknowledgments

Although many individuals have provided background information on the fisheries and oceanography of the Arctic, the author is especially indebted to Drs. D. L. Alverson, F. Favorite and Messrs A. T. Pruter of the Northwest Fisheries Center (National Marine Fisheries Service); to Messrs. D. M. Hickok, E. H. Buck and W. J. Wilson of the Arctic Environmental Information and Data Center (University of Alaska); and to Mr. E. J. Huizer of the Alaska Department of Fish and Game.

REFERENCES

ALT, K. T.

1967   Taxonomy and ecology of the Inconnu, *Stenodus leucichthys nelma*, in Alaska. M.Sc. Thesis, Univ. Alaska, Fairbanks, 106 pp.

ALVERSON, D. L., and N. J. WILIMOVSKY

1966   Fishery investigations of the Chukchi Sea. Chapter 31 in *Environment of the Cape Thompson region, Alaska*, edited by N. J. Wilimovsky and J. N. Wolfe. United States Atomic Energy Commission; pp. 843-860.

ANDRIYASHEV, A. P.

1954   Fishes of the northern seas of the USSR. Zoological Institute, Academy of Sciences of the USSR, No. 53. (Translated by the Israel Program for Scientific Translation, 1964). 617 pp.

ARCTIC ENVIRONMENTAL INFORMATION AND DATA CENTER

1975   Alaska regional profiles: Arctic region. Arctic Environmental Information and Data Center, Univ. Alaska, Anchorage, 217 pp.

ATKINSON, C. E.

1954   A review of research on the salmon fisheries of Alaska. U. S. Bureau of Commercial Fisheries, Biological Laboratory, Seattle, 39 pp. (mimeographed).

BIERI, R.

1962   Zooplankton investigations, oceanographic studies during operation "Wigwam". *Limnol. Oceanogr.* (Supplement) 7: 29-31.

COACHMAN, L. K., and K. AAGAARD

    1974   Physical oceanography of the arctic and subarctic seas. Chapter 1 in *Marine geology and oceanography of the arctic seas*, edited by Y. Herman. Springer-Verlag, New York, pp. 1-72.

ENGLISH, T. S.

    1966   Net plankton volumes in the Chukchi Sea. Chapter 28 in *Environment of the Cape Thompson region, Alaska*, edited by N. J. Wilimovsky and J. N. Wolfe. United States Atomic Energy Commission, pp. 809-815.

KINNEY, P., M. E. ARHELGER, and D. C. BURRELL

    1970   Chemical characteristics of water masses in the Amerasian Basin of the Arctic Ocean. *J. Geophys. Res.* 75(21): 4097-4104.

McROY, C. P., and J. J. GOERING

    1974   The influence of ice on the primary productivity of the Bering Sea. Chapter 21 in *Oceanography of the Bering Sea*, edited by D. W. Hood and E. J. Kelley. Occas. Publ. No. 2, Inst. Mar. Sci., Univ. Alaska, Fairbanks, pp. 403-421.

NIKOL'SKY, G. V.

    1971   Special ichthyology. *Vishaya Shkola*, Moscow, 3rd ed., 471 pp.

PRUTER, A. T., and D. L. ALVERSON

    1962   Abundance, distribution and growth of flounders in the southeastern Chukchi Sea. *J. du Conseil Int'l pour l'Exploration de la Mer* 27(1): 81-99.

QUAST, J. C.

    1974   Density distribution of juvenile Arctic cod, *Boreogadus saida* (Lepechin), in the eastern Chukchi Sea in the fall of 1970. National Marine Fisheries Service. *Fish. Bull.* 72: 1094-1105.

SPARKS, A. K., and W. T. PEREYRA

    1966   Benthic invertebrates of the southeastern Chukchi Sea. Chapter 29 in *Environment of the Cape Thompson region, Alaska*, edited by N. J. Wilimovsky and J. N. Wolfe. United States Atomic Energy Commission, pp. 817-838.

SWARTZ, L. G.

    1966   Sea cliff birds. Chapter 23 in *Environment of the Cape Thompson region, Alaska*, edited by N. J. Wilimovsky and J.

N. Wolfe. United States Atomic Energy Commission, pp. 611-678.

WALTERS, V.

1955 Fishes of western arctic America and eastern arctic Siberia. *Bull. Amer. Museum of Natural History* 106 (art. 5): 259-368.

1961 Winter abundance of *Arctogadus glacialis* in the Polar Basin. *Copeia* 2: 236-237.

WILIMOVSKY, N. J.

1974 Fishes of the Bering Sea: The state of existing knowledge and requirements for future effective effort. Chapter 11 in *Oceanography of the Bering Sea*, edited by D. W. Hood and E. J. Kelley. Occas. Publ. No. 2, Inst. Mar. Sci., Univ. Alaska, Fairbanks, pp. 243-256.

section 6

hydrocarbons in the
arctic marine environment

Assessment of the Arctic Marine Environment: Selected Topics
Copyright © 1976 by Institute of Marine Science, University of Alaska, Fairbanks

CHAPTER **25**

Surveillance of the marine environment for petroleum hydrocarbons

C. D. McAuliffe [1]

Abstract

Large quantities of petroleum hydrocarbons from seeps, along with large amounts of biogenic hydrocarbons, are contributed to the marine environment each year. These hydrocarbons have been contributed to the oceans for long periods of time. This paper will review and evaluate the measured concentrations and amounts of hydrocarbons now in marine waters and sediments, and it will rationalize these concentrations with the mechanisms that destroy hydrocarbons. An estimate of hydrocarbon distribution is given for the North Atlantic Ocean.

INTRODUCTION

Estimates of petroleum hydrocarbons entering the oceans appear in the National Academy of Sciences (NAS) publication, *Petroleum in the marine environment* (1975). Table 25.1, adapted from that publication, lists the best current estimates of hydrocarbons contributed from various sources. The estimates are quite good for inputs from offshore production, transportation, coastal refineries, and municipal and industrial sources. Less confidence can be placed in the estimates for natural seeps, urban runoff, and river runoff. The amount contributed to the oceans from the atmosphere is only an educated guess.

The seepage estimate is based principally on an evaluation by Wilson et al. (1974). These workers suggest a probable range from 0.2 to 6.0 x 10^6

[1] *Chevron Oil Field Research Company, Box 446, La Habra, California 90631.*

TABLE 25.1 Sources of hydrocarbons introduced into the oceans

Source	Best estimate, 10^6 MT/yr	Source	Best estimate, 10^6 MT/yr
Marine transportation	2.13	Urban runoff	0.6
Offshore oil production	0.08	River runoff	1.6
Coastal oil refineries	0.2	Natural seeps	0.6
Industrial waste	0.3	Atmospheric rainout	0.6
Municipal waste	0.3		
		Total	6.11

metric tons (MT) per year for the continental shelf areas of the world, with a best estimate of 0.6 x 10^6 MT/yr. This estimate does not include oil derived from land sources such as breached reservoirs or from subaerial erosion of tar sands, asphalts, and source rocks containing liquid petroleum.

Some of the gas seeps, which have been detected by marine seep surveys and geophysical techniques, probably contain some liquid hydrocarbons. Landes (1973) has documented active seeps and their magnitudes.

If one goes back only one million years, and uses the estimates of Wilson et al. (1974), petroleum additions to the world's oceans from seepage would be approximately 1 x 10^{12} MT.

Currently, man's contributions of petroleum hydrocarbons to the marine environment each year are about 10 times higher than the seepage rate. Hydrocarbon contributions from biogenic sources probably far exceed those from seepage. Bird and Lynch (1974), in a review article, document production of hydrocarbons from all types of plants, animals, and microorganisms. Land plants emit hydrocarbons, including *a*- and *b*-pinene. Rasmussen and Went (1965) estimate that these terpenoids exceed 400 x 10^6 MT/yr worldwide.

Although estimates vary, it is apparent that the marine environment receives currently large amounts of hydrocarbons and has received tremendous quantities in recent geological times.

HYDROCARBONS FOUND IN MARINE WATERS

Methane, the predominant hydrocarbon in the atmosphere and natural waters, is generated by bacteria during decomposition of organic matter. This production occurs in swamps and lakes, as well as in the ocean, in very much larger amounts than for other low-molecular-weight hydrocarbons.

Low-molecular-weight hydrocarbons in waters

The most extensive published measurements of low-molecular-weight hydrocarbons in seawater are by Swinnerton and Linnenbom (1967), Swinnerton et al. (1969), and Swinnerton and Lamontagne (1974). Over seven years, these investigators analyzed approximately 500 surface water samples from the open ocean in the Atlantic, Pacific, Antarctic, Gulf of Mexico, and Caribbean. Their average open-ocean concentrations in nanograms per liter (ng/liter) are: methane, 35.2; ethane, 0.67; ethene, 6.0; propane, 0.67; propene, 2.6; and butanes, 0.13. The surface water concentrations remote from possible sources of contamination were consistent from area to area, suggesting steady-state conditions. The methane content of seawater is about what one would expect from its partial pressure in the atmosphere and its predicted distribution between water and air (McAuliffe 1974).

Higher concentrations of the saturated light gases were observed in areas suspected of contamination either from biogenic or petroleum sources, such as the Mississippi River, and in Potomac and Chesapeake bays. High values in areas of the Mississippi-Gulf coast were probably contributed by subsurface seeps.

Some low-molecular-weight hydrocarbons have been measured in localized areas around oil producing operations (Sackett and Brooks 1974; Brooks et al. 1973).

Swinnerton and Linnenbom (1967) measured methane from the surface down to 500 m, both in the Atlantic Ocean approximately 500 km west of Ireland and in the Gulf of Mexico directly south of Mobile, Alabama, just off the Continental Shelf. The methane concentrations ranged between 30 and 50 ng/liter in the Atlantic and were constant with depth. The concentrations in the Gulf of Mexico were somewhat higher, up to 60 ng/liter at the surface, increasing to 200 at 30 m, and then decreasing with depth. Frank et al. (1970) measured methane, ethane, and propane in depth profiles in the Gulf of Mexico to total depth (some locations to 3600 m). The concentrations observed were fairly uniform with depth and higher than those reported by Swinnerton and Linnenbom (1967). Concentrations (ng/liter) averaged approximately 270 for methane, 7.2 for ethane, and 6.8 for propane.

Kinney et al. (1970) found methane throughout the Cook Inlet, from surface to 170 m in the deepest waters, to be approximately in equilibrium with methane in the atmosphere. Slightly higher concentrations of methane were observed in the northern Cook Inlet, and the authors suggest that this slight increase might be indicative of loss from oil reservoirs. This increase, however, was very small compared to the marked increase shown by Linnenbom and Swinnerton (1970) for the James River-Chesapeake Bay-Atlantic estuarine system and by Swinnerton and Lamontagne (1974) and Brooks et al. (1973) for discharges from the Mississippi River.

Swinnerton and Lamontagne (1974) report relatively uniform ethene and propene concentrations in the open ocean and near shore, even in those areas contaminated with saturated hydrocarbons. In general, the concentrations of ethene and propene decreased with increasing depth in the water column, reaching trace levels at 150 to 200 m. These unsaturated hydrocarbons, however, often showed maximum concentrations, particularly in spring and summer, in water from 30 to 100 m in the Atlantic, Caribbean and Pacific depth profiles.

In oxygenated waters, alkane gases generally tended to decrease with depth. In anoxic areas, however, the situation was reversed: methane increased by three to four orders of magnitude, ethane by one to two orders. Unsaturated hydrocarbons, however, appeared only as traces in truly anoxic environments. These data suggest different mechanisms for producing hydrocarbons in these different environments.

Several oil companies have made a large number of continuous marine seep surveys throughout the world, but the data are proprietary. Jeffery and Zarrella (1970) confirmed the open-ocean concentrations of the low-molecular-weight hydrocarbons reported by Swinnerton and co-investigators. Sigalove and Pearlman (1975) demonstrated the presence of hydrocarbon seeps with a continuously towed hydrocarbon analyzer. Observed anomalies were several hundred times the general background values for methane-through-butane hydrocarbons.

Koons and Brandon (1975) determined the propane-through-butane hydrocarbons in the Coal Oil Point seep area of the Santa Barbara Channel. Concentrations ranged from less than 10 ng/liter in the control areas to a maximum of 290 ng/liter in the seep area surface waters.

During a major oil spill in the Gulf of Mexico, when 1500 barrels per day were being discharged, McAuliffe et al. (1975) measured total C_1 to C_{10} hydrocarbons, obtaining maximum concentrations ranging from less than 1 to 200 μg/liter. Concentrations exceeding 1 μg/liter were restricted to a volume of water approximately 3 m deep by 100 to 300 m wide and about 2 km in length.

Although localized concentrations of low-molecular-weight hydrocarbons are found, the volumes of water involved are rather minor compared to the volume of the world's oceans.

Higher-molecular-weight hydrocarbons

Methods for measuring higher-molecular-weight hydrocarbons in water are not as sensitive as for C_1 to C_{10} hydrocarbons. Fluorescence, the most sensitive, can measure <1 ppb (μg/liter) but is not very diagnostic. Infrared methods can detect about 1 ppb, and gas chromatography and mass spectrometry methods are less sensitive.

Higher-molecular-weight hydrocarbons in the world's oceans have been measured in some areas using several sample collection techniques such as filtering particles of oil with neuston nets; air-sea interface hydrocarbons by screen or teflon discs, and very near-surface waters with slurp bottles; near-surface to about 10 m by opening a submerged bottle or pumping; and wireline sampling to make depth profiles.

Particulate hydrocarbons. Particles of floating tar are variously called tar lumps, tar balls, particulate oil, and pelagic tar. They are collected by skimming with neuston nets having mesh sizes ranging from 150 μm to 1 mm, depending upon the investigator's choice (maximum lump sizes range up to 3 cm in diameter). The amount of tar is most often measured by wet

TABLE 25.2 Summary of pelgic tar measurements by Neuston tows

Location and reference	No. of samples	Tar (mg/m²)	Tar less water (mg/m²)	Mean for area (mg/m²)
NW Atlantic Marginal Sea (Morris 1971)	5	1.0	0.8	
N Atlantic (McGowan et al. 1974)				0.12
Ocean Station Bravo	54?	0.0	0.0	
Ocean Station Charlie	54?	0.12	0.12	
East Coast Continental Shelf (Sherman et al. 1974)	161	0.48	0.38	
(Attaway et al. 1973)	10	0.39	0.31	0.38
Caribbean (Jeffrey et al. 1974)	20	0.74	0.74	
(Sherman et al. 1973)	64	0.16	0.13	0.28
Gulf of Mexico (Jeffrey et al. 1974)	84	1.20	1.20	1.20
Gulf Stream (J. Attaway, J. R. Jadamec and W. McGowan (unpubl. MS, 1973))	18	3.4	2.7	
(Sherman et al. 1968)	0.76	0.60		0.80
Sargasso Sea (Butler et al. 1973)				
Hydrostation S	43	7.50	5.9	
Station S536 (winter)	11	0.92	0.7	
Station S365 (summer)	16	7.54	6.0	
(Morris 1971; Morris and Butler 1973)	25	1.98	1.6	2.1
(McGowan et al. 1974)				
Ocean Station Echo	54?	2.64	2.64	
Ocean Station Delta	54?	1.15	1.15	
(Sherman et al. 1974)				
N. Antilles and Bahamas	86	4.39	3.5	
Mediterranean Sea (Horn et al. 1970)				
High values	37	95	75	
Assumed low values	663?	1	0.79	4.9
Assumed low values	663?	0.1	0.08	4.06
Pacific (Wong et al. 1974) Western,				
Kuroshio System	16	3.8	3.0	3.0
Eastern	17?	0.4	0.3	
Pacific Coast	?	0.0	0.0	0.15

Data sources: National Academy of Sciences (1975) and others.

weight, although Jeffrey et al. (1974) extracted the hydrocarbons for a more quantitative analysis, and McGowan et al. (1974) measured water content.

The National Academy of Sciences publication (1975) summarizes the tar densities observed on about one-third the area of the world's oceans through 1973. Table 25.2, an adaptation from the NAS report, also includes more recent references. It summarizes the mean pelagic tar concentrations as reported by various investigators, mean tar corrected for 21 percent water content (the mean water content measured by McGowan et al. 1974, for 214 samples), and a weighted mean for the various geographic areas.

McGowan et al. (1974) made their 214 neuston tows at four locations in the western mid-Atlantic, ranging from near the southern tip of Greenland to the most southern location between Florida and the Azores. The amount of tar increased from zero at the northern location to an average 2.6 mg/m^2 at the southern station. Sample tar contents were highly variable; the tar concentrations varied by a factor of 12 for eight successive tows at one location during a 48-hr period.

The higher concentrations, averaging 2.6 mg/m^2, found at the southernmost location (Ocean Station Echo), confirm the higher values reported by previous investigators for the Sargasso Sea than for other areas of the north Atlantic. The Sargasso Sea is bordered by a subtropical anticyclonic gyre that moves surface water towards the center, creating a catch-all basin for floating materials.

The Gulf of Mexico might be expected to have a relatively high pelagic tar concentration, because it is a semi-enclosed sea with a weak circulation in the central and western parts and because of its extensive shipping and tanker traffic. Jeffrey et al. (1974), in 104 tows, determined an average concentration of tar in the Gulf of Mexico of 1.2 mg/m^2 and in the Caribbean Sea, 0.7 mg/m^2.

Table 25.3 summarizes the surface areas of the seas in which tar has been measured, the mean tar concentrations, estimates of the total tar, and

TABLE 25.3 Total tar measured on the world's oceans

Location	Area $(10^{12} m^2)$	No. of samples	Mean tar (mg/m^2)	Total tar $(10^3 MT)$	NAS estimates No. of samples	NAS estimates Total $(10^3 MT)$
NW and N Atlantic	10	113	0.12	1.2	5	2
E Coast Cont. Shelf	1	171	0.38	0.37	144	0.2
Caribbean	2	84	0.28	0.56	79?	1.2
Gulf of Mexico	2	84	1.20	2.40	6?	1.6
Gulf Stream	8	186	0.80	6.4	18?	18
Sargasso Sea	7	289	2.1	14.7	95	70
Mediterranean	2.5	700?	4.7	10.1-12.2	700?	50
Kuroshio System	10	16	3.0	30.0	16	38
NE Pacific	40	17?	0.15	6.0	17?	16
	82.5 (23%)			71.7-73.8		197

Total areas of oceans 361

previously reported total tar values (National Academy of Sciences 1975). Although the mean tar concentrations have been corrected for 21 percent water content, they have not been corrected for particulate matter attached or mixed with the tar, and the actual pelagic oil may be lower than that shown. Only in the case of the measurements by Jeffrey et al. (1974), in the Gulf of Mexico and the Caribbean, has true oil been measured.

The more recent and extensive measurements for pelagic tar for some areas have changed the estimates made at the NAS workshop in 1973. Estimates for the northwest and north Atlantic, the east coast continental shelf, Caribbean, and Gulf of Mexico estimates were little changed. However, the amount of tar reported for the Sargasso Sea was reduced from 70,000 to 15,000 MT, and the amount in the Gulf Stream was reduced to about one-third.

As shown in Table 25.2, the difference for the Mediterranean is in the method of calculating the mean tar density. The very high values reported by Horn et al. (1970) appear to be concentrated in areas which may have had abnormally high concentrations of tar and not be typical of the entire Mediterranean. The assumed concentrations shown in Table 25.2 give approximately 11,000 MT of tar (Table 25.3) in the Mediterranean, rather than 50,000.

The large values for the western Pacific (Table 25.2) are based on a relatively small number of samples, so little reliance should be placed on them. Additional measurements in the Kuroshio system could well give lower values, as in the Sargasso Sea.

If the majority of tar found is from tankers and cargo ships (Mommessin and Raia 1975), with a lesser amount from oil seeps, the concentrations should be low in remote oceans such as the Arctic, North Pacific, South Pacific, and Antarctic. Quantities for the Indian Ocean and South Atlantic are unknown.

Hydrocarbons in the water column. As with tar balls collected by neuston nets, most hydrocarbon analyses in the water column have been measured in the Atlantic Ocean, Gulf of Mexico, and the Mediterranean Sea. Table 25.4 summarizes the hydrocarbons measured in the water column for recently reported data and for those studies which involved a number of samples. The National Academy of Sciences Report (1975) or Butler et al. (1973), give data for prior analyses of a few samples for given locations or results reported to Butler as private communications.

Gordon et al. (1974) and Gordon and Keizer (1974) report that very-near-surface samples, obtained with a slurp bottle, contained from 9 to 20 μg/liter (46 and 61 $\mu g/m^2$). Wade and Quinn (1975), using a screen for collection, measured hydrocarbons at the air-sea interface (0.1 to 0.3 mm layer) and found an average concentration of 155 μg/liter (31 $\mu g/m^2$). Ledet and Laseter (1974), using a teflon disc for collection, found somewhat higher values in the Gulf of Mexico. Values in Timbalier Bay averaged 360 $\mu g/m^2$, but average concentrations of only 180 $\mu g/m^2$ were found off Florida. Average values offshore Louisiana varied from 210 to 700 $\mu g/m^2$;

TABLE 25.4 Higher-molecular-weight hydrocarbons in the water column

Location and reference	Depth (m)	No. of samples	Concentration (µg/liter)
Atlantic - tanker routes from	0-0.3	33	8.9 ± 9.6
Caribbean and Gulf Coast to	10	26	3.8 ± 3.5
New York (Brown et al. 1973)			
Near Bermuda (cited by Butler			
et al. 1973, p. 27-28)	>100	?	<1.0
N.W. Atlantic - Nova Scotia	0-3 mm	43	20.4 ±60.7
to Bermuda (Gordon et al.	1	24	0.8 ± 1.3
1974)	5	24	0.4 ± · 0.5
	> 5	?	0.0
(Gordon and Keizer 1974)	1-5 mm	53	9.3 ±18
	1	23	0.6 ± 0.6
	5	24	0.4 ± 0.4
	10-1000	50	0.0
N.W. Atlantic - Nova Scotia	1,10,25	24	4.9 ± 0.9
to the Gulf Stream (Zsolnay			
1974)			
Atlantic - Sargasso Sea	0.1-0.3 mm	17	155.0 ±149
(Wade and Quinn 1975)	0.2-0.3	17	73.0 ±58
Mediterranean (Monaghan	?	?	3.5
et al. 1973)			
Gulf of Mexico - (Ledet and	air-sea	118	$\sim$300? $\mu g/m^2$
Laseter 1974). Alkanes only.	interface		
Gulf Universities Research	?	?	2.0
Consortium, Offshore Ecology			
Investigation (Oppenheimer			
et al. 1974). Alkanes			
N.E. Pacific - Vancouver to	0-0.3	26	0.2
Ocean Station P. (Cretney			
and Wong 1974)			
Santa Barbara Channel - (Koons	1	1	16
and Brandon 1975) Coal Oil	10	1	0.4
Point seep area	55	1	1.0
Non-seep areas	1-400	4	0.3
Cook Inlet - (Kinney et al 1969)	0-170	23	<0.02

the lower value occurred at the same time the samples were collected off Florida.

Brown et al. (1973) report the concentrations of nonvolatile hydrocarbons in surface water samples collected with a pail and at 10 m depth during tanker voyages from the Caribbean and Gulf Coast to New York. Monagham et al. (1973) report data for the Mediterranean Sea.

Near-surface hydrocarbon concentrations in the Pacific are quite low (0.2 μg/liter) as reported by Cretney and Wong (1974). In the Cook Inlet, Kinney et al. (1969) were unable to detect hydrocarbons using a method having a sensitivity of 0.02 μg/liter.

Even in a known oil seep area, Koons and Brandon (1975) report relatively low values in the Santa Barbara Channel at Coal Oil Point. The concentrations varied in the seep area with depth from 0.4 to 16 μg/liter. In nonseep areas, concentrations averaged 0.3 μg/liter.

Table 25.4 shows that hydrocarbon concentrations decrease with increasing water depth. The surface microlayers had the highest concentrations, followed by somewhat lower values measured with slurp bottles. All investigators have found concentrations lower by at least one-half at 5 or 10 m, compared with 0 to 0.3 m and 1 m sampling depths.

Gordon et al. (1974) and Gordon and Keizer (1974) emphasize the difficulties in obtaining water samples without contamination. They document that open-flow subsurface wireline samplers sent down empty and returned not tripped had as high or higher concentrations of nonvolatile hydrocarbons coating the walls than did water samples which had been collected by tripped samplers. Therefore, these workers suggest that many earlier measurements, including their own, are high because of contamination and suggest that even those measured along the tanker routes by Brown et al. (1973) may be too high. The distribution of hydrocarbons in near-surface waters, combined with the low values reported for the northeastern Pacific and the Cook Inlet suggest that Gordon and co-workers are correct — that almost all the higher-molecular-weight hydrocarbons are at less than 10 m.

These facts suggest that the hydrocarbons are probably particulate rather than in true solution. They may be possibly adsorbed on, or absorbed into, particulate or living matter.

HYDROCARBONS IN RECENT MARINE SEDIMENTS

Hydrocarbons in recent marine sediments can have several origins, including subsurface seepage, bacterial generation of methane, and possibly some C_2 to C_4 hydrocarbons, terrestrial and marine plant and animal remains, and particulate matter from erosion of sedimentary rocks in which petroleum hydrocarbons were previously generated.

Recent marine sediments almost always contain higher hydrocarbon concentrations, by three or four orders of magnitude, than does the water column.

Low-molecular-weight hydrocarbons

The biogenic generation of methane with trace amounts of ethane and propane is well documented in the literature (Whelan et al. 1975). In fact, enough gas may form to cause mass movement of sediments (slumping) in offshore areas, such as in the Mississippi River Delta.

Butane-through-heptane hydrocarbons >1 ppm have not been detected in recent sediments, and they are probably not formed biologically. These hydrocarbons appear to form from organic matter subjected to depths of burial exceeding 1000 m at temperatures in excess of 50° C. When present in recent sediments at >1 mg/liter, they are assumed to have migrated from deeper source sediments or petroleum reservoirs (marine seeps).

Hunt (1975) has documented the presence of C_4 to C_7 hydrocarbons in μg/liter concentrations in deep-sea drilling project cores at burial depths from 28 to 800 m. He attributes these hydrocarbons to their generation at lower depths of burial and temperatures than predicted by most geochemists. He has ruled out contamination, because about 5 percent of the samples analyzed were from abyssal plain areas and showed no detectable hydrocarbons, even though the method is sensitive to 5 parts in 10^{12} (ng/liter).

Low-molecular-weight hydrocarbons other than methane are present in seabottom marine sediments in very low concentrations except near seeps. Undoubtedly, numerous petroleum exploration studies have sought such evidence of seeps, but little information is available in the literature. Carlisle et al. (1975) show examples of up to 0.07 mg/liter of C_2 to C_4 hydrocarbons in near-surface marine sediments.

High-molecular-weight hydrocarbons

Table 25.5 gives concentrations of higher-molecular-weight hydrocarbons in recent marine sediments and in some subsurface samples analyzed by Smith (1954).

Local sources of hydrocarbon discharge, such as in bays or rivers, cause locally high concentrations of hydrocarbons in the near-surface sediments. This is shown for a polluted area in coastal France, the Coal Oil Point seep area in the Santa Barbara Channel, San Francisco Bay, Lake Maracaibo, and the discharge areas from the Mississippi River. The areas affected appear to be relatively small.

In remote areas, hydrocarbons in the sediments are low, as measured by Kinney et al. (1969) in the Cook Inlet and by Kinney (1973) in Port Valdez, Alaska. The Cook Inlet samples, however, consisted mainly of sand, gravel, and rocks. One sample, which contained measurable hydrocarbons and was not included in this average, was from Kachemak Bay. The authors do not give a concentration, but they indicate that the hydrocarbons were nonpetroleum.

The data in Table 25.5 suggest a hydrocarbon content of 40 mg/kg for nearshore continental shelf sediments.

Hydrocarbon measurements on deep-sea sediments are limited. J. W. Farrington (personal communication, 1975) reports that the saturate fraction for samples from deep water near Bermuda ranged from 1 to 5 ppm. On a

TABLE 25.5 Higher-molecular-weight hydrocarbons in marine sediments

Location and references	Depth (cm)	No. of samples	Mean concentration
Cook Inlet (Kinney et al. 1969)	0-12	14	<1 µg/kg, wet
Port Valdez (Kinney 1973)	0-12	9	1 mg/kg, wet
France (Tissier and Oudin 1973)			
Polluted	0-30	4	420 mg/kg, dry
Unpolluted	0.30	5	38 mg/kg, dry
Santa Barbara Channel,			
Coal Oil Point seep area,	0-5	24	6,400 mg/1
(Straughan 1974;	5-15?	20	13,600 mg/1
Koons and Brandon 1975)			
Coal Oil Point seep area	0-15?	3	3130 mg/kg, dry
Away from seep area	0-15?	4	94 mg/kg, dry
(Orr and Emery 1956)	?	?	160 mg/kg, dry
San Francisco Bay (DiSalvo and Guard 1975)	Suspended Sediment	18	1175 mg/kg, dry
	0-?	14	1588 mg/kg, dry
Pacific, 60 miles south of Golden Gate	?	1	85 mg/kg, dry
(Pearson et al. 1970)	?	?	$\sim$400 mg/kg, dry
Lake Maracaibo (Battelle Pacific NW Labs 1974; Templeton et al. 1975)	?	10	3125 mg/kg
Gulf of Mexico (McAuliffe et al. 1975) C_{12}-C_{33} fraction near Mississippi River discharges	0-4	9	12 mg/1
Breton-Chandeleur Sounds	0-4	20	1 mg/1
Nearshore and Bays	0-4	21	4 mg/1
(Smith 1954)			
Pelican Island Core	6-51 m	4	41 mg/kg, dry
Grand Isle Core	6 m	3	84 mg/kg, dry
W. Coast of Africa	0-15	4	32 mg/kg, dry
Mississippi Delta	0-15	4	65 mg/kg, dry
Buzzards Bay (Blumer and Sass 1972)	0-10	62	48 mg/kg, dry
Bermuda, deep water (Farrington 1975)	0-15	5	1-5 mg/kg, dry

transect toward New York, the continental slope sediments contained 10 to 20 ppm, and the continental shelf samples were 40 to 60 ppm. Smith (1954) found an average hydrocarbon content of 32 ± 3.7 mg/kg in four samples from beneath 120 to 920 m of water off the west coast of Africa.

Smith (1954) documented the hydrocarbon content of recent sediments from two cores in the Gulf of Mexico off Louisiana. He found fair uniformity in hydrocarbon content down to at least 100 m. These data suggest that the contribution of hydrocarbons to the Mississippi River Delta sediments has been relatively constant in recent geological time. The hydrocarbons apparently originated primarily from erosion of sedimentary rocks containing petroleum hydrocarbons, in addition to terrestrial and marine plant and animal debris. The samples contained several percent organic matter.

DISTRIBUTION OF HYDROCARBONS
IN THE MARINE ENVIRONMENT

Based upon the analysis of water and sediment samples for low- and high-molecular-weight hydrocarbon fractions, it is informative to calculate estimates of the amount of these hydrocarbons in the North Atlantic Ocean for which the most information is available (Table 25.6). The low-molecular-weight gases were calculated from Swinnerton and co-investigators' open-ocean measurements, assuming a uniform concentration throughout the water column for methane, ethane, propane, and butanes. Ethene and propene were calculated for the surface 200 m.

The higher-molecular-weight hydrocarbons at the air-sea interface were determined using the measurements of Gordon et al. (1974), 20.4 µg/liter

TABLE 25.6 Estimated distribution of hydrocarbons in North Atlantic

In water column	Total quantity (MT)
Gases:	
Methane	48,000,000
Ethane, propane, butanes	2,000,000
Ethene, propene	620,000
	(50,620,000)
High-molecular-weight hydrocarbons:	
Air-sea interface	1,500
Near-surface (0-10 m)	132,000
Pelagic tar	24,000
	(157,500)
In sediments	
Gases:	?
High-molecular-weight hydrocarbons (0-1 m):	96,000,000

for a 0 to 3 mm layer (2020 MT); Gordon and Keizer (1974), 9.3 μg/liter for a 0 to 5 mm layer (1530 MT); and Wade and Quinn (1975) 155 μg/liter for a 0 to 0.2 mm layer (1020 MT). These three values average 1520 MT and are very consistent, considering the differences in sampling locations and methods of collection.

The water column amount was calculated by Gordon and Keizer (1974) by integration of their measurements to a depth of 10 m, giving a total of about 4 mg of oil under each square meter of sea surface.

Pelagic tar was calculated by adding to the 15,000 MT for the Sargasso Sea (Table 25.3) an average of 0.4 mg/m^2 for the remaining surface area of the North Atlantic.

Estimates used in calculating the higher-molecular-weight hydrocarbons in the first meter of North Atlantic sediments were a sediment bulk density of 2, a continental shelf area of 5 percent of the ocean, and hydrocarbon concentrations of 40 mg/kg on the shelves and 4 mg/kg elsewhere.

With these assumptions, Table 25.6 reveals that most of the hydrocarbons in the water are gases. They are predominantly, if not exclusively, from natural sources.

The air-sea interface has the highest concentration of higher-molecular-weight hydrocarbons, but it contributes less than 1 percent to the hydrocarbons in the water column. Ledet and Laseter (1974) found this surface layer to be 90 percent saturated hydrocarbons and quite uniform in composition for Louisiana-Florida samples collected at one time. The composition also suggested appreciable contributions from biogenic sources.

The near-surface water hydrocarbons and pelagic tar appear to be particulate material and restricted principally to the first 10 m. The total in the Atlantic is small compared with the hydrocarbon gases. The amount of high-molecular-weight hydrocarbons below 10 m should be left open, however, until diagnostic analyses of uncontaminated samples are obtained.

Good compositional data for hydrocarbons in the water is limited, but the data of Brown et al. (1973) suggest that the hydrocarbons were derived at least in part from petroleum.

No attempt was made to quantify hydrocarbon gases in the first meter of sediments, because data are limited. Methane is probably present in large amounts with lesser amounts of ethane and propane, by at least two orders of magnitude. The C_4 to C_7 hydrocarbons are very low.

Concentrations of high-molecular-weight hydrocarbons and the amount shown for the first meter of sediment (Table 25.6) indicate that both concentrations and amount are higher than for the water column. Areas of petroleum contamination are not sufficiently large, or the concentrations sufficiently high, to measurably affect total hydrocarbons in ocean sediments.

DESTRUCTION OF HYDROCARBONS IN THE MARINE ENVIRONMENT

The destruction of petroleum hydrocarbons by various physical, chemical, and biological mechanisms has been amply documented and summarized in the National Academy of Sciences report (1975). More recent papers include biodegradation (Ladner and Hagstrom 1975; Atlas 1973; ZoBell 1973; and McAuliffe et al. 1975). Klein and Pilpel (1974) have studied additional effects of photooxidation, and Lee (1975) further documented metabolism of hydrocarbons by marine organisms (copepods).

In recent geological times, tremendous quantities of petroleum and biogenic hydrocarbons have been contributed to the marine environment. The measured concentrations and quantities of these hydrocarbons in the oceans and bottom sediments confirm that destructive mechanisms have destroyed the majority of these hydrocarbons. Further, measurements not reported here document no buildup of petroleum hydrocarbons in the atmosphere. Volatile hydrocarbons would reside predominantly in the atmosphere rather than the oceans (McAuliffe 1974), if they were not destroyed.

Most hydrocarbons are present in recent marine sediments, with relatively uniform concentrations, for considerable distances below the sea bottom. This suggests that they have been contributed naturally over geological time and that man's influence has been minor. The data reveal that man's contributions of hydrocarbons have had some influence in populated areas where estuaries or coastal areas have received appreciable quantities from various industrial and municipal wastes.

Pelagic tar in the oceans probably is principally from shipping discharges. It appears to be in lower amounts and have a shorter residence time than originally calculated by Morris and Butler (1973) from a more limited data base. These investigators calculated the half-life for tar in the North Atlantic as between 1.0 and 2.6 yrs. The inclusion of more recent data suggests a half-life between 0.1 and 0.3 yrs. Butler (1975) discussed the evaporative weathering of petroleum residues and the age of pelagic tar. He suggests that pelagic tar degrades and breaks up more rapidly than previously supposed. By assuming a crude-oil residence time of about 1 yr, Butler (1975) requires that the initial masses have a median size of 1 liter, which seems unrealistically large for a water and oil slurry that has passed through a discharge pump. A residence time from 1 to 4 months would better fit the weathering data discussed by Butler.

Biodegradation rates of hydrocarbons decrease in the order of normal alkanes, branched alkanes, aromatics, and cycloalkanes (Perry and Cerniglia 1973). Therefore, if biodegradation were the only mechanism destroying hydrocarbons in the water column (0 to 10 m), this should increase the proportion of aromatic and cycloalkane hydrocarbons relative to n- and to branched alkanes compared with their concentrations found in crude oils and refined products. Brown et al. (1973) document a proportionate increase in cycloalkanes, but they report an apparent decrease in aromatics. This suggests that the aromatic hydrocarbons (known to be photosensitive) are being destroyed by sunlight. Ledet and Laseter (1974) found very low aromatic content in hydrocarbons at the air-sea interface in the Gulf of Mexico.

SUMMARY

Petroleum hydrocarbons and larger amounts of biogenic hydrocarbons have been introduced into the marine environment at least for recent geological times. Current day oil seepage is estimated at 0.6 x 10^6 MT/yr for the continental shelf areas of the world. Man's contributions of petroleum hydrocarbons in recent years have been about 10 times this amount.

Low-molecular-weight hydrocarbons, principally methane with some C_2 to C_4 hydrocarbons, are found throughout the oceans, approximately in equilibrium with methane in the atmosphere. These gases are mostly biogenic, as are ethene and propene found in the top 200 m of water. The concentrations in ng/liter are methane, 35; ethane, 0.7; propane 0.7; butanes, 0.13; ethene, 6; and propene, 2.6.

High-molecular-weight hydrocarbons appear to be restricted to the top 10 m of water, principally as particles, with the concentrations decreasing from the sea-air interface. The approximate concentrations are 0 to 3 mm, 20 μg/liter; 1 m, 1μg/liter; and 5 m, 0.4 μg/liter. The half-life of these hydrocarbons appears to be from 1 to 4 months. Shipping adds some of these hydrocarbons to the Atlantic, Gulf of Mexico, Caribbean, Pacific, Indian and Mediterranean. In more remote seas, petroleum concentrations are probably lower.

Except for methane content, near-bottom marine sediments contain low concentrations of C_2 to C_7 hydrocarbons. Higher-molecular-weight hydrocarbons are generally three orders of magnitude higher in sediments than in water: ~40 mg/kg on the continental shelf, 10 to 20 mg/kg on the slope, and 1 to 5? mg/kg under deep waters. Some sediments in localized areas receiving petroleum hydrocarbons such as seep oil and municipal and industrial wastes contain >500 mg/kg.

An estimated hydrocarbon distribution has been made for the best-studied ocean — the North Atlantic. Concentrations of low-molecular-weight gases in the water column are 5 x 10^7 MT methane through butanes and 6.2 x 10^5 MT ethene and propene. High-molecular-weight hydrocarbons are found at the air-sea interface (1.5 x 10^3 MT), in the upper 10 m (1.3 x 10^5 MT), and in near-surface waters as pelagic tar (2.4 x 10^4 MT). The concentration of high-molecular-weight hydrocarbons in the first meter of sediment is about 1 x 10^8 MT.

The quantity of hydrocarbons in the waters and sediments of the world's oceans is small compared to total additions. This indicates that the destructive mechanisms of physical weathering, biodegradation, and photooxidation are operative. These mechanisms have prevented a build-up of petroleum hydrocarbons in the oceans through geological time and seem to be doing the same for man's recent increased additions.

REFERENCES

ATLAS, R. M.

    1973   Fate and effects of oil pollutants in extremely cold marine environments. Final report, U. S. Defense Documentation Center, A. D. 769895, October 1, pp. 1-15 + tables and figs.

BATELLE PACIFIC NORTHWEST LABORATORIES

    1974   Study of effects of oil discharges and domestic and industrial wastewaters on the fisheries of Lake Maracaibo, Venezuela. Vol. 2. Fate and effects of oil. Richland, Washington, 100 pp.

BIRD, C. W., and J. M. LYNCH

    1974   Formation of hydrocarbons by microorganisms. *Chem. Soc. Rev.* 3: 309-328.

BLUMER, M., and J. SASS

    1972   The West Falmouth oil spill. Tech. Report 72: 19, Woods Hole Oceanographic Institution, Woods Hole, Massachusetts, 125 pp.

BROOKS, J. M., A. D. FREDRICKS, W. M. SACKETT, and J. W. SWINNERTON

    1973   Baseline concentrations of light hydrocarbons in Gulf of Mexico. *Environ. Sci. & Tech.* 7: 639-642.

BROWN, R. A., ET AL.

    1973   Distribution of heavy hydrocarbons in some Atlantic Ocean waters. *In* Proc. Joint Conf. Prevention and Control of Oil Spills. American Petroleum Inst., Washington, D. C., pp. 505-519.

BUTLER, J. N.

    1975   Evaporative weathering of petroleum residues: The age of pelagic tar. *Mar. Chem.* 3: 9-21.

BUTLER, J. N., B. F. MORRIS, and J. SASS

    1973   *Pelagic tar from Bermuda and the Sargasso Sea.* Spec. Publ. 10, Bermuda Biological Station for Research. Harvard University Printing Office, Cambridge, Massachusetts, 346 pp.

CARLISLE, C. T., G. G S. BAYLISS, and D. G. VANDELINDER

    1975   Distribution of light hydrocarbon in seafloor sediments: Correlations between geochemistry, seismic structure, and possible reservoired oil and gas. Proc. 1975 Offshore Tech. Conf., Vol. 3, pp. 65-70.

CRETNEY, W. J., and C. S. WONG

1974 Fluorescence monitoring study at ocean weather station "P." In *Marine pollution monitoring (Petroleum)*. Spec. Publ. 409, Nat'l Bur. Standards, U. S. Government Printing Office, Washington, D. C., pp. 175-177.

DISALVO, L. H., and H. E. GUARD

1975 Hydrocarbons associated with suspended particulate water in San Francisco Bay waters. *In* Proc. 1975 Conf. Prevention and Control of Oil Pollution. American Petroleum Inst., Washington, D. C., pp. 169-173.

FRANK, D. J., W. SACKETT, R. HALL, and A. FREDERICKS

1970 Methane, ethane, and propane concentrations in Gulf of Mexico. *Bull. Amer. Assoc. Petrol. Geol.* 54: 1933-1938.

GORDON, D. C., and P. D. KEIZER

1975 Hydrocarbon concentrations in seawater along the Halifax-Bermuda Section: Lessons learned regarding sampling and some results. In *Marine pollution monitoring (Petroleum)*. Spec. Publ. 409, Nat'l Bur. Standards, U. S. Government Printing Office, Washington, D. C., pp. 113-115.

GORDON, D. C., P. D. KEIZER, and J. DALE

1974 Estimates using fluorescence spectroscopy of the present state of petroleum hydrocarbon contamination in the water column of the northwest Atlantic Ocean. *Mar. Chem.* 2: 251-261.

HORN, M. H., J. M. TEAL, and R. H. BACKUS

1970 Petroleum lumps on the surface of the sea. *Science* 168: 245-246.

HUNT, J. M.

1975 Origin of gasoline range alkanes in the deep sea. *Nature* 254: 411-413.

JEFFERY, D. A., and W. M. ZARRELLA

1970 Geochemical prospecting at sea. *Bull. Amer. Assoc. Petrol. Geol.* 54: 853-854.

JEFFREY, L. M., ET AL.

1974 Pelagic tar in the Gulf of Mexico and Caribbean Sea. In *Marine pollution monitoring (Petroleum)*. Spec. Publ. 409, Nat'l Bur. Standards, U. S. Government Printing Office, Washington, D. C., pp. 233-235.

KINNEY, P. J.

    1973   Baseline hydrocarbon concentrations. In *Environmental studies of Port Valdez*, edited by D. W. Hood, W. E. Shiels, and E. J. Kelley. Occas. Publ. 3, Inst. Mar. Sci., Univ. Alaska, Fairbanks, pp. 397—409.

KINNEY, P. J., D. K. BUTTON, and D. M. SCHELL

    1969   Kinetics of dissipation and biodegradation of crude oil in Alaska's Cook Inlet. *In* Proc. Joint Conf. Prevention and Control of Oil Spills. American Petroleum Inst., New York, pp. 333—340.

KINNEY, P. J., D. M. SCHELL, and D. K. BUTTON

    1970   Quantitative assessment of oil pollution problems in Alaska's Cook Inlet. Tech. Report R69—16, Inst. Mar. Sci., Univ. Alaska, Fairbanks.

KLEIN, A. E., and N. PILPEL

    1974   The effects of artificial sunlight upon floating oils. *Water Res.* 8: 79-83.

KOONS, C. B., and D. E. BRANDON

    1975   Hydrocarbons in water and sediment samples from Coal Oil Point area, offshore California. Proc. 1975 Offshore Tech. Conf., Vol. 3, pp. 5133—521.

LADNER, L., and A. HAGSTROM

    1975   Oil spill protection in the Baltic Sea. *J. Water Pollution Control Fed.* 47: 796-809.

LANDES, K. K.

    1973   Mother Nature as an oil polluter. *Bull. Amer. Assoc. Petrol. Geol.* 57: 637—641.

LEDET, E. J., and J. L. LASETER

    1974   Alkanes at the air-sea interface from offshore Louisiana and Florida. *Science* 186: 261—263.

LEE, R. F.

    1975   Fate of petroleum hydrocarbons in marine zooplankton. *In* Proc. 1975 Conf. Prevention and Control of Oil Pollution. American Petroleum Inst., Washington, D. C., pp. 549—553.

LINNENBOM, V. J., and J. W. SWINNERTON

1970 Low-molecular-weight hydrocarbons and carbon monoxide in sea water. In *Organic matter in natural waters*, edited by D. W. Hood. Occas. Publ. 1, Inst. Mar. Sci., Univ. Alaska, Fairbanks, pp. 455—467.

McAULIFFE, C. D.

1974 Determination of C_1 to C_{10} hydrocarbons in water. In *Marine pollution monitoring (Petroleum)*. Spec. Publ. 409, Nat'l Bur. Standards, U. S. Government Printing Office, Washington, D. C., pp. 121-125.

McAULIFFE, C. D., ET AL.

1975 Chevron Main Pass Block 41 oil spill: Chemical and biological investigations. *In* Proc. 1975 Conf. Prevention and Control of Oil Pollution. American Petroleum Inst., Washington, D. C., pp. 555—565.

McGOWAN, W. E., W. A. SANER, and G. L. HUFFORD

1974 Tar ball sampling in the western North Atlantic. In *Marine pollution monitoring (Petroleum)*. Spec. Publ. 409, Nat'l Bur. Standards, U. S. Government Printing Office, Washington, D. C., pp. 83-84.

MOMMESSIN, P. R., and J. C. RAIA

1975 Chemical and physical characterization of tar samples from the marine environment. *In* Proc. 1975 Conf. Prevention and Control of Oil Pollution. American Petroleum Inst., Washington, D. C., pp. 155—167.

MONAGHAN, P. H., J. H. SEELINGER, and R. A. BROWN

1973 The persistent hydrocarbon content of the sea along certain tanker routes. Prelim. report, 18th Annual Tanker Conf., American Petroleum Inst., Washington, D. C., 298 pp.

MORRIS, B. F.

1971 Petroleum: Tar quantities floating in the northwestern Atlantic taken with a new quantitative neuston net. *Science* 173: 430-432.

MORRIS, B. F., and J. N. BUTLER

1973 Petroleum residues in the Sargasso Sea and on Bermuda beaches. *In* Proc. Joint Conf. Prevention and Control of Oil Spills. American Petroleum Institute, Washington, D. C., pp. 521—529.

NATIONAL ACADEMY OF SCIENCES

1975   *Petroleum in the marine environment.* (Proc. workshop held
        21—25 May 1973), Airlie House, Va.). NAS-NRC Ocean
        Affairs Board, National Academy of Sciences, Washington, D.
        C., 107 pp.

OPPENHEIMER, C. H., R. MIGET, and H. KATOR

1974   Hydrocarbons in seawater and organisms and microbiological
        investigations. Final report submitted to Gulf Universities
        Research Consortium for Offshore Ecology Investigations,
        Project OE73HJM.

ORR, W. L., and K. O. EMERY

1956   Composition of organic matter in marine sediments: Prelimi-
        nary data on hydrocarbon distribution in basins of southern
        California. *Bull. Geol. Soc. of Amer.* 67: 1247—1258.

PEARSON, E. A., P. N. STORRS, and R. E. SELLECK

1970   A comprehensive study of San Francisco Bay, Vol. 7. Sum-
        mary, conclusions and recommendations. Report 67—5, Sani-
        tary Engineering Research Laboratory, Berkeley, California, 85
        pp.

PERRY, J. J., and C. E. CERNIGLIA

1973   Studies on the degradation of petroleum by filamentous fungi.
        In *The microbial degradation of oil pollutants,* edited by D.
        C. Ahearn and S. P. Myers. Publ. LSU—SG—73—01, Center for
        Wetland Resources, Louisiana State Univ., Baton Rouge, Louisi-
        ana, pp. 89—94.

RASMUSSEN, R. A., F. W. WENT

1965   Volatile organic materials of plant origin in atmosphere. In
        Proc. Nat'l Academy of Sciences 153: 215-220.

SACKETT, W. M., and J. M. BROOKS

1974   Use of low-molecular-weight hydrocarbon concentrations as
        indicators of marine pollution. In *Marine pollution monitoring
        (Petroleum).* Spec. Publ. 409, Nat'l Bur. Standards, U. S.
        Government Printing Office, Washington, D. C. pp. 243—244.

SHERMAN, K., ET AL.

1974   Distribution of tar balls and neuston sampling in the Gulf
        Stream. In *Marine pollution monitoring (Petroleum).* Spec.
        Publ. 409, Nat'l Bur. Standards, U. S. Government Printing
        Office, Washington, D. C., pp. 243—244.

SIGALOVE, J. J., and M. D. PEARLMAN

    1975   Geochemical seep detection for offshore oil and gas exploration. Proc. 1975 Offshore Tech. Conf., Vol. 3, pp. 95—102.

SMITH, P. V., JR.

    1954   Studies on origin of petroleum: Occurrence of hydrocarbons in recent sediments. *Bull. Amer. Assoc. Petrol. Geol.* 38: 377—404.

STRAUGHAN, D.

    1974   Field sampling methods and techniques for marine organisms and sediments. In *Marine pollution monitoring (Petroleum).* Spec. Publ. 409, Nat'l Bur. Standards, U. S. Government Printing Office, Washington, D. C., pp. 183-187.

SWINNERTON, J. W., and R. A. LAMONTAGNE

    1974   Oceanic distribution of low-molecular-weight hydrocarbons: Baseline measurements. *Envir. Sci. & Tech.* 8: 657—663.

SWINNERTON, J. W., and V. J. LINNENBOM

    1967   Gaseous hydrocarbons in seawater: Determination. *Science* 156: 1119—1120.

SWINNERTON, J. W., V. J. LINNENBOM, and C. H. CHEEK

    1969   Distribution of methane and carbon monoxide between the atmosphere and natural waters. *Envir. Sci. & Tech.* 3: 836—838.

TEMPLETON, W. L., ET AL.

    1975   Oil pollution studies on Lake Maracaibo, Venezuela. *In* Proc. 1975 Conf. Prevention and Control of Oil Pollution. American Petroleum Inst., Washington, D. C., pp. 489—496.

TISSIER, M., and J. L. OUDIN

    1973   Characteristics of naturally occurring and pollutant hydrocarbons in marine sediments. *In* Proc. Joint Conf. Prevention and Control of Oil Spills. American Petroleum Inst., Washington, D. C., pp. 205-214.

WADE, T. L., and J. G. QUINN

    1975   Hydrocarbons in the Sargasso Sea surface microlayer. *Mar. Poll. Bull.* 6: 54—57.

WHELAN, T., J. M. COLEMAN, J. N. SUHAYDA, and L. E. GARRISON

  1975   The geochemistry of recent Mississippi River Delta sediments: Gas concentration and sediment stability. Proc. Offshore Tech. Conf., Vol. 3, pp. 71-84.

WILSON, R. D., ET AL.

  1974   Natural marine oil seepage. Science 184: pp. 857—865.

WONG, C. S., D. R. GREEN, and W. J. CRETNEY

  1974   Quantitative tar and plastic waste distribution in the Pacific Ocean. Nature 247: 30—32.

ZoBELL, C. E.

  1973   Bacterial degradation of mineral oils at low temperatures. In The microbial degradation of oil pollutants, edited by D. C. Ahearn and S. P. Myers. Publ. LSU—SG—73—01, Center for Wetland Resources, Louisiana State Univ., Baton Rouge, Louisiana, pp. 153—161.

ZSOLNAY, A.

  1974   Hydrocarbon content and chlorophyll correlation in the waters between Nova Scotia and the Gulf Stream. In Marine pollution monitoring (Petroleum). Spec. Publ. 409, Nat'l Bur. Standards, U. S. Government Printing Office, Washington, D. C., pp. 255—256.

CHAPTER **26**

Hydrocarbon studies in the benthic environment at Prudhoe Bay

D. G. SHAW *and* L. M. CHEEK[1]

Abstract

Measurements of aliphatic hydrocarbons are reported for sediments and biota collected in and around Prudhoe Bay, Alaska, during the summer of 1974. The results of these measurements indicate that these hydrocarbons are largely, or totally, of biogenic origin. Special problems associated with carrying out the field work of this project are also discussed.

INTRODUCTION

Work in north temperate regions has suggested that petroleum hydrocarbons may be taken up by marine species. To understand the movement of petroleum hydrocarbons through food-webs in the arctic marine environment at Prudhoe Bay, we must have knowledge of food habits of benthic species to determine trophic interactions to be expected there. Such understanding also requires direct measurement of the kinds and amounts of hydrocarbons now present in various elements of the food-webs. The hydrocarbon measurements described here serve a double purpose, since they also provide baseline information on hydrocarbons in Prudhoe Bay organisms. It is reasonable to expect that some arctic marine organisms accumulate unusually large amounts of fats and oils for use during the long, sunless winter. Such animals might also have an enhanced tendency to retain petroleum following exposure.

In the summer of 1974, a continuing program of environmental studies at Prudhoe Bay, Alaska, was begun. These studies include investigations of

---

[1] *Institute of Marine Science, University of Alaska, Fairbanks, Alaska 99701.*

benthic biota, sediment characteristics, and the ambient levels of hydrocarbons in and around Prudhoe Bay. Like many embayments of the Beaufort Sea, Prudhoe Bay is very shallow, with depths seldom exceeding 7 ft. In this setting, a dock to receive seagoing barges required a gravel causeway extending approximately 4400 ft into the bay. The effects of this dock on the benthic environment of Prudhoe Bay are the immediate subject of the study reported here. Understanding these effects is particularly important, since similar gravel structures are contemplated for the extension of petroleum exploration and development into the Beaufort Sea.

LOGISTICS AND SAMPLING

Prudhoe Bay largely freezes to the bottom each winter. This has limited our sampling activities to a single collection during the summer. Sediment samples for hydrocarbon analysis were collected before the causeway was built on two transects approximately 500 ft on either side of its projected location. Beginning 500 ft from shore, samples were collected every 1000 to 4500 ft from shore. In this way, a loop of 10 sediment samples for hydrocarbon analysis was collected around the causeway site. Shorter transects were sampled about one mile on either side of the causeway site. Additional sediment samples were obtained at other nearshore locations within Prudhoe Bay and in the vicinity of Gull, Niakuk and Argo islands. Fish collected by gill-net at the causeway site were made available to us for hydrocarbon analysis by the Alaska Department of Fish and Game. Our sampling stations are shown in Figure 26.1 and Table 26.1.

The difficulty of winter sampling has already been mentioned. Even in summer, logistic problems of this work, like most arctic undertakings, are considerable. A small boat that can operate in less than two feet of water is essential for work within Prudhoe Bay. Thus, many standard oceanographic and marine biological sampling techniques that require large platforms and winches are impractical. The principle collecting procedure used was divers with scuba gear. Operating out of an open, shallow draft boat also limited sampling activities to near-perfect weather and ice conditions. To collect a sediment sample for hydrocarbon analysis, a diver was handed a pre-cleaned glass jar which he filled directly at the bottom. The jar was returned to the boat and recapped. Ashore, the jar was frozen until analysis. The sediments were generally sandy and contained organic material which appeared to be tundra fragments.

ANALYTICAL METHODS

Analyses of sediment and biota for hydrocarbons have been carried out using a procedure similar to that of Farrington et al. (1972). Samples were extracted in a Soxhlet apparatus using equal parts of benzene and methanol. The lipids thus extracted were partitioned into hexane, then dried and concentrated. Following saponification or removal of elemental sulfur with copper, as necessary, the extract was subjected to column chromatography

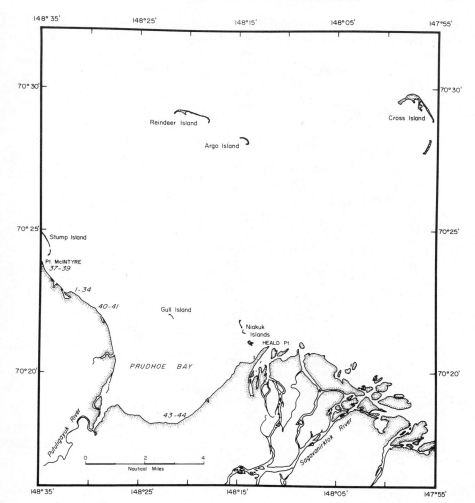

Fig. 26.1 Map of sampling locations at Prudhoe Bay.

on a mixed bed of silica and alumina. The aliphatic hydrocarbon containing eluate was concentrated and analyzed by gas chromatography. The concentration of total extractable hydrocarbons was determined gravimetrically.

RESULTS AND DISCUSSION

Weights of hydrocarbons extracted from sediments are presented in Table 26.2. These weights were obtained gravimetrically. We now believe that this procedure gives values that may be too high, and these weights are probably best regarded as upper limits. The sediments from Prudhoe Bay were general-

TABLE 26.1 Hydrocarbon sampling locations in Prudhoe Bay

Station	Location	Distance from shore (ft)
	1000 ft SE of causeway	
2		500
4		1500
6		2500
8		3500
10		4500
	100 ft NW of causeway	
26		500
28		1500
30		2500
32		3500
	1 mi NW of causeway	
37		1500
38		2500
39		3500
	1 mi SE of causeway	
40		1500
41		2500
43	Adjacent to old dock	

TABLE 26.2 Hydrocarbon concentrations in Prudhoe Bay sediments

Station	Extractable hydrocarbons (mg/kg wet-wt of sample)
2	7.1
4	5.3
6	23
8	43
10	27
26	26
28	26
30	10
32	18
37	35
38	6.8
39	5.2
40	5.2
41	5.6
43	2.2
Argo	6.5
Gull	1.8
Niakuk	11

ly sandy and contained some organic material which appeared to be tundra fragments. None of the samples shows the regular pattern of *n*-alkanes associated with petroleum.

There is considerable variability among the hydrocarbons extracted from the 18 sampling locations in the Prudhoe Bay area. However, no clear trends in the data are apparent. The lowest concentration of hydrocarbons in sediment was found at Gull Island. The sediment collected at that location was coarser than the others and did not appear to contain tundra fragments.

(Second-year sampling during August 1975 included acquisition of fragments from on shore and replicate sampling at one location to provide information about the degree of variance at a given point. This should aid in our understanding of the first-year sediment hydrocarbon data presented here).

Table 26.3 gives concentrations of hydrocarbons from fish collected at Prudhoe Bay; selected chromatograms are presented in Figure 26.2. All of the species analyzed are highly mobile; the individuals probably spent considerable time outside Prudhoe Bay itself. It is possible to interpret these results to a greater extent than for the sediments, because in some cases more than one individual of a given species was analyzed, and also, better resolution was obtained in gas chromatography by the use of 0V-101 liquid phase rather than Apiezon L which was used for the sediments.

Flesh from two individuals of the species *Coregonus autumnalis* (Arctic cisco) were analyzed separately. In both the total concentrations of hydrocarbons (Table 26.2) and in the array of hydrocarbons shown in the gas chromatograms (Fig. 26.2 a and b), these two individuals were similar but not identical. The total concentrations are within experimental error and are thus equivocal; however, the gas chromatograms, despite their qualitative similarity, showed several differences in intensities of smaller peaks. The only peak identified is pristane (marked: Pr). None of the other major peaks has the retention time of an *n*-alkane. These peaks are probably terpenoid hydrocarbons. Phytane, a hydrocarbon characteristic of petroleum, is absent.

Skin from the two Arctic ciscos was pooled and analyzed. The gas chromatogram (Fig. 26.2, c) is qualitatively similar to those of the flesh (Fig. 26.2, A and B). The total concentration of hydrocarbons in the skin

TABLE 26.3 Hydrocarbon concentrations in Prudhoe Bay fish

Sample	Extractable hydrocarbons (mg/kg wet-wt of sample)
Arctic cisco (1), flesh	10
Arctic cisco (2), flesh	13
Arctic cisco (1 + 2), skin	50
Four-horned sculpin (1) flesh	3.8
Four-horned sculpin (2) flesh	4.2
Arctic char, flesh	3.6
Arctic flounder, dorsal flesh	5.2
Arctic flounder, ventral flesh	9.6

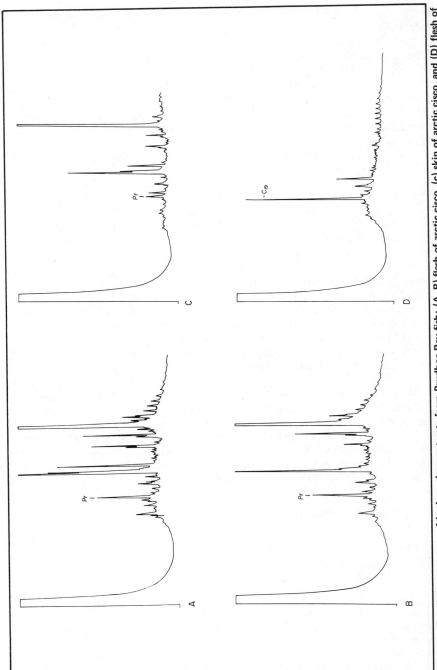

Fig. 26.2 Gas chromatograms of hydrocarbon extracts from Prudhoe Bay fish: (A, B) flesh of arctic cisco, (c) skin of arctic cisco, and (D) flesh of arctic char.

(Table 26.2) is significantly higher than that in the flesh. This observation is in keeping with the fact that these fish store oil subcutaneously. In the skinning process, these oils go with the skin.

Table 26.2 also shows the results of the analysis of two individuals of the species *Myoxocephalus quadricornis* (four-horned sculpin). As with the Arctic cisco, the total hydrocarbon concentrations for the two individuals show similar hydrocarbon contents. Also the gas chromatograms are generally similar (to each other, but not to the ciscos).

A single individual of the species *Salvelinus alpinus* (Arctic char) was analyzed. The gas chromatogram (Fig. 26.2, D) of the hydrocarbons from the Arctic char show a quite different pattern. In the char, nonadecane is the major peak. Again, phytane is absent.

Flesh of a single individual of the species *Liopsetta glacialis* (Arctic flounder) was divided into dorsal and ventral portions, which were analyzed separately. The total concentrations by hydrocarbons were 9.6 and 5.2 mg/kg, respectively. We are aware of no metabolic or structural feature of the flounder that would lead to this nearly twofold enrichment of the dorsal flesh in hydrocarbons. Additional analyses of *Liopsetta glacialis* will be required to substantiate or disprove this finding. The dorsal and ventral gas chromatograms are qualitatively quite similar.

In summary, it can be stated that the kinds and amounts of hydrocarbons present in fish and sediment taken from Prudhoe Bay during 1974 indicate that these hydrocarbons are largely or totally biogenic in origin.

Note added in proof. Analysis of sediment samples collected in 1975 and re-analysis of some 1974 samples by gas chromatography and combined gas chromatography mass spectroscopy shows the presence of an array of hydrocarbons characteristic of petroleum at concentrations similar to those of simultaneously observed biogenic hydrocarbons. It is not known at this writing (April 1976) whether this petroleum is the result of natural seeps or development activities or whether the petroleum has caused ecological upset.

Acknowledgments

The authors wish to thank the Atlantic Richfield Company and the Alaska Sea Grant Program for fianancial support and the Alaska Department of Fish and Game for assistance in sample collection. Contribution no. 285, Institute of Marine Science, University of Alaska.

REFERENCE

FARRINGTON, J., C. S. GIAM, G. R. HARVEY, P. L. PARKER, and J. TEAL

1972   Analytical techniques for selected organic compounds. In *Marine pollution monitoring: Strategies for a national program*, edited by E. D. Goldberg. Allan Hancock Foundation, Los Angeles, Calif., pp. 152-176.

Assessment of the Arctic Marine Environment: Selected Topics
Copyright © 1976 by Institute of Marine Science, University of Alaska, Fairbanks

CHAPTER **27**

Sample size in environmental assessments

T. A. GOSINK [1]

Abstract

Of the many errors that can occur during the analysis of a sample, some of the most serious can happen at the collection site and are fundamental in character. The critical question is not that of contamination, but whether samples taken are either large enough or numerous enough to be representative of the matrix. Data are presented which show that a 1-liter seawater sample is probably inadequate, even though a usable analytical detector response can be obtained from a 10-ml sample. Evidence also shows that analysis of replicate samples from the same station at the same time which display poor precision could very well be accurate. It is suggested that for trace element or particulate analyses, multiple samples of 1 to 10 liters each, or at least a composite of that size, should be analyzed for each component of interest. Other analytical errors such as time factors between sample collection and analysis or preservation, surface contamination factors, and filtration are briefly discussed in this chapter.

INTRODUCTION

Differences in reported trace element concentrations in apparently similar types of environmental and biological samples are frequently encountered. Several interlaboratory calibration studies have revealed very large variations in values for oxygen (Carritt and Carpenter 1966), nutrients (Palmork 1969), various trace metals (Brewer and Spencer 1970), and lead (Patterson 1974) in analyses on subsamples from the same parent samples. A significant number of these variations may be attributed to poor analytical techniques;

---

[1] *Institute of Marine Science, University of Alaska, Fairbanks, Alaska 99701.*

however, it is possible that some analytical differences may be valid. New data for analyses of aluminum, chromium, and selenium, together with reference to the existing literature, are used in support of the following statements regarding sampling, subsampling, sample size, and storage.

PROCEDURES

Procedures for the gas chromatographic analysis of aluminum and chromium (Gosink 1975) and for selenium (Gosink and Reynolds 1975) have been described elsewhere.

For studies on the effect of surface contamination, two types of samples were employed. For the Atlantic coastal water samples, two NIO bottles were thoroughly cleaned with detergent, rinsed, filled with dilute (~1 M) HCl for about 1 hr and then thoroughly rinsed with deionized water. These bottles were kept clean and closed throughout the first portion of the sampling period. Sampling was accomplished from a small boat about 2 km offshore from Rudee Inlet, Virginia Beach, Virginia, while drifting about 1 km north over a 45-min period. A sampler was tied to a weighted line and lowered, while still clean and closed, to a depth where the top of the container was a little more than 1 m below the surface. Another line, attached to the top cover of the sampler, was pulled repeatedly so as to open momentarily, thus filling the sampler. The water was transferred immediately to clean polyethylene bottles and mildly acidified (pH ~4). The same sampler was then opened, passed through the surface of the ocean four times, lowered to the same depth for about 1 min, and then triggered closed in the normal fashion with a messenger. This water sample was also immediately transferred and stored as described above. A second, replicate, set of samples was then taken with another clean sampler.

The Elizabeth River (Norfolk, Virginia) samples were taken from a helicopter platform using a specially constructed small glass and rubber sampling device. Glass tubes were mounted on a plexiglass block with a small weighted line below it in such a way that one tube was open when it entered the water and the other remained sealed and clean. The tube ends were sealed with small coated rubber balls which were held together through the length of the tubes with rubber bands. The device was rigged in such a way that the closed (protected) tube opened briefly when the open (exposed) tube was closed after a line was pulled to free a retaining pin. In this way, small-volume (35 to 40 ml) samples of "protected and exposed" water were obtained simultaneously from a depth of about 0.5 m. The glass tubes used in obtaining the samples had been cleaned with NOCHROMIX and silanized with HMDS. The constructed tubes and rubber were washed in detergent, rinsed, soaked for several hours in dilute HCl solution, and then thoroughly rinsed with deionized water before storage in the closed position. A clean duplicate, device was used to obtain a second set of samples within 5 min. These samples were transferred immediately to small clean polyethylene storage bottles containing one drop of dilute HCl (pH ~4).

For other Al and Cr analyses, some samples were either pipetted or expressed through a 0.5-μ, filter-equipped, graduated plastic syringe into precleaned tubes or bottles containing the buffered chelating solutions used for the gas chromatographic analysis. Other samples were drawn from a standard Niskin bottle.

RESULTS AND DISCUSSION

Sampling

The first and most critical step in any analysis is the time-consuming task of taking the sample. In this limited discussion on analytical errors, it is assumed that the samples are properly secured without contamination or loss. For trace metal studies, the nearly universal choice is a sampler constructed of plastic material. Such problems as the density or porosity of these plastics, trace element contaminants in the plastics, and surface loss characteristics have been discussed elsewhere (Spencer and Brewer 1970).

Sample containers present a surface resulting from aging the sampler in seawater. The theory is that this practice will equilibrate any surface-active binding sites with the water of interest, thus reducing the chance of trace element exchange during the short, yet critical, time that the sample is in the container (Lewin and Chen 1973). Very possibly, a surface-active or oil film constitutes the real surface in contact with the water in the sampler. Most samplers available on the market are opened when mounted on the hydrowire and then passed through a surface known to be enriched with organic and surface-active materials and trace elements associated with these materials (e.g., Duce et al. 1972). To test the contaminating effect of this surface film on subsurface analyses, two sets of replicate samples were taken with clean, unaged samplers. Table 27.1 presents these results and clearly shows a contamination trend in the open or exposed samplers, which have average values for aluminum and chromium in the raw water up to four times as high as the samples taken from the surface-protected devices.

The wide variation in the results for a single sample are attributable to a large extent to the fact that individual subsamples (five or more in each case) were too small (<10 ml each) to be representative, a point which is discussed below.

TABLE 27.1 Trend of the effect of the water surface on the analysis of aluminum and chromium in subsurface raw water samples

Source	Date	Number of samples analyzed	Type	Concentration[1]	
				Al	Cr
Elizabeth River,	8/25/73	8	protected	55 ±65/20-225	7.5 ±10/nd*-50
Norfolk, Va.	8/25/73	8	open	72 ±30/20-210	25 ±22/ng-80
Atlantic coastal,	9/12/73	5	protected	75 ±25/35-100	4 ±2/1.5-7
Virginia Beach, Va.	9/12/73	8	open	100 ±45/55-185	8.5 ±8.0/nd-19

*nd = not detected above blank level by individual analysis.

Number and size of samples and subsamples

Because of time and volume of work factors, the analyst usually accepts many assumptions in an analytical scheme. Often the first mistake is to overlook the fact that the oceans at any given level really are not well mixed over relatively large areas. Evidence of water nonuniformity is well documented in at least three sources. The first is the well-established fact that Langmuir cycles cause distinct and severe differences in the concentrations of flora, fauna, and debris at the surface and for a considerable distance below, along with the accompanying inherent concentration factors on and within these components of the sea (Sutcliffe et al. 1963; Williams 1967; Parker and Barson 1970; Duce et al. 1972; George and Edwards 1973). To obtain representative surface samples from a region affected by Langmuir cycles, at least two precisely taken casts are required at the maximum and minimum points of concentration. In practice, this is an impossible task because of drift and because the cycles are frequently difficult or impossible to discern visually.

Evidence that surface waters, at least, are far from uniform (even a matter of <10 m) is further seen in high altitude photography and the recent Earth Resources Technology Satellite (ERTS) images which show extensive fine structure in many plumes reaching far out to sea. Figure 27.1 is a fourth or fifth generation reproduction of some randomly selected satellite pictures of coastal oceans. Relatively few of these photometrically sharply delineated boundaries (at least in third generation pictures) are discernible from a surface vessel. Within these larger plumes in coastal areas, as much as 20 percent real differences in the transmittance properties of the water occur within a few meters, as measured by a towed continuous transmissometer, implying, on an absolute minimum, the same amount of horizontal variance in the trace metals associated with the particulates (Bowker et al. 1973; Berg et al. 1975).

The report showing large variations in particulate iron concentrations for repeated casts in the region of Panama by Schaeffer and Bishop (1958) can be interpreted as showing these plume variations. The highly variable results of Lewis and Goldberg (1954) for several single casts in the Pacific Ocean, using a triplicate cluster of bottles 5 m apart, perhaps reflect some vertical variations in the water column.

In our analysis of some Alaska surface waters by repeated sampling, variations such as those of Schaeffer and Bishop (1958) were observed. These results are discussed in the next section.

Wangersky (1974) commented on significant variability for multiple cluster sampling for particulate organic material and on the value obtained as a function of the sample size. That is to say, larger samples tend to normalize the results.

The third area of evidence demonstrating the nonuniformity of seawater comes from relatively recent reports in the literature based on continuous vertical measurements of temperature or oxygen. If a marked discontinuity for 5 to 50 m depth within a water column for these parameters can be demonstrated, it is reasonable to assume that trace elements within that

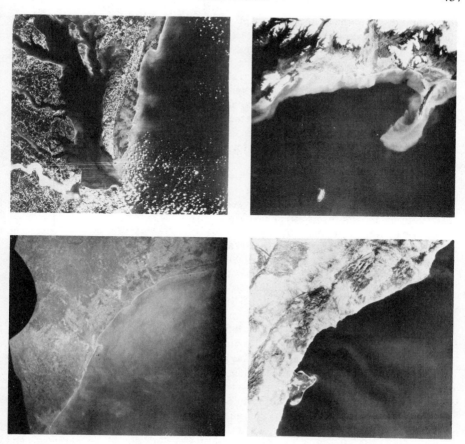

Fig. 27.1 High altitude pictures showing sediment plumes extending far out to sea. Pictures (a), (b), and (d) are ERTS MSS Band 5 (green wavelengths), approximately 125 miles across. Picture (c), smaller in scope, is from the Gemini series. Locations are (a) Chesapeake Bay, (b) Cape St. Elias—Middleton Island in the Gulf of Alaska (c) Gulf of Mexico, Galveston—Houston area, and (d) Mediterranean near Barcelona, Spain.

band can also vary. Speculatively, an oxygen discontinuity could very well be due to the presence or recent past presence of a deep scattering layer such as one containing shrimp which are respiring in that layer and causing chemical change. More concretely, some of these oxygen variations have been demonstrated to be caused by eddies of water from neighboring different water masses diffusing into the sampling area (Lambert 1974).

Subsampling

The errors involved in subsampling become more important as the element of interest diminishes in concentration. In mineral analysis of rocks, it quite often becomes the major experimental error (Ingamells and Switzer 1973). It

is commonly presumed that a sample is well mixed, that the particulates
(0.1 to 100 μ) are numerous, at least randomly distributed, and that equal
volume subsamples contain the same number of particulates. Such assumptions are often wrong.

An accepted practice in rock analysis is to crush and grind a large
sample, in which case a subsample taken will contain at least 10^6 particles.
If fewer particles are taken for analysis, statistical theory indicates that
deviations in individual minerals taken may become significant (Kleeman
1967). In estuarine and coastal water, samples containing 10^6 particles would
correspond to subsamples of 1 liter and probably around 10 liters for
clearer water (Sheldon et al. 1972; Bowker et al. 1973). The data in Table
27.1 and subsequent small-volume sample data amply demonstrate this
discrepancy. One of the results of a recent IDOE workshop on lead analyses
was the recommendation that the subsample for individual analysis be at
least 2 liters in volume (Patterson 1974).

Although the vast majority of particles in seawater are on the order of
10 μ in diameter or smaller, there are small but significant numbers of
particles of 10 μ diameter and larger which are a very real part of the
system and must be included in the total analysis. Concentration factors of
trace elements on and in particles or microorganisms commonly are on the
order of 10^2 to 10^4, and those of 10^6 are not unknown. A single particle
weighing 10^{-3}g (such as a clump of microorganisms or algal cells, or perhaps
an iron floc), and having a concentration factor of 10^6 for the element of
interest, raises the concentration of that element by a factor of two in a
1-liter sample, or usually within experimental error for <1 ppb level work.
But if that 1-liter sample were divided into 10 subsamples of 100 ml each,
the analyst, not having a prior knowledge of the single enriched particle,
would reasonably be tempted to assume that he had good precision for nine
analyses, but that on one of the analyses he must have contaminated the
subsample because it was anomalistically high (20 times the average of the
other nine). In actuality, the high or outlier value should be reported and
averaged with the others, since this is characteristic of a Poisson (asymmetric) or bimodal distribution which trace elements may follow, as opposed
to Gaussian (symmetric) distribution (Ingamells and Switzer 1973). High
precision is not a measure of accuracy. Particulates in the ocean do follow
Poisson or bimodal size distributions (Sheldon et al. 1972; Bowker et al.
1973), and one or two such greatly enriched particles per liter or at least
per m^3 should be expected from a variety of sources, including a minor
plankter species not in bloom or a sediment floc.

When relatively small numbers of enriched particles are present in a
system, the analyst working with 10-ml raw water subsamples is more likely
to be presenting erroneous low background results with good precision, with
an occasional extraordinarily high value, whereas the person analyzing large
(>500 ml) samples will have poor precision but better accuracy. If the
number of enriched particles is larger but the analysis subsample is still too
small to take a representative number of particles each time, then the
analyst will obtain high values with only an occasional low value. Wangersky
(1974) presents similar arguments for particulate organic matter analysis

values and the size of the sample. Again, high precision is not a measure of good accuracy.

We have seen Poisson-type distribution for some trace metal samples taken in Prince William Sound, Valdez Arm, and in Resurrection Bay, both in Alaska. When small subsamples were used (<10 ml), most tubes containing the subsample and chelating agent were colorless and had reasonably similar analysis, but those tubes showing a yellowish to light brown coloration, presumably due to the presence of one or more iron-rich particles, almost invariably were the ones which had substantially higher aluminum and particularly chromium values. The coloration usually appeared in about 5 to 10 percent of the tubes, but for some water sources it amounted to as high as 60 to 80 percent of the tubes. The variations in the small number of small volume samples reported in Table 27.1 can be attributed in part to the fact that the samples used were too small for any one sample to be representative, requiring large numbers of samples to be statistically accurate with poor precision, and thus were presented as a trend, and to demonstrate the subsample variability phenomenon.

Table 27.2 and Figure 27.2 show the results of larger numbers of 10-ml and 160-ml raw water and 0.5-μ filtered water samples pipetted directly from various sources, or within 30 min from Niskin samplers, into containers with the reagents used to prepare the samples for gas chromatographic analysis. One can see the asymmetric distribution of results in the histograms (Fig. 27.2) of some of these analyses, whereas the field blank values for the Chena River and Valdez samples on the other hand show a reasonably smooth standard Gaussian distribution of results for the 11 analyses performed. Field blank values for aluminum (that is, the blank tubes and reagents are exposed to the same environment as the samples) tend to be two to four times larger than blanks prepared in the laboratory (Gosink 1975). Field blank chromium values, on the other hand, show little variation, if any, from the low to undetectable laboratory blanks.

The Chena River was calm and clear with only tannins visible. The Valdez samples were taken from very calm waters which contained considerable amounts of glacial scour. Analyses 4 to 6 were from a region closer to the main glacial scour input to Valdez Arm, the Lowe River, as compared to a station about 2 km further north which yielded analyses 7 to 12 (Table 27.2). There was no visually discernible difference in the high level of scour in the water between those two stations; yet there is about a 20 to 30 percent variation in analytical values (lower for the station farther from the source) between the two stations for both raw and filtered samples. The variation is also evident in the filtered water samples which were taken from Resurrection Bay, (analyses 20 and 21, Table 27.2) which had no visible glacial scour; yet the values for Al at the station closer to Bear Glacier (20) is twice that found at the station 21 located 16 km further inland.

Joyner (1964), like others (e.g., Feely et al. 1971), who employed a spectrophotometric technique to follow particulate aluminum and iron, saw similar results in that the concentrations of these elements increase as one approaches a coastal source. He has shown that, at least for coastal water, the method can be used to track runoff plumes.

The values for aluminum and chromium in the 20 small samples and the 3 large samples (analyses 2 and 3, Table 27.2) are within experimental error. There was no coloration in any of the 20 tubes nor in any of the larger volume samples which had an analytical range of 13.1 to 13.8. The larger standard deviation shown in Table 27.2 is the result of the uncertainty in the blanks. On the other hand, there does not seem to be any correlation

TABLE 27.2   Aluminum and chromium analyses of various surface water samples as a function of sample size and color

Sample source and analysis numbers	Al (ppb ±σ)	Cr (ppb ±σ)	Sample size (ml)	Number of samples analyzed	Color
Field blanks for the Chena and Valdez samples 1	7.7± 2.6	nd	(container and reagents)	11	none
Chena River, 2 August 1974:					
2 Raw	15.2± 3.1	nd[1]	10	20	none
3 Raw	13.5± 2.7	nd	160	3	none
Valdez Arm, 25 July 1974: 61°5'45"N, 145°21'W					
4 Raw	82.6±16.5	nd	10	10	none
5 Raw	103	1.5	160	1	slight
6 Filtered (0.5 μ)	32.4± 7.1	nd	10	6	none
61°6'34"N, 145°20'W					
7 Raw	54.6± 8.0	nd	10	10	none
8 Raw	71.8	1.3	160	1	very slight
9 Filtered (0.5 μ)	24.6± 8.6	nd	10	5	none
26 July 1974:					
10 Raw	58.1±19.1	nd	10	5	none
11 Raw	88	na[2]	160	1	slight
12 Filtered (0.5μ)	20.0± 9.3	nd	10	4	none
End of Scripps Pier, 9 April 1974:					
13 Raw	14 ± 3	nd	8	4	none
14 Raw	93 ± 10	∿1.5	8	10	none
15 Raw	120 ± 20	2.3±0.6	8	4	intense
Prince William Sound, 8 May 1974:					
16 60°40'N, 146°45'W	20.2± 11.7	nd	8	10	none
17 15 min later	49 ± 30	nd[3]	8	5	1 colored
18 60°50'N, 147°01'W	64.3± 30	∿1.5	8	10	4 colored
19 15 min later	12.4±	nd	8	5	none
Resurrection Bay, November 1973:					
20 59°52'N, 149°27'W, 0.5 μ filtered	81 ± 22	1.2±0.3	2	3	—
21 60°3'N, 149°27'W 0.5 μ filtered	44 ± 19.6	1.5±0.4	2	3	—

[1] nd = not detected above blank level (see statement on page 14).
[2] na = not analyzed.
[3] In the one colored sample, the chromium content was 5 ppb, thus giving an average of 1.0 ± 2.2 ppb.

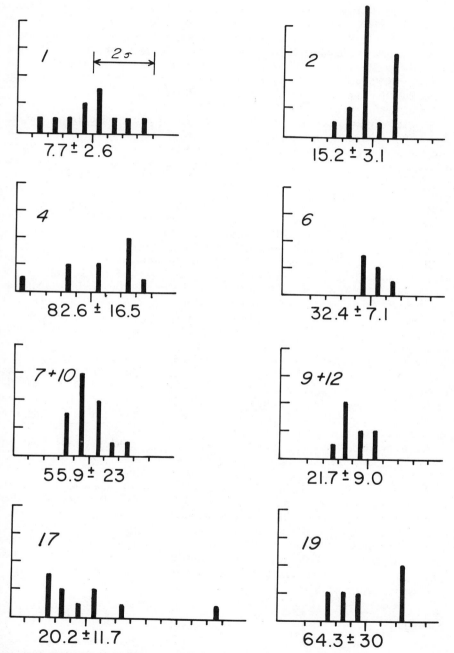

Fig. 27.2 Histograms of individual aluminum analyses for some of the stations shown in Table 27.2.

between the analyses by the small volume method (nos. 4, 7, and 10) as compared to the larger volume samples taken during the same sampling periods (nos. 5, 8, and 11, respectively) which required about 45 min each to accomplish while drifting not more than 300 m (wind calm). Although all of the small volume samples in these cases were colorless, however, all of the larger volume samples (equivalent to the variability in 16 10-ml samples) must have picked up enriched particles, in that each developed some color and had higher analytical values.

A most striking example of this color phenomenon came from samples taken from the end of the Scripps Pier (nos. 13 to 15). A series of four samples were pipetted directly from surface water into reagent tubes, developed no color, and had reasonably low analytical values. However, two additional samples were collected in clean, rinsed 250-ml bottles about 15 min later for return to the lab for other studies. One bottle had some visible particulate matter in it, whereas the other did not. However, subsamples drawn from both of these bottles had markedly higher analytical values, especially the one with the particulates in it (15) — which showed a pronounced increase in the chromium content.

The values for the Prince William Sound Stations (nos. 16 to 19) also showed temporal variations. Analyses for 16 and 17 and for 18 and 19 are for subsamples taken from the same Niskin sample bottle used for replicate near-surface casts for the same stations about 15 min apart in time. Ten subsamples were taken from the first casts (16 and 18) and five subsamples each from the replicate casts (17 and 19). Three of the analytical values comprising the average value reported in no. 17 are very similar to the ones comprising analysis 16, but two of the analyses from no. 17 were high, one with slight color, and thus the average is higher. The variations for the other station were markedly greater. For one cast, tubes used in four of the analyses developed color and high analytical results, and the other six were in fairly good agreement although higher than all of the values for the five subsamples from the replicate cast (no. 19).

Atkinson and Stefanson (1969) also show as much as an order of magnitude variation in some of their replicate near-surface analysis of aluminum and iron in southeastern coastal U. S. waters.

The data shown in Table 27.2 are in the standard Gaussian format. If, however, the data are treated so as to accommodate the typical outlier values of Poisson distributions by utilizing a log transform (Natrella 1963), the mean values are altered in the direction of the skew. For example, entry 2 takes on a mean of 13.3 (note the closer approximation to the large-volume sample 3) when the transform $y = 1/2 \ln \left| \frac{1+x}{1-x} \right|$ is used. For skews in the other direction (e.g., no. 6), the transform equation $y = \ln x$ produced a mean of 32.8.

Storage

Much or all information about speciation is lost, and there is a possibility that the sample itself may become useless if a sample or subsample must go through a period of storage. Lewin and Chen (1973) have recently demonstrated how rapidly soluble iron changes to particulate iron in raw samples in containers, showing measurable changes within 30 min. The data for aluminum and chromium shown in Table 27.3 support Lewin and Chen's findings.

Five small (2-ml) samples of clear water from Resurrection Bay were pipetted directly into tubes with the buffer and gas chromatography preparation reagents. Then a 2-liter Niskin bottle was rinsed and filled with water from the same place that subsequent subsamples had been taken. Raw water subsamples taken from the Niskin bottle as soon as it was returned to the lab (within 30 min) had analytical values (no. 2) which were similar to those taken directly from the Bay (no. 1). Two hours later, samples were taken from the Niskin bottle after the water in the bottle had been shaken. These samples had substantially lower analytical values for aluminum and chromium for both the raw and 0.5-μ filtered water (nos. 3 and 6). Samples of the water in the Niskin bottle were then transferred to a clean 250-ml glass sample storage bottle and mildly acidified with HC1. Another portion was used to rinse a 0.5-μ filter apparatus, the first two 100-ml portions were discarded, and the subsequent filtrate was used to rinse another clean sample bottle before it was filled and also mildly acidified. Subsamples from these two bottles were taken 2 days later, and the results are given as analyses 4 and 7 in Table 27.3.

TABLE 27.3   Effect of storage time on the aluminum and chromium contents of a seawater sample from Resurrection Bay (Lowell Point), Alaska

	Al (ppb ±σ/orange)	Cr (ppb ±σ/orange)	Number of samples analyzed
Raw water			
1 Direct from bay, NIO bottle	31 ±30/nd-52	11 ±10/nd-26.3	5
2 1/2-hr storage	25 ±28/nd-52	8.5±3.3/nd-37	5
3 2-hr storage	3.2±19/nd-16	nd	5
4 2 days, pH 4, glass bottle	3.3±28/nd-33	nd	5
Filtered water (0.5 μ) from NIO Bottle			
5 1/2-hr storage	14 ±21/nd-29	nd	5
6 2-hr storage	4 ±27/nd-35	nd	5
7 2 days, pH 4, glass bottle	nd	nd	5

nd = not detected above average blank level.

To be fully accurate, it is not possible to specify general storage conditions which are suitable for all trace elements. In general, a low pH ($<$2) is desirable. Not only must trace metal solutions be acid, but they frequently need to be oxidative as well. Nitric acid usually serves well to both acidify and maintain an oxidizing medium, except that lead can be leached from glass by nitric acid and will oxidize hydrocarbon plastic bottle walls to produce metal binding sites.

In the case of mercury, even nitric acid apparently does not stop losses, but only slows them (Coyne and Collins 1972; Rosain and Wai 1973); and, in our work with selenium, it has been shown that unless the sample is fully 1 M in HNO_3, serious losses can occur in a short period of time. We have reported this elsewhere (Gosink and Reynolds 1975), and some of the data are presented in Table 27.4.

TABLE 27.4 Effect of storage time on selenium*

Substance analyzed	Se (ppm)	n	Remarks
Zostera marina (eelgrass)	2.0-2.3	2	Fresh, live, HNO_3, preserved
	0.8	1	Fresh, live, from aquarium
	0.1-0.43	4	Stored dry
Reduced mud from estuary in Southampton, England	0.9-1	4	After 1 hr
	<0.5	2	After 21 hrs, storage in closed flasks
Southampton estuary water	$1.4-4.1(x10^{-3})$	10	Prepared immediately
	$0.2^† (x10^{-3})$	1	3-day-old unpreserved sample

*Source: Gosink and Reynolds 1975.
†Our abiological, HCl-acidified standards proved stable for longer periods.

Museum samples of biological specimens are frequently taken for trace metal analysis, but a recent publication confirms suspicions that their trace element contents are markedly changed after 1 month's storage despite precautions to prevent contamination (Gibbs et al. 1974).

Filtration and acidification

A commonly accepted practice is to filter seawater samples through membranes rated at 0.45-μ and to accept arbitrarily the material which passes as being in solution. For many purposes, and as a very practical beginning, this is a fine procedure. Like most other arbitrarily accepted assumptions, however, it is not without its faults. Such a filtration procedure will remove all living organisms and will allow only those particles of colloidal size and smaller to pass. However, in itself, as a manipulation of the sample, it is a possible source of contamination or a means of element loss (Korenman 1968; Marvin et al. 1970). It is a necessary operation to prevent interferences in many cases and is absolutely necessary in gross speciation work in order to distinguish between particulate, adsorbed, and dissolved states of the element.

Although a 0.45-μ membrane will stop all biological organisms, it will not stop all large metal organic compounds. We observed in one case that nearly all of the aluminum content would pass a 0.2-μ filter. This water, taken from Valdez, Alaska, contained very fine glacial scour particles (Table 27.5). Similar results were reported by Hood (1967), who showed that the major copper fraction was filterable at >0.01 μ, and zinc and manganese were filterable at <0.002 μ.

Histograms of filtrate subsample analyses (Fig. 27.2; nos. 6, 9, and 12 from Table 27.2) reveal an asymmetry in those results just as the raw water samples did. Unfortunately, the time factor during this study did not allow for the development of this potentially very useful method if distinguishing among the above-mentioned gross speciation fractions of aluminum.

Other acidification results are also shown in Table 27.5. The Valdez samples from the summer months (no. 5), when a larger load of glacial scour was present, show high total aluminum content and an analyzable amount of chromium.

TABLE 27.5   Aluminum and chromium values for acid-treated raw and filtered water samples

Source	Al (ppb ±σ)	Cr (ppb ±σ)	Number of samples analyzed	Sample size (ml)
Valdez, 18 December 1973				
1 Raw	98 ±20	na	3	2
2 Raw, acid-treated	173 ±65	na	3	2
3 Filtrate (0.2 μ)	27 ±13	na	3	2
4 Filtrate, acid-treated	156 ± 6	na	3	2
Valdez, 25 July 1974				
5 Raw, acid-treated	500 ±60	1.1 ±0.2	10	10
Scripps Pier, 9 April 1974*				
6 Two acid-treated	29 ± 3	∿0.1	4	8
7 Raw water samples	136 ±68	2.8 ±2.4	4	8
Resurrection Bay, November 1973				
8 Raw, acid-treated	450 ±50	13	3	2
Prince William Sound, May 1974				
9 60°40′N, 146°45′W	110 ±23	1.7	4	8

na = not analyzed.
*see text for explanation.

Acidified sample results from the drastically different Scripps Pier samples show corresponding results. The low values (no. 6) were for samples taken at the same time as the low raw values (Table 27.2, no. 13). These were pipetted directly from the ocean into tubes containing HNO_3. The high value (no. 7) corresponds to sample 14 in Table 27.2; the high analysis for the samples did not show color and were taken from a 250-ml bottle.

The pressure of the vacuum used in the filtration must be controlled, since microorganism cells will rupture if the flow shear velocity is too great, thus releasing to the filtrate any elements for which they have a concentration factor. To alleviate this possible error, membrane vacuum filtrations should always be accomplished, regardless of bloom conditions, at no more than 4 psi (<1/3 atm) vacuum.

CONCLUSIONS

Although all of the new data here reported are for coastal waters, the arguments concerning errors and analytical variations can be extended to open ocean situations. Indeed, similar variations in deep ocean data have been clearly presented by Lewis and Goldberg (1954) and by Wangersky

(1974). It appears that variations in replicate samples can very well be real and not due entirely to analytical error.

It is obvious that subsamples of at least 1 liter should be used for each analysis and that multiple replicates are most desirable. If at all possible, the analysis or, at least, proper storage, should be started within one hour of the sample bottle closing. Sampling methods should be modified so that the contaminated surface is avoided and containers are opened preferably only at the sampling depth. The pressure rupture disc sampling method should be investigated for this purpose.

Outlier values should be included in the statistical treatment of the data and not discarded. When necessary, Poisson rather than Gaussian treatment of the data should be applied.

Conference Discussions

Korringa — Phytoplankton in the sea often show clear-cut concentration ridges perpendicular to the direction of the wind — the Langmuir effect. Should it not be advisable to collect samples always sailing in the direction of the wind to ensure that the samples cover both the ridges and the water between? We should take the distance between two ridges into account.

Gosink — Yes, one should pay attention to these Langmuir cycles, as you suggest. In the case of particulates this is easy, but it is difficult to get a representative water sample.

Korringa — Is there a way out by taking a larger sample when particles are included? A 1 m^3 sample may contain a ctenophore or a little fish, or a 1 km^3 sample might include a school of fish or a dolphin.

Gosink — It is a never ending problem, but take larger samples — those at least 2 to 10 liters in volume. One criterion that may be used is based on some good statistical work done by geologists on approaching the asymmetric distribution of elements in rocks — i.e., every subsample must contain 10^6 particles. For nearshore water, this would correspond to about 1 liter and for clear offshore water about 10 liters; therefore, 2 to 10 liters is a good range for sampling.

McAuliffe — Do you recommend taking large samples and then analyzing subsamples — i.e., take 10 analyses of a given volume? Or should one take 10 samples and analyze these 10 samples or at least pool them to one sample?

Gosink — The latter solutions are statistically satisfactory, but then you have to worry about contamination due to handling. I would prefer to process one large sample — or better still, several large samples.

Short — Is the distribution of trace element concentrations log-normal in any of the cases you examined?

Gosink — Yes, I have treated some of the data with log-normal statistics, and this does tend to normalize things.

Lamoureux — Have you done any work on the variability of common nutrients such as nitrates and phosphates? If so, what type of variability would you expect?

Gosink — Very little. It is reasonable to assume that variations should be expected due to the presence of various blooms which tend to stay in confined areas.

Acknowledgments

This work was sponsored by a grant from the National Science Foundation (GA37740). Contribution no. 290, Institute of Marine Science, University of Alaska.

REFERENCES

ATKINSON, L. P., and V. STEFANSON

    1969   Particulate aluminum and iron in seawater off the southeastern coast of the United States. *Geochim. Cosmochim. Acta* 38: 1449-1453.

BERG. W. W., JR., P. FLEISCHER, G. R. FREITAG, and E. L. BRYANT

    1975   A continuous, turbidity monitoring system for coastal surface waters. *Limnol. Oceanogr.* 20: 137-141.

BOWKER, D. E., ET AL.

    1973   Correlation of ERTS multispectral imagery with suspended matter and chlorophyll in lower Chesapeake Bay. *In* A report to the symposium on significant results obtained for the Earth Resources Technology Satellite-1, Vol. 1, Sec. B, pp. 1291-1298.

BREWER, D. G., and D. W. SPENCER

    1970   Trace element intercalibration study. Tech. Rep. No. 70-62, Woods Hole Oceanographic Institution, Woods Hole, Massachusetts.

CARRITT, D. E., and J. H. CARPENTER

    1966   Comparison and evaluation of currently employed modification of the Winkler Method for determining dissolved oxygen in seawater; A NASCO report. *J. Mar. Sci.* 24: 286-317.

COYNE, R. V., and J. A. COLLINS

1972 Loss of mercury from water during storage. *Anal. Chem.* 44: 1093-1096.

DUCE, R. A., ET AL.

1972 Enrichment of heavy metals and organic compounds in the surface microlayer of Narragansett Bay, Rhode Island. *Science* 176: 161-163.

FEELY, R. A., W. M. SACKETT, and J. E. HARRIS

1971 Distribution of particulate aluminum in the Gulf of Mexico. *J. Geophys. Res.* 76: 5893-5902.

GEORGE, D. G., and R. W. EDWARDS

1973 Daphnia distribution within Langmuir circulations. *Limnol. Oceanogr.* 18: 798-800.

GIBBS, R. H., JR., E. JAROSEWICH, and H. L. WINDOM

1974 Heavy metal concentrations in museum fish specimens: Effects of preservatives and time. *Science* 184: 475-477.

GOSINK, T. A.

1975 Rapid simultaneous determination of picogram quantities of aluminum and chromium from water by gas phase chromatography. *Anal. Chem.* 47: 165-168.

GOSINK, T. A., and D. REYNOLDS

1975 Selenium analysis of the marine environment. Gas chromatography and some results. *Marine Science Communications* 1: 101-114.

HOOD, D. W.

1967 Chemistry of the oceans: Some trace metal organic associations and parameter differences in top one meter of surface. *Environ. Sci. and Tech.* 1: 303-305.

INGAMELLS, C. O., and P. SWITZER

1973 A proposed sampling constant for use in geochemical analysis. *Talanta* 20: 547-568.

JOYNER, T.

1964 The determination of particulate aluminum and iron in coastal waters of the Pacific northwest. *J. Marine Res.* 22: 259-268.

KLEEMAN, A. W.

1967 Sampling error in chemical analysis of rocks. *J. Geol. Soc.* 14: 43-47, Australia.

KOREMAN, I. M.

1968 *Analytical chemistry of low concentrations.* Israel Program for Scientific Translation, J. Schmorak, translator, Jerusaleum.

LAMBERT, R. B., JR.

1974 Small scale dissolved oxygen variation and the dynamics of Gulf Stream eddies. *Deep Sea Res.* 21: 529-546.

LEWIN, J., and C-H. CHEN

1973 Changes in the concentration of soluble and particulate iron in sea water enclosed in containers. *Limnol. Oceanogr.* 18: 590-596.

LEWIS, G. J., JR., and E. D. GOLDBERG

1954 Iron in marine waters. *J. Mar. Res.* 13: 183-197.

MARVIN, K. T., R. R. PROCTOR, and R. A. NEAL

1970 Some effects on filtration on the determination of copper in fresh water and salt water. *Limnol. Oceanogr.* 15: 320-325.

NATRELLA, M. G.

1963 The use of transformations. In *Experimental statistics.* Handbook 91, Chapter 20, National Bureau of Standards.

PALMORK, K. H.

1969 Determination of reliability in marine chemistry by means of intercalibration and statistics. In *Chemical oceanography: An introduction* edited by R. Lange, Universitetsforlagert, Oslo, Norway, pp. 91-103.

PARKER, B., and G. BARSON

1970 Biological and chemical significance of surface microlayers in aquatic ecosystems. *Bioscience* 20: 87-93.

PATTERSON, C. C.

1974 Lead in seawater. *Science* 183: 533-534.

ROSAIN, R. M., and C. M. WAI

1973 Rate of loss of mercury from aqueous solution when stored in various containers. *Anal. Chem. Acta* 65: 279-284.

SCHAEFFER, M. B., and Y. M. M. BISHOP

    1958   Particulate iron in offshore waters in the Panama Bight and in the Gulf of Panama. *Limnol. Oceanogr.* 3: 137-149.

SHELDON, R. W., A. PRAKASH, and W. H. SUTCLIFFE, JR.

    1972   The size distribution of particles in the ocean. *Limnol. Oceanogr.* 17: 327-340.

SPENCER, D. W., and D. G. BREWER

    1970   Analytical methods in oceanography. *In* CRC critical reviews in solid state sciences, September, pp. 409-478.

SUTCLIFFE, W. H., E. R. BAYLOR, and D. W. MENZEL

    1963   Sea surface chemistry and Langmuir circulation. *Deep Sea Res.* 10: 223-243.

WANGERSKY, P. J.

    1974   Particulate organic carbon: sampling variability. *Limnol. Oceanogr.* 19: 980-984.

WILLIAMS, P. M.

    1967   Sea surface chemistry: Organic carbon, organic and inorganic nitrogen and phosphorous in surface film and subsurface water. *Deep Sea Res.* 14: 791-800.

Assessment of the Arctic Marine Environment: Selected Topics
Copyright © 1976 by Institute of Marine Science, University of Alaska, Fairbanks

CHAPTER **28**

Comparison of two methods for oil and grease determination

J. W. SHORT, S. D. RICE, *and* D. L. CHEATHAM[1]

Abstract

A gravimetric method is used by government regulatory agencies for determining levels of oil pollutants in discharge waters. This method involves extraction with an organic solvent, evaporation at elevated temperatures, and gravimetric determination of the residue. The authors compare oil content determined by the gravimetric method with oil content determined by infrared spectrophotometry for toxic water-soluble fractions of Prudhoe Bay and Cook Inlet crude oils and a No. 2 fuel oil.

The gravimetric method is adequate for grease but not for the oils. Recovery of a synthetic grease standard was 98 percent, whereas the recovery of the three pure oils ranged from 52 to 65 percent by the gravimetric method. Recovery of all the oils and the grease standard was essentially 100 percent by the infrared method. The differences between the two methods are ascribed to significant losses of volatile compounds from the oils during the evaporation step of the gravimetric method.

Gravimetric estimates of oil in toxic concentrations of water-soluble fractions ranged from 0 to 36 percent of those determined by the infrared method. Oil content of the No. 2 fuel oil water-soluble fractions was below detectable limits of the gravimetric method (1.5 mg/liter). Four-day median tolerance limits of shrimp *(Eualus fabricii)* and scallops *(Chlamys rubida)*, as evaluated by the infrared water-sol-

[1] *National Marine Fisheries Service (NOAA), Auke Bay Fisheries Laboratory, Box 155, Auke Bay, Alaska 99821.*

uble fractions of the three oils, were between 0.25 mg/liter (No. 2 fuel oil) and 3.82 mg/liter (crude oils).

It is concluded that the gravimetric method is sensitive to heavier compounds, but these have only a casual relationship to acute toxicity. Concentrations of oil in water known to have adverse effects are much lower than can be detected by the standard gravimetric method. When oil concentrations in water are to be measured and correlated with chemical toxicity, the gravimetric procedure should be supplemented with a method specific for the more soluble and volatile components.

INTRODUCTION

A gravimetric method (Taras et al. 1971, Section 137) for the determination of oil and grease has been used by the Federal Environmental Protection Agency as a method for monitoring oil concentrations in waste-water effluents (U.S. Environmental Protection Agency 1973). The petroleum industry is rapidly expanding and developing in Alaska and is responsible for oil-contaminated waste water in activities varying from drilling and production platform operations to tanker transport. The waste water is generally treated before release, but treatment usually entails a physical process that only removes the gross surface contamination rather than the dissolved and dispersed components remaining in the water. In contaminated ballast water on tankers, the petroleum (crude or refined) may be in contact with water for weeks before treatment.

In all of these cases, soluble components of petroleum approach saturation in water before treatment and release. Whether saturation occurs depends on several factors, including crude oil composition, quantity of oil, quantity of water, length of mixing time, and amount of mixing energy. The composition of oil components dissolved and dispersed in the water is certainly different from the composition in the parent oil. Indeed, both Anderson et al. (1974) and Bean et al. (1974) observed aromatic enrichment of water-soluble extracts of various oils. The data of Bean et al. are especially relevant to Alaska, because their studies included water-soluble fractions (WSFs) of Prudhoe Bay crude oil.

The lower boiling aromatic compounds (such as benzene, toluene, and naphthalene), which are ubiquitously associated with petroleum products, are among the most water-soluble components (McAuliffe 1966) of these products. They are also among the most volatile. The solubility of these compounds increases with decreasing temperature, so that the carrying capacity for aromatic compounds is highest in arctic waters. These same aromatic compounds are also among the most toxic components of petroleum products. Benzene is known to be toxic at 5 to 10 ppm; naphthalene is about 10 times more toxic to fish than benzene or gasoline (Boylan and Tripp 1971). Anderson et al. (1974), using water-soluble extracts of different oils as toxicants, associated greater toxicity with these extracts than with oil-water dispersions and ascribed the difference to aromatic enrichment of the extracts. Our own laboratory has observed similar enrichment and increased toxicity.

It seems likely that the gravimetric method might be inappropriate for setting water quality standards for oil-water solutions, because WSFs contain toxic aromatics that would probably be underestimated when measured by this technique. The gravimetric method involves organic extraction of the water sample and subsequent distillation at temperatures as high as 60° C. It measures the higher-molecular-weight residues that may be present in a variety of associations with water (emulsions, soap solutions, adsorptions on particulate matter floating at the surface, etc.). The gravimetric method was not intended for components with appreciable vapor pressures at the distillation temperatures, because the low-molecular-weight compounds tend to be lost during the distillation phase.

To evaluate the efficacy of the gravimetric method for measuring petroleum hydrocarbon concentrations in toxic oil-water solutions, we have determined the recovery efficiency of the method for Prudhoe Bay and Cook Inlet crude oil and No. 2 fuel oil, the relationship between the gravimetric method and an infrared (IR) spectrophotometric method for these oils, and the toxicity of these oil-water solutions as measured by bioassays with shrimp (*Eualus fabricii*) and scallops (*Chlamys rubida*).

MATERIALS AND METHODS

Standard method for oil and grease determination

Quantities of oil in seawater were estimated by the gravimetric method (Taras et al. 1971, Section 137), except that aliquots of 750 ml rather than 1 liter were used. Distillation temperatures never exceeded 50° C. Petroleum ether was used as the organic extractant.

Infrared determinations

Quantities of oil in seawater were also estimated by Gruenfeld's method (1973). Carbon tetrachloride extracts of an acidified seawater sample were analyzed by IR spectrophotometry at 2930 cm^{-1}. This is the analytical wavelength for paraffinic hydrocarbons. Concentrations in milligrams per liter were determined by reference to known dilutions of weighted quantities of whole oil in carbon tetrachloride (CCl_4).

Relative aromatic hydrocarbon concentrations in the three oils and their WSFs were measured by comparing absorbances at 3040 cm^{-1}. Aromatic C-H bonds absorb at this wavelength. The ratio of absorbance at the wavelength to absorbances at 2930 cm^{-1} was used to compare differences in composition between pure oils and WSFs of those oils (Table 28.3).

Recovery studies

To test the recovery efficiency of the gravimetric procedure, we dissolved known quantities of the three test oils in 40 ml of petroleum ether. These 40-ml aliquots were then added to 750 ml of seawater and shaken together in a separatory funnel for 2 min. The organic phase was separated and combined with a second extraction. These combined extracts were then distilled and evaporated and the residues weighed in accordance with the gravimetric method. A similar experiment was performed using a known quantity of Crisco[2] grease as a standard. When using Crisco, it has been observed that only slight losses occur with the gravimetric method (Taras et al. 1971, Section 209). The recovery efficiency of the Crisco grease standard was determined on a known concentration of 17.34 mg/40-ml aliquot of petroleum ether. This corresponds to a grease concentration of 23.12 mg/liter of seawater.

The IR method was tested in the same manner except that known quantities of oil or Crisco were added to CCl_4, an aliquot of which was shaken with seawater for 30 sec in a separatory funnel. The CCl_4 was then separated and analyzed by IR spectrophotometry.

Appropriate blanks were used to correct for naturally occurring organic extractables in seawater.

Detection limits

The estimated lower detection limit of the gravimetric method is twice the sample standard error associated with the gravimetric determinations of oil in the Crisco grease standard and the three oils tested for recovery efficiencies.

The lower detection limit of the IR method varies with the ratio of solvent (CCl_4) volume to sample volume used during extraction. We have not found the method reliable for oil concentrations below about 0.30 mg/liter of seawater.

Preparation of water-soluble fractions of oils

Toxic solutions were prepared by mixing 1 percent oil-water solutions (10 ml oil/liter seawater) of each of the three oils for 20 hrs with a paint stirring propeller. The speed of the propeller was adjusted so that the vortex never exceeded 25 percent of the depth of the solution. During the stirring process, the solutions were kept in constant temperature baths at 6° C. Salinity of the seawater was 30°/₀₀. After the 18 liters of mix were allowed to stand for 3 hrs, the WSF was siphoned off. Water aliquots were taken for hydrocarbon concentration measurements: five aliquots of 750 ml each for gravimetric analysis and one aliquot of 100 ml for IR analysis. The remaining volume was used in bioassays to determine toxicity.

Bioassays

Static bioassays were conducted for up to 4 days in 19-liter glass jars aerated at approximately 100 bubbles per minute. Temperatures were maintained at ambient seawater levels (4 to 5° C) during the exposure. Tissue-to-volume ratios never exceeded 1 g tissue (wet weight) per liter of seawater. To obtain data suitable for computerized probit analysis, we exposed 10 test animals to each dose level for 96 hrs. At the end of that

[2] *Use of a brand name does not imply endorsement by the National Marine Fisheries Services, NOAA.*

time the fraction of living animals was recorded to determine the 96-hr TLm (median tolerance limit).

RESULTS AND DISCUSSION

Recovery studies

The purpose of these recovery experiments was twofold. First, we wished to determine only the losses of organic extractables that occurred during the analytical steps after organic extraction in both methods. The total recovery efficiencies for standard oils and greases have already been studied for both methods (gravimetric method, Taras et al. 1971 — Section 209; IR method, Gruenfeld 1973). Second, we had to develop a technique that separated the organic from the aqueous phase in the gravimetric procedure in such a way that no water remained in the organic phase; still, we needed a technique that collected sufficient organic solvent so that the remaining solvent contained an insignificant quantity of organic extractables. This was because even 0.1 ml of seawater contains about 3 mg salt, which would lead to a very significant error.

Recovery efficiency of the gravimetric method for Crisco grease was 97.5 percent, and recovery efficiency of the IR method was 101.5 percent (Table 28.1). These percentage recoveries indicate that our technique in performing both the gravimetric and IR methods is satisfactory. Recovery efficiency of the gravimetric method for the oils tested was between 46 and 71 percent (Table 28.1), and the recovery efficiency of the IR method for these oils was essentially 100 percent. These results indicate that significant losses occur during the gravimetric procedure with the test oils.

TABLE 28.1 Percentage recovery of Crisco grease and three oils determined by the Standard Methods gravimetric procedure and the infrared procedure; recovery (reported in percent) is the mean of five determinations.

| | Percentage recovery | | | |
| | Infrared | | Gravimetric | |
Type of material tested	Mean	95% confidence interval	Mean	95% confidence interval
Crisco	101.5	± 2.6	97.5	± 3.5
Cook Inlet crude oil	102.8	± 3.2	52.1	± 5.9
Prudhoe Bay crude oil	102.3	± 2.7	65.1	± 5.5
No. 2 fuel oil	100.6	± 2.7	53.5	± 6.5

The losses occurring with the gravimetric method are probably associated with the solvent distillation-evaporation step. Crisco grease has a negligible vapor pressure at 50° C (it melts at about 45° C), so that no volatility losses would be expected at this temperature. Benzene has a vapor pressure of 100 mm at 26.1° C, and octane has a vapor pressure of 40 mm at 45.1° C (Weast and Selby 1967). These and similar compounds comprise a significant fraction of the crude oils. Number 2 fuel oil contains few mononuclear aromatics such as benzene, but it does contain appreciably volatile paraffinic and aromatic compounds. Thus, volatility losses would be expected of these compounds at the temperatures encountered during the solvent distillation-evaporation step of the gravimetric method. That these compounds comprise a significant portion of the test oils is confirmed by the severity of the losses encountered during the gravimetric procedure.

Analysis of water-soluble fractions

The gravimetric analyses were even less satisfactory for estimating oil content in WSFs than in pure oils. Gravimetric estimates were approximately 29 to 36 percent of the crude oil concentrations estimated by IR analysis (Table 28.2). The concentration of the fuel oil in water was below the limits of detection of the gravimetric method.

These gravimetric estimates of WSFs are lower than those that would be expected from the data presented in Table 28.1. If the composition of oil in the WSFs were the same as in the pure oil, then one would expect gravimetric estimates of oil content to be lower than IR estimates by the percentages indicated in Table 28.1. The fact that the gravimetric estimate of a given oil in the WSF is an even smaller percentage of the IR estimate (Table 28.2) indicates that the composition of oil in the WSF is different from the pure oil.

TABLE 28.2 Analysis of oil concentrations in 100% water-soluble fractions by the infrared and gravimetric procedures. Analytical wavelength of the IR method here is 2930 cm^{-1}. Results are means of five determinations; the IR method gave no significant variance between five replicates of the same oil.

Type of oil tested	Concentration of oil (mg/liter)			Ratio of gravimetric and infrared values	
	Infrared	Gravimetric			
	Mean	Mean	95% confidence interval	Percentage	95% confidence interval
Cook Inlet crude oil	13.72	4.92	± 1.19	25.9	± 0.10
Prudhoe Bay crude oil	17.43	5.10	± 1.15	29.3	± 0.07
No. 2 fuel oil	0.82	(0)*		—	

*Indistinguishable from seawater blank values.

This difference can be explained by assuming that the oil in the WSF is enriched in volatile paraffins and aromatics from the pure crude oil. For example, 52.1 percent of a pure sample of Cook Inlet crude oil was recovered by the gravimetric method (Table 28.1); the WSF of Cook Inlet oil had 4.92 mg/liter, as analyzed by the gravimetric method (Table 28.2). If the composition of the WSF were identical to the pure crude oil, then one would have expected the IR method to show the presence of

$$\frac{4.92 \text{ mg/liter}}{0.521} = 9.44 \text{ mg/liter}$$

Actually, the IR method showed the presence of 13.72 mg/liter. We ascribe the difference (4.28 mg/liter) as being mainly due to the enrichment of the WSF with volatile paraffins from the pure crude oil. These are the most soluble of the paraffins (McAuliffe 1969). They are efficiently measured by the IR method, but because of the volatility of the more soluble paraffins, losses of these compounds would occur during the distillation step of the gravimetric method.

The WSFs also contained significantly greater quantities of aromatic hydrocarbons than the parent oils, and these aromatic hydrocarbons were not efficiently estimated by either the IR or the gravimetric methods. Our method of preparing WSFs involved prolonged contact of oil with water and relatively gentle mixing. Under these conditions, one would expect that the WSF would be enriched in soluble components and that the formation of dispersed droplets (which requires a high mixing energy) would be reduced. Aromatic enrichment in the WSF from the parent oil is demonstrated in Table 28.3, where the absorbance of 3040 cm^{-1} is compared to the absorbance at 2930 cm^{-1}. The 3040 cm^{-1} wavelength is sensitive to aromatic compounds, and increased absorbance indicates increased concentrations in the WSF. The wavelength of 2930 cm^{-1} is specific for paraffinic hydrocarbons. The absorbance at 3040 cm^{-1} relative to that at 2930 cm^{-1} was greater for the WSFs than for the pure oils. Anderson et al. (1974) produced WSFs with crude and refined oils and observed similar enrichment of aromatic hydrocarbons. Thus, the WSFs contain relatively more aromatics and paraffins than do the parent oils; and since the aromatics and paraffins can have significant vapor pressures at $50°$ C, they are susceptible to losses in the distillation step of the gravimetric method.

Because of this aromatic enrichment and the volatility of aromatics, the efficiency of recovery of oil in WSFs by the gravimetric method relative to

TABLE 28.3 Ratio of aromatic absorbance to paraffinic absorbance in the IR method for pure oil and water-soluble fractions. Number of determinations is in parentheses.

Type of oil tested	$\dfrac{OD \text{ at } 3040 \text{ cm}^{-1}}{OD \text{ at } 2930 \text{ cm}^{-1}}$ Oil	$\dfrac{OD \text{ at } 3040 \text{ cm}^{-1}}{OD \text{ at } 2930 \text{ cm}^{-1}}$ WSF
Cook Inlet crude oil	= 0.0342 (5)	0.328 ± 0.067 (16)
Prudhoe Bay crude oil	0.0365 (5)	0.469 ± 0.105 (8)
No. 2 fuel oil	0.0360 (5)	0.0526 (2)

the IR method are even lower than those reported in Table 28.2. Since the IR wavelength (2930 cm^{-1}) used to determine oil concentrations in Table 28.2 is not sensitive to aromatic compounds, more oil components were present than were estimated by the IR procedure. The gravimetric method therefore actually recovered a somewhat smaller percentage of the compounds present than Table 28.2 indicates.

Bioassays

The WSFs produced from Prudhoe Bay and Cook Inlet crude oils and No. 2 fuel oil were all significantly toxic to shrimp and scallops (Table 28.4). The toxicity data with oil concentrations measured by IR at 2930 cm^{-1} indicate that fuel oil was most toxic (i.e., least quantities killed half the shrimp or scallops in 96 hrs), and Prudhoe Bay crude oil was slightly more toxic than Cook Inlet crude oil, and that differences in sensitivity between shrimp and scallops were negligible. Our bioassay data indicate that dilutions of WSFs prepared from the crude oils and the fuel oil are toxic to both scallops and shrimp at concentrations measured by IR spectrophotometry that are near or below the detection limits of the gravimetric method (1.50 mg/liter organic extractables).

TABLE 28.4   Toxicity of water-soluble fractions of three types of oil to two test animals as determined by IR method. Milligrams per liter values are the 96 hrs TLm together with the lower and upper bounds of the 95% confidence interval. TLm's were calculated by probit analysis.

Test animal and type of oil	Median tolerance limit (TLm) at 96 hrs	95% confidence interval
Shrimp *(Eualus fabricii)*		
Cook Inlet crude oil	3.17	2.85–3.54
Prudhoe Bay crude oil	1.94	1.77–2.12
No. 2 fuel oil	0.25	0.13–0.49
Scallops *(Chlamys rubida)*		
Cook Inlet crude oil	3.15	2.94–3.35
Prudhoe Bay crude oil	2.06	1.90–2.18
No. 2 fuel oil	0.49	0.39–0.61

The water-soluble extracts as prepared in this study are probably a reasonable approximation of petroleum components in waste water effluent. These effluent waters are characterized by prolonged contact between oil and water, so that solubility effects are probably significant. Neff (see Anderson et al. 1974) observed that oil components in WSFs increased nearly linearly with time over 30 hrs, when bacterial action became important. The same study noted that WSFs that had been mixed for 20 hrs were highly enriched in aromatic components. Because of the possible influence of bacterial action, the 20-hr mixing time was chosen by Anderson et al. as a standard. The prolonged approach to equilibrium (which is probably never actually achieved) of their WSFs indicates that the slow solubilization of petroleum components is an important effect associated with prolonged contact between oil and water. Experiments in our own laboratory support this view: we observed that with time naphthalene concentrations beneath an oil slick increased in the presence of an air bubbler, presumably because of the mixing effect of the air bubbles.

Applications of the standard gravimetric method

Rather than quantitating hydrocarbons in WSFs, the gravimetric method is better suited for quantitating heavier hydrocarbon compounds like tars, which have low vapor pressures and low water solubilities and typically float at the surface. Harmful effects from these compounds are more frequently associated with physical coating than with chemical toxicity. The effects of physical coating are obvious on seabirds, marine organisms, beaches, and fishing gear.

It is quite possible that oil could be present in the water column (mainly in the form of dispersed droplets), so that better recoveries than those reported in Table 28.2 would be obtained. The composition of droplets would then be similar to the pure oil, and recoveries would be similar to those presented in Table 28.1. If the history of the oil-contaminated water were such that prolonged weathering would be significant, then recoveries approaching 100 percent of the oil contaminants present might be expected because compounds subject to volatility losses would presumably be absent. Nevertheless, evidence from this study indicates that compounds amenable to analysis by the gravimetric method have only a vague relationship to toxicity. The 100 percent stock solution of the WSF prepared from fuel oil showed zero organic extractables by the gravimetric method; yet it was more toxic than WSFs prepared from the two crude oils.

We believe that this lack of detectability occurs because the toxicity is due to aromatic compounds, which are not amenable to analysis by the gravimetric method and which are among the more readily water-soluble of the compounds in oil. However, recent evidence (Lysyj and Russell 1974) indicates that water in contact with oil tends to become enriched in polar compounds present in the oil. These compounds are not amenable to analysis by any technique that relies on extraction of oil-derived compounds into a low polarity organic solvent. Thus, it is possible that the toxicity of WSFs is due to the presence of these polar compounds.

Organic extraction gravimetric procedures that quantitate volatile compounds subsequent to extraction and gravimetric analysis of the residues do exist (American Society for Testing and Materials 1970). However, these methods, which involve volumetric determination of the volatile compounds in samples as large as 4 liters, have a sensitivity limit of about 5 ppm.

There are several analytical methods for measuring oil in the water column that may have better correlations with chemical toxicity: IR (Gruenfeld 1973), ultraviolet (Levy 1971), flourescence spectroscopy (Gordon et al. 1974), gas-liquid chromatographic analyses (Bean et al. 1974; Jeltes and den Tonkelaar 1972), gas stripping-volumetric determination (Webber and Burks 1952), and total organic carbon (Lysyj and Russell 1974), among others. Each method has the ability to measure oil-derived hydrocarbons in water but with varying specificities, sensitivities, ease, and costs.

Oil associated with water is not amenable to exhaustive quantitative description by some simple chemical test. Only under certain conditions will the composition of oil in water approach the composition of the contaminating oil. More commonly, the composition of oil in water will be very different because of a wide variety of physical and chemical factors — as observed in the WSFs prepared in this study. The concept of a quantity of oil dissolved in water is not sufficiently specific to establish a monitoring program. A more fruitful approach may be to establish those qualities associated with oil-contaminated water that require monitoring and then to select a chemical technique (or combination of techniques) most appropriate. For example, the gravimetric method might be quite acceptable for monitoring grease on surface waters; such an analysis would be indicated if the grease were clogging a filtration system. On the other hand, marine phytoplankton photosynthesis is affected by very low levels of oil-derived hydrocarbons, as measured by flourescence spectroscopy (Gordon and Prouse 1973).

In summary, the WSFs prepared as we have described are enriched in those compounds likely to escape detection by the gravimetric method. The gravimetric method lacks the sensitivity to detect toxic concentrations of hydrocarbons in WSFs of crude and refined oils, and it is sensitive to compounds that have only a vague relation to the most readily toxic components of oil. Thus, where it is desirable to quantitate oil concentrations in water for correlation with chemical toxicity, alternative methods of analysis seem indicated.

SUMMARY

The solvent evaporation step of the gravimetric procedure for oil and grease causes losses of up to 45 percent of the pure oil known to be present. The water-soluble fractions (WSFs) of oil are enriched in those volatile components likely to be lost during the solvent evaporation step of the gravimetric oil and grease procedure. Acute toxicity of WSFs of oil to scallops and shrimp, as measured by median tolerance limit (TLm) at 96 hrs, are well below the sensitivity limits of the gravimetric procedure. In addition, compounds amenable to analysis by the gravimetric procedure have only a vague relationship to acute toxicity. Where it is desirable to avoid effects associated with the chemically toxic fractions of oils, we recommend that analytical techniques more sensitive than the gravimetric procedure to the toxic fractions of oils be used.

Acknowledgments

The authors wish to acknowledge the Shell Oil Company for supplying the Cook Inlet crude oil, Linda Swenson of the State of Alaska Department of Environmental Conservation, and the editorial staff at the Auke Bay Fisheries Laboratory for their assistance in preparing the manuscript.

REFERENCES

AMERICAN SOCIETY FOR TESTING AND MATERIALS

1970 Standard method of test for oily matter in industrial water. In *Annual Book of ASTM Standards*, Part 23. American Society for Testing and Materials, Philadelphia, Pennsylvania, pp. 295-301.

ANDERSON, J. W., ET AL.

1974 Characteristics of dispersions and water-soluble extracts of crude and refined oils and their toxicity to estuarine crustaceans and fish. *Mar. Biol.* 27: 75-88.

BEAN, R. M., J. R. VANDERHORST, and P. WILKINSON

1974 Interdisciplinary study of the toxicity of petroleum to marine organisms. Battelle Pacific Northwest Laboratories, Richland, Washington, 31 pp. + appendix.

BOYLAN, D. B., and B. W. TRIPP

1971 Determination of hydrocarbons in seawater extracts of crude oil and crude oil fractions. *Nature* 230 (March): 44-47.

GORDON, D. C., JR., P. D. KEIZER, and J. DALE

1974 Estimates using fluorescence spectroscopy of the present state of petroleum hydrocarbon contamination in the water column of the northwest Atlantic Ocean. *Mar. Chem.* 2: 251-261.

GORDON, D. C., JR., and N. J. PROUSE

1973 The effects of three oils on marine phytoplankton photosynthesis. *Mar. Biol.* 22: 329-333.

GRUENFELD, M.

1973 Extraction of dispersed oils from water for quantitative analysis by infrared spectrophotometry. *Environ. Sci. Technol.* 7: 636-639.

JELTES, R., and W. A. M. den TONKELAAR

1972 Gas chromatography versus infrared spectrometry for determination of mineral oil dissolved in water. *Water Res.* 6: 71-278.

LEVY, E. M.

1971 The presence of petroleum residues off the east coast of Novia Scotia, in the Gulf of St. Lawrence, and the St. Lawrence River. *Water Res.* 5: 723-733.

LYSYJ, J., and E. C. RUSSELL

    1974   Dissolution of petroleum-derived products in water. *Water Res.* 8(11): 863-868.

MCAULIFFE, C.

    1966   Solubility in water of paraffin, cycloparaffin, olefin, acetylene, cycloolefin, and aromatic hydrocarbons. *J. Phys. Chem.* 4: 1267-1275.

    1969   Determination of dissolved hydrocarbons on subsurface brines. *Chem. Geol.* 4(1969): 225-233.

TARAS, M. J., A. E. GREENBERG, R. D. HOAK, and M. C. RAND (Eds.)

    1971   *Standard methods for the examination of waste and wastewater.* 13th Ed. American Public Health Association, Washington, D. C., 874 pp.

U. S. ENVIRONMENTAL PROTECTION AGENCY

    1973   Guidelines establishing test procedures of analysis of pollutants. *Federal Register* 2(199): 28758-28760.

WEAST, R. C., and S. M. SELBY (Eds.)

    1967   *Handbook of chemistry and physics.* The Chemical Rubber Company, Cleveland, Ohio.

WEBBER, L. A., and C. E. BURKS

    1952   Determination of light hydrocarbons on water. *Anal. Chem.* 24(7): 1086-1087.

arctic epilog

Bearing north to the future

D. W. HOOD

As indicated throughout this book, the North Country is besieged by the powerful forces of industrialization, exploitation, and population demands for living, recreational and resource space. As in any frontier development, there are fierce antagonists for or against any major move on the part of those who are affected by the action. More often than not, the motivation for conflicts is based on personal bias and does not stand in the best democratic tradition of what is best for the whole of society. In fact, even the question of what is best for society in the North has been in itself a serious subject for discussion throughout times past, and its solution will probably continue to be pursued for decades to come. Unfortunately, most discussions on this topic lack the benefit of pertinent documentation. Opinions are far too often based on a body of information developed for temperate or tropical regions, and application is attempted without the proper and necessary modifications.

Experience has shown that most deductions based on experience outside the northern regions have been in error. As examples, most biologists would have thought that light was the primary limiting factor to photosynthesis in the North as elsewhere. But the evidence is otherwise: during the summer months in the North, the concentration of a nitrogen-containing fertilizer component is in nearly total control of the rate of plant growth. In other seasons, such factors as the nature of the ice substrate or the stability of the near-surface water column are of more direct influence to plant growth than is light. Few would have guessed that such an apparently inert substance as ice can be a site of remarkable biological activity. And how many would have estimated the magnitude of permafrost importance to engineering of the trans-Alaska pipeline as has ultimately been demanded.

The need for information on how the northern ocean functions is probably of higher priority than any other oceanographic data sought in the world at the present time. The rationale for this statement rests on the importance of the North as a supply of food resources and in support of a community of biological ecosystems which are either partially or wholly dependent on the sequence of well-timed events in the Arctic. Because of the lack of experience of the main body of workers in this scientifically

465

intriguing region — and therefore a less than enthusiastic support for northern studies by the scientific community, the understanding of the oceanographic processes of the region is far behind that of the lower latitudes. A major deterrent has been the rigorous conditions of the study area and the almost total lack of suitable platforms for work, at least by United States scientists, in the Arctic other than during the summer months. Scientific interest is shifting, however, and programs such as PROBES — which has the potential for becoming an unusually productive ecosystem study of the Bering Sea — and baseline studies of the outer continental shelves of Alaska, as well as Canadian work in the Beaufort Sea and development of the North Sea, will undoubtedly lead the way to more concentrated effort and better logistic support for this region.

A problem not exclusively related to the Arctic, but generally relevant to the whole of science, is the ever-increasing costs of keeping a productive worker at the so-called laboratory bench collecting new scientific data. A host of bureaucratic syndromes, which are very effective in enlarging the management role in research but of doubtful value to the overall effort, have run the cost of experimentation, whether in the field or in the laboratory, to dangerously high levels. Partly as inflation, partly through social benefits, bureaucracy acts to overcontrol scientific investigation. Although many bureaucratic innovations have value, few can be successfully argued philosophically. For the very nature of bureaucracy is to enhance itself, to preserve and feed on itself, while always enlarging the body on which to feed — until the output or work done by the system is minimal, and yet the body may be ever-increasing its activities. We seem to be more caught up in this non-productive engulfing tendency in the North than elsewhere, because it becomes the obligation of the few expert workers in the Arctic to inform the would-be northern experts of the bureaucracy from areas south about the program and conditions under their control — thus further diluting the efforts of those in the position to obtain new needed information. It is doubtful if much data-collecting in the North is presently facing better than about a three-to-one ratio of cost to the taxpayer or sponsor to cost of the man at the bench in the environmental field. If logistic costs are added, then the ratio worsens.

Since field data in the North are so costly to obtain, a sharp focus on the kinds of data needed in response to a particular question becomes even more important. As recommended earlier in this book, the assessment of the arctic environment to evaluate the effects of oil and gas extraction, in particular, should focus on nearshore processes which are critical steps in the life cycle of commercial fish or of the so-called sensitive species. The adults of the system are in most cases already under stress from commercial fishing, and the likelihood of a catastrophic oil spill significantly reducing the large adult population seems somewhat remote. Every precaution should be made to prevent and retain such spills, but this concern lies in the area of good engineering practice and not in environmental assessment. The assessment studies should concentrate in understanding the oceanographic process in those environments where likelihood of significant damage to the biota is greatest. The nearshore environment is certainly the most critical of the ocean environments with respect to impact from oil and gas development.

Some of the greatest scientific unknowns are associated with the active nearshore environment. In the open ocean, it is often possible to combine data obtained over a period of several years to describe oceanographic features, because the changes which occur are on a very long time-scale. On the continental shelf, changes are more rapid, and surveys may be required as often as twice a year to describe the processes. Nearshore events occur rapidly, and observations must be made at least quarterly; in some cases, almost continuous monitoring is required of the nearshore environment.

In addition to staging rapidly changing events, the nature of the nearshore environment itself is a transition zone between land and water. Here is where ocean tides and waves dissipate their energy against the beach — where fresh water meets the sea. The coastal waters are the major site of waste discharge from cities and industry; the nearshore zone is the area most critically affected by spillage of oil or other contaminants. It is the scene of anadromous fish migration — where many species of ocean fish spend their early lives. And, in the Arctic, shorefast ice covers the nearshore waters each year.

Studies of the nearshore marine environment include consideration of those associated land events which directly influence the processes of the water. Little would be understood of an estuary if the nature of the incoming river water were not carefully documented. The change in precipitation in the drainage basin of the river may have profound effects on the juvenile forms of organisms living in the relatively quiet waters of the estuary. Changes in the amounts of incoming water may alter the circulation pattern and flushing characteristics; through the sediments transported, these changes may in turn limit light penetration and therefore photosynthesis. Other processes can also be affected — nutrient cycling, trace-metal chemistry, microbial growth.

Faced with so complex a situation, many oceanographic scientists have good reason for a hesitancy to work in the nearshore environment. This is particularly true in the physical sciences. More difficult to handle than the intricate scientific problem is the high cost of deploying necessary current-meter moorings and occupying hydrographic stations frequently enough for meaningful results. Other disciplines have similar problems. Although the daily cost of suitable platforms for inshore work may be reduced because of the size of the ship required, the frequency of use compensates to offer no real saving in total costs for a given study program.

Inshore studies, to be worthwhile, must be highly interdisciplinary. The days are gone when a single scientist can sample an estuary and produce a definitive statement on the quality of the water or condition of life there. He might be able to determine whether a catastrophe has indeed occurred, but he may well be unqualified to determine the cause, the path of approach, other damage to the system, or how long the effects will last. If he were able to know whether such a catastrophe would likely occur again, there might be nothing he could do to prevent it or, in most cases, to know the extent of change to the biota. A chemist working alone would not have the benefit of knowing the effect of changes in water quality on the biota. The geologist would be in great difficulty indeed to describe sedimentation processes associated with a hazardous situation without the benefit of a chemist concerned with absorption and exchange — and a physicist to determine the circulation patterns. The physicist can measure dispersion rates, circulation patterns and flushing rates — but these data are of little value in themselves, if the purpose of assessment is to determine the effect of perturbation on the biota of the region. The biologist working alone can determine the numbers and kinds of organisms and changes in the population that may occur, but he will be hard pressed to explain the changes without the benefit of data from the chemist on substances to which organisms may have been exposed — without information from the geologist on changes that may have occurred in the substrate — and without contribution from the physicist on how the substances got there and how long they will stay.

From the above discussion, it can be seen that assessment of the arctic marine environment should be a well-focused, highly interdisciplinary study of nearshore processes in order to be most effective in evaluating and preserving the important ecosystem of the North. Because inshore studies cost much in time and facilities, it is not possible to study all the areas which are potentially exposed to development. It is therefore necessary to select "type" areas for detailed study until a reasonable understanding is obtained of the events that keep these systems working. Sufficient evidence should also be obtained in the whole of the area to permit comfortable extrapolations from a studied type area to a non-study area.

The recent surge of industrial activity in the Arctic will hopefully be accompanied by a genuine interest and intelligent resources for pursuing scientific understanding of how this extraordinary region of the earth operates. The need for available facilities is clear. A single oceanographic research vessel capable of working in the rigor of storms and ice in the high latitudes would possibly cost as much as $10 million to build, and up to 20 percent of that again for annual operating funds. Yet, is this not a small price to pay to gain understanding and therefore give means for protecting a multi-billion dollar fishery. In the Bering Sea alone, the value of the annual catch taken by several nations totals well over a billion dollars.

The oil, gas and hard mineral resources of this region are largely unknown, but their extraction is likely imminent in order to maintain the economy of the industrialized world. There is no fundamental reason why the northern seas cannot be used for both fisheries and mineral extraction. It will require discipline, diligence and perseverance. But most of all, it will require adequate resources to obtain the necessary information so that the hazards of development to the endemic biota can be avoided. To do less can only lead to further disappointment in the ability of the human species to live in its environment compatible with other elements of nature.

Clearly, we can no longer afford to err.

a perduring symbol of arctic achievement

Fram   'forward'

Norwegian iceship
sharing history with explorers Nansen and Sverdrup
in discovery of north polar ocean (1893-1896)

The *Fram* marked a hard course during that period
— over a thousand days adrift
           in the grip of the Arctic ice pack —
and four continuous years (1898-1902)
of *New Land* explorations

No easier are the challenges of present-day goals
in the Arctic
Onward . . .
to resource assessment and rational exploitation
with respect for the future of a unique environment